Cytotoxicity - Understanding Cellular Damage and Response

Edited by Anil Sukumaran
and Mahmoud Ahmed Mansour

Published in London, United Kingdom

Cytotoxicity - Understanding Cellular Damage and Response
http://dx.doi.org/10.5772/intechopen.100956
Edited by Anil Sukumaran and Mahmoud Ahmed Mansour

Contributors
Eslin Ustun Karatop, Ayla Melisa Aksu, Cagla Eren Cimenci, Emusani Ramya, D. Narayana Rao, Leisheng Zhang, Hui Cai, Arinzechukwu Ude, Emmanuel Ogbodo, Kelechi Okeke, Kaiyven Afi-Leslie, Özlem Dağdeviren Özsöylemez, Gül Özcan Arican, Margaret A. Agnes Jordan, Annie M.L. Willson, Upninder Kaur, Devyani Sharma, Rakesh Sehgal, Nanik Siti Aminah, Khun Nay Win Tun, Alfinda Novi Kristanti, Hnin Thanda Aung, Yoshiaki Takaya, Chakrabhavi D. Mohan, Kereyagalahally H Narasimhamurthy, Nichhapurada Kallesha, Kanchugarakoppal S. Rangappa, Areg Hovhannisyan, Hovhannes Ghazaryan, Usha Singaravelu, Muthusamy Velusamy Nivedha, Molabanti Vasupriya, Santhanakrishnan Jayasree, Subramaniam Mahalakshmi, Gunasekaran Sinduja, Saravanan Priyadharshini, Nebu G. George Thomas, Nandakumar Kalarikkal, Saumya John, Mridula Sreedharan, Valliaveettil Thomas George, Nibu Varghese, Sabu Thomas, Mekha Grace Grace Varghese, Sherin Sara George, Anil Sukumaran, Vishnupriya K. Sweety, Betsy Joseph, Biba Vikas

First published in London, United Kingdom, 2023 by IntechOpen
IntechOpen is the global imprint of INTECHOPEN LIMITED, registered in England and Wales, registration number: 11086078, 5 Princes Gate Court, London, SW7 2QJ, United Kingdom

British Library Cataloguing-in-Publication Data
A catalogue record for this book is available from the British Library

Additional hard and PDF copies can be obtained from orders@intechopen.com

Cytotoxicity - Understanding Cellular Damage and Response
Edited by Anil Sukumaran and Mahmoud Ahmed Mansour
p. cm.
Print ISBN 978-1-80356-245-2
Online ISBN 978-1-80356-246-9
eBook (PDF) ISBN 978-1-80356-247-6

For EU product safety concerns:
IN TECH d.o.o., Prolaz Marije Krucifikse Kozulić 3, 51000 Rijeka, Croatia,
info@intechopen.com or visit our website at intechopen.com.

Meet the editors

Prof. Anil Sukumaran is a distinguished dental professional and academician with an extensive career spanning more than 36 years. He is Senior Consultant Professor and Chairman of Hospital Research at the Hamad Medical Corporation, Doha, Qatar. An alumnus of the University of Hong Kong, Prof. Anil earned his Ph.D. in Dentistry in 2002. His commitment to excellence led him to undertake advanced training at the UK Center for Oral HIV Research, Guy's and St. Thomas's Hospital in London. He boasts an impressive array of fellowships including from the Royal College of Pathologists (FRCPath), Royal College of Physicians and Surgeons in Glasgow (FDS, RCPS), International College of Dentist (FICD), and the Pierre Fauchard Academy (FPFA). His contributions to the dental field have earned him numerous accolades. These include the Gold Medal in BDS from India, the Ratan H. Doctor National Award from India, the Wang Gungwu Scholarship for Research Students in Hong Kong, and the Golden Quill Award for research excellence from King Saud University, Riyadh, Saudi Arabia. Prof. Anil has a rich history of academic engagements. He has mentored undergraduate, postgraduate, and doctoral students across diverse nations like India, Hong Kong, Saudi Arabia, Qatar, the United Kingdom, the Netherlands, South Korea, and the United States. His decade-long tenure at King Saud University saw him don the roles of a full professor, chairman, and member of various accreditation boards, cementing his mark in the global dental education sector. Furthermore, Prof. Anil established a medical research facility focusing on biomaterials and dental at the Ministry of Defense, Saudi Arabia. A prolific writer, he has contributed more than 400 research papers to renowned journals and authored several chapters and books in his domain. Prof. Anil Sukumaran is a beacon of excellence, expertise, and dedication in dentistry and academia.

Prof. Mahmoud Mansour is a Professor of Biochemistry at the College of Pharmacy, King Saud bin Abdulaziz University for Health Sciences, Saudi Arabia. He received a pharmacy degree from Al-Azhar University Cairo, Egypt, in 1984, and a Ph.D. in Clinical Biochemistry from the Karolinska Institute, Stockholm, Sweden in 1992. As a fellow, Prof. Mansour contributed significantly to defining some aspects of the molecular biology of tissue injury and induction of lever toxicity and renal toxicity, with a focus on the role of antioxidants in the prevention and treatment of different toxic effects. In addition, he contributed significantly to developing a new strategy for tumors, especially those in the liver. His research has shown that proteasome inhibitors play an important role in preventing carcinogenesis of the liver by metastatic tumors. At King Saud bin Abdulaziz University for Health Sciences, Prof. Mansour continued to work on the pathogenesis of cancer and on issues concerning the cellular and molecular mechanisms of cancer biology and the metastatic process. He is currently conducting studies on liver cancer using the oncogene Gankyrin.

Contents

Preface

In the constantly evolving domain of cellular biology, few topics possess the gravitas and ubiquity of cytotoxicity. This deleterious phenomenon, where cells respond adversely to external agents or intrinsic imbalances, stands at the nexus of various biological processes and pathological states. Hence, a comprehensive understanding of cytotoxicity is paramount, not only for seasoned researchers but also for newcomers to the field. This book, *Cytotoxicity – Understanding Cellular Damage and Response*, is an authoritative source of knowledge, laying out the landscape of current research, understanding, and advancements in this crucial area.

The chapters enclosed within these covers are not mere literary entries but rather a curated selection of methodical explorations, each echoing the dedication of its contributors. From detailing the molecular mechanisms underpinning cytotoxic reactions to understanding the myriad of external agents triggering such responses, this book traverses the broad spectrum of cytotoxicity. It delves into contemporary methodologies employed in the study of cellular damage, elucidating techniques that enable researchers to discern subtle nuances in cell responses. Furthermore, the chapters accentuate the translational potential of understanding cytotoxicity, particularly in drug development and disease therapeutics.

A notable characteristic of this volume is its inclusivity. We have encompassed established principles and pioneering discoveries, fostering an environment where tradition and innovation coalesce. The result is a body of work that is both foundational and forward looking.

For the seasoned researcher, this compendium harmonizes the diverse facets of cytotoxicity, presenting an opportunity to witness the field's trajectory and identify promising frontiers. For the neophyte, it serves as a reliable torchbearer, illuminating a path through the intricate maze of cellular responses and consequences.

As the editor of this volume, I am profoundly grateful to each contributor for their rigorous scholarship and commitment to advancing the boundaries of knowledge. Their collective wisdom, distilled in these pages, cements this book's status as a keystone resource in cytotoxicity.

I hope that the readers, be they students, scholars, or professionals, will benefit from the diligence and expertise of these pages, advancing the noble cause of biomedical science.

Anil Sukumaran
Hamad Medical Corporation,
Doha, Qatar

Mahmoud Ahmed Mansour
King Saud bin Abdulaziz University for Health Sciences,
Riyadh, Saudi Arabia

Chapter 1

Cytotoxicity and Cell Viability Assessment of Biomaterials

Anil Sukumaran, Vishnupriya K. Sweety, Biba Vikas and Betsy Joseph

Abstract

Biocompatibility testing is essential for medical devices and pharmaceutical agents, regardless of their mechanical, physical, and chemical properties. These tests assess cytotoxic effects and acute systemic toxicity to ensure safety and effectiveness before clinical use. Cell viability, indicating the number of healthy cells in a sample, is determined through various assays that measure live-to-dead cell ratios. Cytotoxicity measures a substance's potential for cell damage or death, and is evaluated through numerous assay methods based on different cell functions. Ensuring biocompatibility is crucial for the successful integration of medical devices and pharmaceuticals into clinical practice. As part of the evaluation process, researchers utilize a range of cell viability assays and cytotoxicity tests to assess the potential impact of these products on living cells. The results of these tests inform the optimization of cell culture conditions and drug candidates, as well as guide the development of safer, more effective medical devices. By thoroughly examining the interactions between devices, drugs, and biological systems, researchers aim to minimize the risk of adverse reactions and improve patient outcomes.

Keywords: cell viability, cytotoxicity, biocompatibility, trypan blue dye exclusion assay, ATP assay, MTT assay, DNA synthesis cell proliferation assays, Raman micro-spectroscopy, MTT assay

1. Introduction

Biocompatibility testing is a crucial aspect of the development and evaluation of medical devices and pharmaceutical agents. Ensuring that these products are safe and effective for human use is of paramount importance, as they can directly interact with the body's tissues and cells. The primary goal of biocompatibility testing is to assess the safety and effectiveness of medical devices and pharmaceutical agents before they come into contact with the human body. These products may possess a variety of mechanical, physical, and chemical properties that can potentially impact their interactions with biological systems. Biocompatibility testing helps identify any potential risks associated with these interactions, allowing for the optimization of product design and formulation to minimize the likelihood of adverse effects. The components of biocompatibility or the tissue response of the clinically relevant performance of biomaterials are cytotoxicity, genotoxicity, mutagenicity, carcinogenicity, and immunogenicity. Medical devices

IntechOpen

and pharmaceutical agents must undergo a series of tests to determine their cytotoxic effects and acute systemic toxicity. These tests are essential in ensuring that the products are safe for clinical use and do not pose any undue risks to patients.

1.1 Cell viability and cytotoxicity assessments

A key component of biocompatibility testing involves the evaluation of cell viability and cytotoxicity. Cell viability refers to the number of healthy, functioning cells in a sample, while cytotoxicity measures the potential of a substance to cause cell damage or death. By assessing these parameters, researchers can gain insights into how medical devices and pharmaceutical agents may affect the body at the cellular level [1]. Various assays can be employed to measure cell viability and cytotoxicity, including dye exclusion methods, metabolic activity-based methods, ATP assays, and DNA synthesis cell proliferation assays, among others. These tests can help determine whether a substance exhibits direct cytotoxic effects or impacts cell proliferation, providing valuable information on its safety profile.

1.2 Biocompatibility testing for medical devices

Medical devices, which encompass a wide range of products used in various clinical disciplines, must demonstrate good biocompatibility to be deemed safe for use. This is particularly important for devices that come into direct contact with the body's tissues and cells, such as implants, prosthetics, and surgical instruments. Biocompatibility testing for medical devices involves assessing the compatibility of these products with the biological systems they will encounter during use. This includes evaluating the interaction between the device and the living tissues and cells it will come into contact with, as well as examining the potential for adverse reactions, such as inflammation, infection, or rejection. By thoroughly evaluating the biocompatibility of medical devices, researchers can develop products that are more likely to integrate successfully with the body and promote positive patient outcomes.

2. Parameters used in cell-based assays

Cell-based toxicological assays are designed to evaluate the potential toxic effects of various substances, including drugs, chemicals, and environmental pollutants, on living cells. These assays measure a range of cellular parameters that can be affected by toxic agents. By examining these factors in cell-based toxicological assays, researchers can gain a comprehensive understanding of the effects of various substances on cells and identify potential therapeutic targets, mechanisms of action, and potential side effects [2]. This information is crucial for the development of safer and more effective drugs, chemicals, and other products. Here are some common parameters used in cell-based toxicological assays:

2.1 Cell viability

Cell viability is a measure of the number of living cells in a sample. Cell viability assays are crucial for determining the overall health and survival of cells in response to various treatments. These assays are based on cellular functions that are specific to living cells, such as metabolic activity or membrane integrity [3]. A decrease in viability indicates a toxic effect, whereas an increase may indicate a protective or stimulatory effect. Several assays, such as MTT, XTT, WST-1, Neutral Red, and Alamar Blue, can be used to assess cell viability based on metabolic activity, dye uptake, or ATP levels. Damage to the cell membrane can be identified by the presence of intracellular substances, like lactate dehydrogenase (LDH), in the suspension medium. This can be assessed when cells are in contact with materials or material extracts in cell cultures [4]. A decline in metabolic activity might signal cell death before the breakdown of the membrane occurs. The MTT assay, introduced by Mosmann, is a widely used technique for evaluating cell viability [5].

2.2 Cell proliferation

Cell proliferation refers to the rate at which cells divide and increase in number. Assessing cell proliferation provides insights into how toxic agents affect cell division and growth. Inhibition of cell proliferation can indicate the potential anti-cancer effects of a drug, while excessive inhibition may indicate general toxicity. Cell proliferation is a crucial parameter in evaluating cytotoxicity. The cytotoxic effects of a biomaterial can threaten cell viability by compromising its structural or metabolic integrity and affecting its regenerative capacity [6]. Toxic agents can inhibit cell proliferation, which can be assessed using assays such as BrdU (bromodeoxyuridine) incorporation, EdU (5-ethynyl-2′-deoxyuridine) incorporation, or Ki-67 staining. A simple approach to measure cell proliferation involves comparing cell counts in cultures exposed to test material extracts for varying durations with control cultures. This method typically requires trypsinizing the cell culture and counting individual cells with a microscope or electronic cell counter [7]. Evaluating the protein content of cell cultures is a practical test for assessing the toxicity of biomaterials [8]. Cell proliferation assays can be used to determine the optimal concentration of a drug or compound for further testing.

2.3 Cytotoxicity

Cytotoxicity is a measure of cell damage or death caused by toxic agents. Cytotoxicity assays focus on detecting the extent of cell damage or death caused by toxic agents. These assays provide information on the direct effects of a substance on cells, which can be useful for identifying potential therapeutic targets or understanding the mechanisms underlying toxicity. Common cytotoxicity assays include LDH (lactate dehydrogenase) release assay, which measures the release of LDH from damaged cells, and the trypan blue dye exclusion assay, which measures membrane integrity. Cytotoxicity assays can also be used to determine the selectivity of a drug or compound for different cell types.

2.4 Apoptosis and necrosis

Apoptosis (programmed cell death) and necrosis (uncontrolled cell death) are distinct mechanisms of cell death that can be triggered by toxic agents.: Understanding the mechanisms of cell death induced by toxic agents is essential for developing targeted therapies and identifying potential side effects. Apoptosis and necrosis are distinct types of cell death, and differentiating between them can provide insights into the mode of action of a toxic agent and its potential therapeutic value. Assays such as annexin V/propidium iodide staining, TUNEL (terminal deoxynucleotidyl transferase dUTP nick end labeling) assay, and caspase activity assays can be used to assess apoptosis and necrosis.

2.5 Oxidative stress

Toxic agents can cause oxidative stress by inducing the production of reactive oxygen species (ROS) and/or impairing cellular antioxidant defenses. Assays such as DCFDA (2′,7′-dichlorofluorescin diacetate) fluorescence, lipid peroxidation assays, and glutathione assays can be used to evaluate oxidative stress. Evaluating oxidative stress is crucial for understanding how toxic agents affect cellular redox balance, which plays a vital role in maintaining cellular homeostasis. Oxidative stress can lead to cellular damage, dysfunction, and eventually cell death. Identifying agents that induce or prevent oxidative stress can help in the development of novel therapeutic strategies.

2.6 Genotoxicity

Genotoxic agents can cause DNA damage, which may lead to mutations, chromosomal aberrations, or DNA strand breaks. Genotoxicity assays assess the potential of a substance to damage DNA, which can lead to mutations, chromosomal aberrations, or other genomic changes. These assays can help identify potential carcinogens, mutagens, or teratogens and provide insights into the mechanisms of DNA damage and repair. Genotoxicity can be assessed using assays such as the comet assay (single-cell gel electrophoresis), micronucleus assay, and γ-H2AX (phosphorylated histone H2AX) staining.

2.7 Cellular morphology

Toxic agents can induce changes in cellular morphology, such as cell shrinkage, membrane blebbing, or cytoplasmic vacuolization. Examining cellular morphology provides a visual assessment of the effects of toxic agents on cell structure and organization. Changes in cellular morphology can indicate alterations in cellular functions, such as cell adhesion, migration, or differentiation, which can help elucidate the mechanisms underlying toxicity. These morphological changes can be visualized using light microscopy, phase-contrast microscopy, or fluorescence microscopy.

Studies examining morphological changes caused by cell adhesion to compatible and incompatible material surfaces have shown that human fibroblasts proliferate extensively on glass, but experience inhibition on hydrophobic biomaterials. Furthermore, cell rounding, detachment, and decreased proliferation have been observed [9]. This incompatibility of the biomaterial surface is known as intrinsic toxicity. The rounding of cells and other morphological alterations often occur before the loss of cell viability, which is accompanied by the disconnection of cells from the substrate [10]. Another characteristic of cell morphological changes is the increased vacuolation of the cytoplasm, often involving the formation of autophagosomes. Cytoplasmic vacuolation has been recognized as a dependable indicator of toxicity [11].

2.8 Protein synthesis and enzyme activity

Toxic agents can affect protein synthesis or the activity of specific enzymes, which can be assessed using assays such as western blotting, enzyme-linked immunosorbent assay (ELISA), or enzymatic activity assays. Assessing protein synthesis and enzyme activity can reveal how toxic agents affect specific cellular processes or signaling pathways. Identifying the proteins or enzymes affected by a toxic agent can help in understanding its mode of action and potential therapeutic applications.

2.9 Calcium signaling

Calcium signaling plays a crucial role in various cellular processes, and alterations in intracellular calcium levels can be indicative of toxic effects. Calcium signaling is involved in various cellular processes, such as cell division, migration, and apoptosis. Disruptions in calcium signaling can result in cellular dysfunction or death. Evaluating calcium signaling can provide insights into the effects of toxic agents on cellular communication and function. Calcium signaling can be assessed using fluorescent calcium indicators such as Fluo-4 or Fura-2.

2.10 Mitochondrial function

Toxic agents can affect mitochondrial function, leading to changes in mitochondrial membrane potential, respiration, or biogenesis. Mitochondria play a central role in cellular energy production and metabolism, and disruptions in mitochondrial function can have significant consequences for cellular health. Assessing mitochondrial function can reveal the potential effects of toxic agents on cellular bioenergetics and provide insights into the mechanisms of mitochondrial dysfunction and related diseases. Assays such as JC-1 staining (for membrane potential) and Seahorse XF analysis can be used to evaluate mitochondrial function.

3. Cell viability assessment

Cell viability assessment is an indispensable aspect of biocompatibility testing for medical devices and pharmaceutical agents. It plays a critical role in evaluating the safety and effectiveness of these products, ensuring that they do not cause undue harm to living cells when they come into contact with the human body. Cell viability is a fundamental parameter in the evaluation of the biocompatibility of medical devices and pharmaceutical agents. It is defined as the number of healthy, functioning cells in a sample and serves as an essential indicator of the impact of a particular substance on cellular health. By assessing cell viability, researchers can gain insights into the potential toxic effects of medical devices and pharmaceuticals, enabling them to optimize their design and formulation to minimize the risk of adverse effects. Moreover, cell viability assessment is crucial in various other aspects of biomedical research, such as understanding the mechanisms of action of specific genes, proteins, and signaling pathways involved in cell survival or death. Furthermore, it plays a significant role in the development of novel therapeutic strategies, as researchers must ensure that new drugs or treatments do not cause unacceptable levels of cell damage or death.

3.1 Applications of cell viability assessment

Cell viability assessment has a wide range of applications in biomedical research and drug development. Some key applications include:

- Evaluating the effect of drug candidates on cells: Cell viability assays can be used to determine the cytotoxic potential of new drug candidates, providing valuable information on their safety profile and aiding in the selection of promising candidates for further development.

- Optimizing cell culture conditions: Cell viability assessment can help researchers optimize the culture conditions for various cell types, ensuring that cells remain healthy and functional during in vitro experiments.

- Investigating the mechanisms of cell death: Cell viability assays can be employed to study the molecular mechanisms underlying cell death, providing insights into the biological processes involved in various diseases and enabling the development of targeted therapies.

- Screening for potential therapeutic agents: High-throughput cell viability assays can be used to screen large libraries of compounds for their potential to enhance cell survival or promote cell death, facilitating the identification of novel therapeutic agents.

3.2 Methods for determining cell viability

There are numerous methods available for determining cell viability, each with its advantages and limitations. Some of the most commonly used methods include:

Dilution: A simple technique in which the number of viable cells is estimated based on their ability to proliferate in a diluted environment. The dilution method involves serially diluting a cell suspension and then assessing cell growth in each dilution. Viable cells will continue to grow and divide, whereas non-viable cells will not proliferate. By comparing cell growth in each dilution, researchers can estimate the percentage of viable cells in the original sample. Although this method is simple and relatively easy to perform, it may not be suitable for all cell types or experimental conditions and may be less sensitive than other methods.

Surface viable count: This method involves counting the number of viable cells on a solid surface, such as a culture dish, after exposure to the test substance. The surface viable count method is based on the ability of viable cells to grow and form colonies on a solid surface, such as a culture dish or agar plate. After incubating the cells with the test substance, the viable cells are plated onto the surface and allowed to grow for a specified period. Researchers then count the number of colonies formed, which is proportional to the number of viable cells in the sample. This method is relatively simple and can provide accurate results, but it may not be suitable for non-adherent cell types or slow-growing cells.

Roll tube: A technique in which viable cells are embedded in a semi-solid agar medium and incubated for a specified period, allowing for the observation of cell growth and viability. In the roll tube method, viable cells are mixed with a semi-solid agar medium and poured into a glass tube, which is then rolled to create a thin layer of agar containing the cells. The tube is incubated, allowing viable cells to grow and form visible colonies within the agar. By counting the colonies, researchers can determine the number of viable cells in the sample. This method is useful for detecting slow-growing or fastidious cells but can be more labor-intensive and time-consuming than other methods.

Nalidixic acid: This method uses nalidixic acid to selectively inhibit the growth of nonviable cells, enabling the determination of viable cell counts. The nalidixic acid method involves selectively inhibiting the growth of nonviable cells by incorporating the antibiotic nalidixic acid into the culture medium. Viable cells will continue to grow in the presence of the antibiotic, whereas nonviable cells will not. By comparing cell growth in the presence and absence of nalidixic acid, researchers can estimate the percentage of viable cells in the sample. This method can be highly specific but may not be suitable for all cell types, as some cells may be resistant or sensitive to nalidixic acid.

Fluorogenic dye: Fluorescent dyes, such as calcein-AM or propidium iodide, can be used to stain live or dead cells, respectively, allowing for the quantification of viable cells using fluorescence microscopy or flow cytometry. Fluorogenic dyes are molecules that emit fluorescence when bound to specific cellular structures or molecules. For cell viability assessment, researchers often use two different dyes: one that selectively stains live cells and another that selectively stains dead cells. By measuring the fluorescence intensity of each dye in the sample, researchers can determine the proportion of live and dead cells. This method is highly sensitive and can provide rapid results but may be affected by factors such as dye penetration, cell autofluorescence, and photobleaching.

Trypan Blue cell viability assay: A widely used method that involves the use of the Trypan Blue dye, which selectively stains dead cells, allowing for the estimation of viable cell counts using a hemocytometer or automated cell counter. The Trypan Blue cell viability assay is a widely used and straightforward method for determining cell viability. The Trypan Blue dye selectively stains dead cells with compromised membrane integrity, while live cells with intact membranes remain unstained. After incubating the cells with the dye, researchers count the number of stained (dead) and unstained (viable) cells using a hemocytometer or automated cell counter. This method is relatively simple and quick, but it may not provide accurate results for certain cell types or experimental conditions, such as when cell membrane integrity is temporarily altered or when cell autofluorescence interferes with the detection of the dye.

The diverse methods available for cell viability assessment offer a range of options to evaluate the impact of various substances on cellular health. Each method has its advantages and limitations, and the choice of the most suitable method depends on factors such as the cell type, experimental conditions, and desired level of sensitivity and specificity.

4. Cytotoxicity assessment

Cytotoxicity assessment is an essential component of the evaluation process for pharmaceutical agents, medical devices, and other substances that may come into contact with living cells or tissues. It involves the study of the potential harmful effects of these substances on cells, including cell damage and cell death. Cytotoxicity is the degree to which a substance can cause damage to cells. Assessing cytotoxicity is crucial for ensuring the safety and effectiveness of medical devices and pharmaceutical agents, as well as other substances that may come into contact with living cells or tissues. Understanding the cytotoxic effects of these substances helps researchers identify potential hazards and optimize the design and formulation of products to minimize the risk of adverse effects on the human body.

Moreover, cytotoxicity assessment is a valuable tool in various other aspects of biomedical research, such as the investigation of the mechanisms of cell death and the identification of novel therapeutic targets. By studying the cytotoxic effects of specific compounds or treatments, researchers can gain insights into the biological processes involved in cell damage and death, which may contribute to the development of new therapeutic strategies for a wide range of diseases and conditions.

4.1 Applications of cytotoxicity assessment

Cytotoxicity assessment has a wide range of applications in biomedical research and drug development. Some key applications include:

Evaluating the safety of medical devices and pharmaceutical agents: Cytotoxicity assays can be used to determine the potential harmful effects of medical devices and pharmaceuticals on living cells, ensuring that these products do not cause unacceptable levels of cell damage or death.

Screening for potential therapeutic agents: High-throughput cytotoxicity assays can be used to screen large libraries of compounds for their ability to selectively kill cancer cells or other target cell populations, facilitating the identification of novel therapeutic agents.

Investigating the mechanisms of cell death: By studying the cytotoxic effects of specific compounds or treatments, researchers can gain insights into the biological processes and signaling pathways involved in cell damage and death. This knowledge can contribute to a deeper understanding of the mechanisms underlying various diseases and conditions, as well as the development of new therapeutic strategies.

Evaluating the efficacy of therapeutic interventions: Cytotoxicity assessment can be used to measure the effectiveness of therapeutic interventions, such as chemotherapy or radiation therapy, in inducing cell death in target cell populations. This information is crucial for optimizing treatment regimens and developing more effective therapeutic strategies.

Assessing the potential toxicity of environmental contaminants: Cytotoxicity assays can be employed to evaluate the potential harmful effects of environmental contaminants, such as pollutants, pesticides, and industrial chemicals, on living cells. This information is vital for understanding the risks associated with exposure to these substances and developing strategies to minimize their impact on human health and the environment.

Cytotoxicity assessment is an indispensable tool in the evaluation of medical devices, pharmaceutical agents, and other substances that may come into contact with living cells or tissues. The various methods available for measuring cytotoxicity offer researchers a range of options for assessing the potential harmful effects of these substances on cells and for investigating the mechanisms of cell damage and death.

4.2 Methods for measuring cytotoxicity

There are numerous methods available for measuring cytotoxicity, each with its advantages and limitations. Some of the most commonly used methods include:

1. *MTT assay:* The MTT assay is a colorimetric method that measures the reduction of a yellow tetrazolium salt (MTT) to a purple formazan product by metabolically active cells. The amount of formazan produced is proportional to the number of viable cells, providing an indirect measure of cytotoxicity.

2. *LDH release assay:* Lactate dehydrogenase (LDH) is a cytoplasmic enzyme that is released into the culture medium when cells undergo damage or death. The LDH release assay measures the activity of LDH in the culture medium as an indicator of cytotoxicity.

3. *ATP assay:* The ATP assay quantifies the amount of intracellular adenosine triphosphate (ATP) as a measure of cell viability. A decrease in ATP levels can indicate cytotoxicity, as damaged or dead cells typically have reduced ATP content.

4. *Annexin V/propidium iodide staining:* This method uses fluorescent dyes to stain cells undergoing apoptosis (annexin V) or with compromised membrane integrity (propidium iodide). By analyzing the fluorescence intensity of each dye, researchers can determine the proportion of apoptotic and necrotic cells, providing insights into the cytotoxic effects of a test substance.

5. *Colony formation assay:* The colony formation assay measures the ability of single cells to grow and form colonies in the presence of a test substance. A reduction in the number of colonies formed can indicate cytotoxicity.

4.3 MTT assay

The MTT assay, also known as the 3-(4,5-dimethylthiazol-2-yl)-2,5-diphenyltetrazolium bromide assay, is a widely used colorimetric method for assessing cell viability and cytotoxicity [12]. This assay measures cellular metabolic changes using colorimetric shifts. It is based on the conversion of the purple tetrazolium dye MTT into insoluble formazan by the nicotinamide adenine dinucleotide phosphate (NADPH)-dependent cellular oxidoreductase enzymes. The reductive activity occurring in the mitochondria of living cells is employed to assess cell viability [6]. The MTT assay quantifies live cells by gauging mitochondrial activity, as it correlates with the number of formazan crystals [13]. Despite being the gold standard for cytotoxicity testing, the conversion to formazan crystals is influenced by various factors like metabolic rate and the number of mitochondria [14].

The MTT assay is based on the idea that proliferating cells exhibit a higher rate of MTT conversion, while nonviable or slow-growing cells have reduced metabolism and lower MTT reduction levels. After MTT application, formazan crystals are dissolved in a solution containing dimethyl sulfoxide or sodium dodecyl sulfate. Formazan concentrations can be measured using a spectrophotometer between 540 and 720 nm [15]. This method can provide an accurate dose-response curve for small cell numbers, test multiple parameters simultaneously, and is straightforward and highly replicable. The MTT assay is primarily used for in vitro testing of cytotoxic effects of various novel drugs at different concentrations and evaluating drug resistance in cell lines. It also assesses in vitro drug effects and their potential clinical applications. Due to its simplicity, the MTT assay is widely used to determine the toxicities of polymers, alloys, and ceramics [16]. However, the assay does not differentiate between cytostatic and cytotoxic effects, and results may be inaccurate if the cell population is low [17, 18]. Additionally, the MTT test is cell-specific and requires solubilization. Although highly sensitive, it works only with adherent cell targets. Since all cells must be killed during the protocol, this assay cannot be used for follow-up studies.

The MTT assay is relatively easy to perform, requiring only the addition of the MTT reagent to cell cultures, incubation, and subsequent solubilization and quantification of the formazan product. Non-destructive: As the MTT assay measures cell viability indirectly through the reduction of the MTT reagent, it does not require the destruction of cells or the use of invasive techniques.

4.4 AlamarBlue assay

The AlamarBlue assay offers a straightforward and dependable approach to measuring cell viability. It employs a fluorometric technique to detect cellular metabolic activity. Mitochondrial enzymes with diaphorase activity, such as NADPH dehydrogenase, reduce resazurin (oxidized form; 7-hydroxy-3H-phenoxazin-3-1-10-oxide) to resorufin (reduced form) [19]. The AlamarBlue assay has been utilized to examine trophoblast viability, migration, and invasion [20].

Cell viability can be assessed in 96-well plates after exposure to the biomaterial being studied. This test offers several advantages [19]. It is a straightforward method that uses a water-soluble substance, applicable to both suspended and attached cells, and features a fluorometric and colorimetric growth indicator [21]. Furthermore, the reagents are harmless to both cells and technicians. This test eliminates the necessity for washing and extraction steps, allowing for easy differentiation of endothelial cell viability and cell concentrations. It is a cost-effective test that enables continuous monitoring of endothelial cell metabolism and viability [22]. However, the reduction process may reverse with high cell numbers and extended culture times. One limitation of this assay is that it is not a direct cell counting technique. The assay relies on metabolic pathways that can be influenced by various factors, such as individual cell-reducing capacity and agents that affect mitochondrial activity or directly reduce resazurin [23].

4.5 LDH (lactate dehydrogenase) release assay

The LDH (lactate dehydrogenase) release assay is a widely used method for evaluating cell viability, cytotoxicity, and membrane integrity in various cell types. LDH is an intracellular enzyme that is released into the extracellular environment upon cell membrane damage or cell lysis. Lactate dehydrogenase release assay is a two-step rapid colorimetric test to assess the quantification of cell numbers in vitro [24]. The principle of the LDH release assay is based on the conversion of lactate to pyruvate by LDH in the presence of a cofactor, NAD+ (nicotinamide adenine dinucleotide). During this process, NAD+ is reduced to NADH, which then reacts with a specific tetrazolium salt, producing a colored formazan product. The absorbance of the formazan product can be measured using a spectrophotometer or microplate reader, with the intensity of the color directly proportional to the amount of LDH released and, consequently, the number of damaged or non-viable cells. Lactate dehydrogenase (LDH) is a stable cytoplasmic enzyme in every cell. Cytotoxicity is assessed by the activity of cytoplasmic enzymes released by damaged cells [25]. LDH is released into the cell culture when there is apoptosis, necrosis, or other cellular destruction in the membrane [26]. This test can detect the cytotoxic effects of various agents or environmental factors [27]. In the first stage, LDH catalyzes the conversion of lactate to pyruvate by reducing NAD+ to NADH. Following this, diasphorase enzymes reduce the tetrazolium salt to a red formazan in the presence of NADH.

Colorimetric lactate dehydrogenase (LDH) assay has also been used for the evaluation of antiviral activity against bovine viral diarrhea virus in vitro. Using the NADH produced during the conversion of lactate to pyruvate to reduce a second compound in a coupled reaction into a product with easily quantifiable properties makes it simple to quantify LDH activity. In this assay, the reduction of a yellow tetrazolium salt, INT, by NADH into a red, water-soluble dye of the formazan class is measured using absorbance at 492 nm. The amount of formazan is proportional to the amount of LDH in the culture, which is proportional to the number of dead or damaged cells. The advantages are that LDH assay reflects the membrane integrity, and the reagent does not damage viable cells. The drawback is that it is not super sensitive. Despite these advantages, there are some limitations to the LDH release assay.

4.6 MTS assay

The MTS assay is utilized to evaluate cell proliferation, cell viability, and cytotoxicity. It can determine cell viability after exposure to various cytokines, growth hormones, cytotoxic drugs, and anticancer agents [28]. The MTS assay can also be used to assess the effects of chemical and physical treatments on the biocompatibility of human bone and tendon tissues for clinical applications [12]. The test's principle is that a colored formazan dye is generated when MTS tetrazolium molecules are reduced by live mammalian cells and other species' cells.

Living cells' mitochondrial reductase enzymes convert MTS to formazan crystals in the presence of phenazine methosulfate, an electron-coupling agent. These reduced formazan crystals are water-soluble, eliminating the need for an additional solution or washing step to dissolve them in the cell culture medium [29]. A spectrophotometer can measure these formazan crystals at 490–500 nm. The MTS reagent solution has better storage stability compared to MTT or XTT molecules. One advantage of tetrazolium assays that yield a water-soluble formazan is the ability to periodically measure absorbance from the test plates during the initial incubation stages. Multiple readings may be useful during assay development, but it is crucial to keep the plates in the incubator between readings to maintain a relatively constant environment [30].

4.7 XTT assay

The XTT (2,3-bis-(2- methoxy-4- nitro-5-sulfophenyl) -2H-tetrazolium –5-carboxanilide) assay is similar to the MTT assay but relies on the reduction of the XTT reagent to a soluble orange formazan product. This allows for a simpler and faster assay, as the formazan product can be directly quantified in the cell culture medium without the need for solubilization. The colorimetric change indicates cell viability, proliferation, and cytotoxicity through a nonradioactive test [31]. The biomaterial to be evaluated is placed in 96-well microplates and an adherent or suspension cell culture. Metabolically active cells reduce yellow tetrazolium salt (sodium 3′-[1- (phenylaminocarbonyl)- 3,4- tetrazolium]-bis (4-methoxy6-nitro) benzene sulfonic acid hydrate or XTT) into an orange formazan dye.

A scanning multiwell spectrophotometer (ELISA reader) measures formazan dye. XTT assay can assess cell proliferation when exposed to growth factors, cytokines, and nutrients. It can measure the increase in the overall activity of mitochondrial dehydrogenases that corresponds to the increase in the number of living cells and the amount of orange formazan formed. Cytotoxicity can also be measured using XTT assay by measuring the cytotoxic or growth-inhibiting agents such as inhibitory antibodies [32]. This assay is compatible with resorbable and nonresorbable guided tissue regeneration membranes in cultures of primary human periodontal ligament fibroblasts and human osteoblast-like cells [33]. Unlike MTT, where cells must be lysed to solubilize the formazan salt before absorbance measurement, XTT does not require cell lysis. This allows easier monitoring of the same samples at different time intervals [32].

4.8 WST-1 assay

The WST-1 assay (Water-Soluble Tetrazolium Salt-1) is a colorimetric method used to evaluate cell viability, cytotoxicity, and proliferation. It is based on the reduction of a tetrazolium salt, WST-1, to formazan by cellular dehydrogenase enzymes present in metabolically active cells. The formazan dye produced is directly proportional to the number of viable cells, and its absorbance can be measured using a spectrophotometer [34]. WST-1 assay is an improved version of the MTT assay and offers several advantages. Unlike the MTT assay, which requires solubilization of the formazan crystals in a separate step, the WST-1 assay generates a water-soluble formazan product. It is water-soluble, quick and sensitive [35]. This feature simplifies the experimental procedure and allows for the absorbance measurement without the need for additional steps or cell lysis. As a result, it is possible to measure cell viability more quickly and monitor the same cell samples at multiple time points.

The WST-1 assay is widely used in various applications, such as assessing the cytotoxic effects of drugs, chemicals, or nanoparticles on cell lines, screening for potential anticancer agents, and testing the biocompatibility of biomaterials. The assay is performed by adding the WST-1 reagent to cell cultures in a 96-well plate format, followed by incubation for a specific period, usually ranging from 1 to 4 hours. The absorbance of the formazan dye produced is then measured at a wavelength of around 450 nm using a microplate reader. One of the limitations of the WST-1 assay is its sensitivity to environmental factors and culture conditions, such as pH and serum concentration [35]. WST-1 assay is unsuitable for assessing the cell toxicity of Mn-containing materials in vitro [35]. It is essential to optimize the experimental conditions and maintain a consistent environment during the assay to obtain reliable and accurate results. Additionally, the WST-1 assay measures metabolic activity rather than directly counting the number of viable cells, so it may not always accurately reflect the true cell viability, especially in cases where metabolic activity is altered by the experimental treatments. Despite these limitations, the WST-1 assay remains a popular and convenient method for assessing cell viability, cytotoxicity, and proliferation due to its simplicity, speed, and compatibility with various cell types and experimental conditions.

4.9 ATP assay

The ATP assay is a sensitive and reliable method for evaluating cell viability, metabolic activity, and cytotoxicity in various fields of biomedical research. It is based on the quantification of adenosine triphosphate (ATP), the primary energy currency of living cells, which serves as an indicator of cell health and metabolic activity. The ATP assay is centered on the detection and quantification of intracellular ATP levels, which directly correlate with the number of viable, metabolically active cells in a given sample. The most common method of ATP quantification involves the use of a bioluminescent enzyme, luciferase, which catalyzes the oxidation of luciferin in the presence of ATP, producing light as a byproduct. The emitted light is then measured using a luminometer, with the intensity of the luminescent signal being proportional to the ATP concentration and, consequently, the number of viable cells [36].

The ATP assay is highly sensitive, capable of detecting even small changes in ATP levels and cell viability, making it suitable for assessing the effects of various substances on cellular metabolism and health. The ATP assay provides rapid results, often within minutes, allowing researchers to quickly assess cell viability and metabolic activity in response to various experimental conditions. The ATP assay is a powerful tool for assessing cell viability and metabolic activity in a wide range of applications, from drug discovery and toxicology studies to basic cell biology research. Its sensitivity, speed, and adaptability make it a popular choice among researchers in various fields of biomedical science. However, the limitations of the ATP assay, such as assay interference and extracellular ATP contamination, should be carefully considered when interpreting results and selecting the most appropriate method for a specific research question or experimental condition. The evaluation of biomaterials' cytocompatibility can be performed through a variety of cytotoxicity tests, as demonstrated in **Table 1**.

Assay	Merits	Demerits
MTT Assay	Considered as the gold standard for cytotoxicity testing	The conversion to formazan crystals depends on metabolic rate and number of mitochondria resulting in many known interferences
AlamarBlue Assay	This test eliminates the need for washing and extraction steps. It is a simple method based on a water-soluble substance. It can be used for both suspended and attached cells.	This is not a direct cell counting technique. It relies on metabolic pathways that can be affected by various factors.
LDH release Assay	LDH assay reflects the membrane integrity, and the reagent does not damage viable cells.	This assay lacks sensitivity.
MTS Assay	Superior storage stability of MTS reagent solution. Water-soluble reduced formazan crystals eliminate the need for the washing step.	The testing environment should be kept relatively constant for consistent results
XTT Assay	Measure the increase in the overall activity of mitochondrial dehydrogenases that corresponds to the increase in the number of living cells. Does not require cell lysis.	Time-consuming and not very sensitive
WST-1 Assay	It is water-soluble, quick, and sensitive	Particles such as carbon nanotubes and magnesium particles can interfere with the results
ATP Assay	Highly sensitive, as it measures the adenosine triphosphate (ATP) content in cells, which correlates well with the number of metabolically active and viable cells	ATP assay is vulnerable to interference from certain factors, such as the presence of extracellular ATP, ATP-degrading enzymes, or substances that affect the bioluminescent reaction.
Sulforhodamine B Assay	Simple, rapid, and cost-effective method for measuring cell viability and cytotoxicity based on the cellular protein content.	It measures cellular protein content rather than directly assessing cell viability or metabolic activity and the changes in protein content due to factors other than cytotoxicity could affect the assay results.
Neutral Red Assay	Cheaper and more sensitive than many other cytotoxicity tests. Does not require unstable reagents like in the case of tests using tetrazolium salts.	Once started, it must be completed in less than 3 hours. The accuracy of the absorbance readings of this assay is affected by the visible needle-like crystals precipitates of the neutral red dye.
Trypan blue dye exclusion Assay	It is a simple and rapid technique dye exclusion test	The viability is indirectly accessed based on the cell membrane integrity. Very time-consuming.

Assay	Merits	Demerits
GSH Assay	The assay's sensitivity is increased by the enzyme glutathione reductase, which causes enzymatic recycling of GSH.	High polarity and limited stability
Protease viability marker Assay	Highly specific for measuring cell viability, as they detect the activity of specific proteases that are released from cells when they undergo necrosis or apoptosis.	Have a limited dynamic range, which means that they may not be able to detect subtle changes in cell viability.
Clonogenic cell survival Assay	Highly sensitive, as they can detect small changes in cell survival and growth.	Time-consuming, as they require cells to grow and form colonies over a period of several days or weeks.
DNA synthesis cell proliferation assays (e.g., BrdU or EdU)	Highly specific for measuring cell proliferation, as they directly measure DNA replication, which is a key component of cell division.	DNA synthesis cell proliferation assays can lack precision, as they only measure one aspect of cell proliferation.
Agar Diffusion Assay	Cost-effective and straightforward cytotoxicity screening test	It cannot be used for biomaterials that do not dissolve through agar.
Raman Micro-Spectroscopy	High sensitivity; Raman micro-spectroscopy can provide quantitative data at a molecular level.	Raman micro-spectroscopy can be a complex technique to perform and analyze, requiring specialized equipment and expertise.

Table 1.
Cytotoxicity assays as part of the cytocompatibility assessment of biomaterials.

4.10 Sulforhodamine B assay

The Sulforhodamine B (SRB) assay is a colorimetric method used for measuring cell viability, proliferation, and cytotoxicity. It has been widely employed to evaluate the efficacy of anticancer agents and to determine the cytotoxic effects of various substances on cell cultures [37]. The SRB assay is based on the ability of the protein-binding dye, sulforhodamine B, to interact with the basic amino acid residues of cellular proteins under mild acidic conditions. Upon fixation, the dye binds to the proteins in the cells, and the amount of bound dye is proportional to the cell mass or protein content [38]. One of the advantages of the SRB assay is its sensitivity and accuracy, as well as its low cost compared to other cell viability assays. Additionally, the SRB assay is relatively simple and can be performed in a high-throughput manner, making it suitable for large-scale screening studies. Furthermore, the SRB assay is compatible with various cell types, including adherent and suspension cells, and does not require cell lysis or radioactive reagents [39].

However, the SRB assay has some limitations. For instance, the assay may not be suitable for measuring cell viability in certain situations, such as when cells produce high amounts of extracellular matrix, which can interfere with the dye binding to intracellular proteins. Additionally, the SRB assay is not a direct measure of cell number, and it relies on the assumption that protein content is proportional to the number of viable cells. Therefore, factors that affect protein synthesis or degradation may influence the results of the assay [39].

4.11 Neutral red assay

The neutral red dye easily penetrates nonionic cell membranes and accumulates in lysosomes. The structural integrity of these lysosomes serves as an indicator of cell viability, which is the foundation of the neutral red uptake assay. This method can quantitatively measure live cells [40]. The neutral red assay technique is based on the degree of absorption and binding of the dye by living cells [41]. This cell viability assay aids in the in vitro evaluation of biomaterials. The underlying principle is that dying cells, due to altered membrane properties, can no longer take up neutral red. As a result, living cells can be distinguished from dead or dying cells based on differences in neutral red uptake. After being exposed to the dye for 3 hours, the cells are briefly rinsed with a phosphate buffer solution. Cells are seeded on a 96-well plate and then exposed to test material or control substances in a nutritive cell culture medium for 24 hours. Live cells take up neutral red into their lysosomes after 24 hours of exposure, but as cells begin to die, their ability to incorporate neutral red decreases. The cells are subsequently treated with an acidified ethanol solution to release the incorporated dye. Neutral red that has been released is measured at 540 nm and correlated to the number of viable cells. The viability of unexposed cells measured at 540 nm is set at 100%. The lysosomal capacity for dye incorporation, the foundation of the neutral red dye assay, can be employed to differentiate between living, injured, and dead cells. Viability curves can be generated based on absorption data to determine the concentration of the test chemical needed to inhibit neutral red dye uptake by 50%. Lysosomal swelling agents have been demonstrated to cause an increase in neutral red uptake, potentially leading to an underestimation of cytotoxicity [42]. However, the neutral red assay can produce false-positive or false-negative results. The neutral red assay is more affordable and sensitive than many other cytotoxicity tests [43]. However, being a sensitive test, it must be completed immediately once initiated, usually within 3 hours after the cells have been treated with the dye [41]. This assay does not require unstable reagents like tests using tetrazolium salts. Another limitation is that the absorbance readings' accuracy is affected by the visible needle-like crystal precipitates of the neutral red dye [44].

4.12 Trypan blue dye exclusion assay

The trypan blue assay is a dye exclusion test that provides a straightforward and quick method for assessing cell viability. It is based on the principle that live cells possess intact cell membranes, which exclude the Trypan Blue dye, while dead or damaged cells take up the dye due to compromised membrane integrity. Viable cells have intact cell membranes, allowing them to exclude certain dyes (such as trypan blue, Eosin, and propidium), while dead cells cannot [45]. In this test, a cell suspension is mixed with trypan blue dye. The dye's absorption or exclusion is then visually examined, as viable cells will display clear cytoplasm, while nonviable cells will exhibit blue cytoplasm. A significant drawback of this technique is that it indirectly assesses viability based on cell membrane integrity. It is possible for a cell to be nonviable while still having an intact membrane. Conversely, cells with compromised membranes might recover and become fully viable. Another limitation is the subjective evaluation of dye uptake, which may cause small amounts of dye uptake to go undetected, potentially indicating cell damage.

One solution to this issue is to evaluate dye exclusion using a fluorescent dye and a fluorescence microscope instead of using trypan blue with a transmission microscope. However, determining dye uptake and cell viability using the cell's light scatter properties can be quite complex. A notable limitation of this method is that it is time-consuming, although some protocols show that the trypan blue exclusion assay can be performed in under 10 minutes [46].

4.13 GSH assay

In human cells, the majority of glutathione (90–95%) is present in its reduced form (GSH). It plays a role in numerous regulatory processes, such as signal transduction, gene expression, DNA and protein synthesis, proteolysis, cell growth and apoptosis, cytokine and immune responses, protein glutathionylation, and the maintenance of mitochondrial function and integrity [47]. The glutathione assay is a colorimetric test that identifies alterations in GSH and GSSG levels during oxidative stress [48] using the enzymatic recycling technique with glutathione reductase and Ellman's reagent. This assay can measure reduced glutathione (GSH), oxidized glutathione (GSSG), total glutathione (GSH + GSSG) concentrations, and their ratio in various samples, including blood, plasma, serum, cultured cells, and tissues.

The glutathione reductase enzyme converts GSSG to GSH, generating a yellow chromophore that can be detected spectroscopically at 415 nm. Consequently, the concentration in an unknown sample is determined by evaluating the absorbance at 415 nm and comparing it to the standard curve for GSSG. This curve is plotted each time glutathione quantification is performed. The assay's sensitivity is enhanced by the enzyme glutathione reductase, which facilitates the enzymatic recycling of GSH. However, some drawbacks of the glutathione assay include its high polarity, limited stability, and the aliphatic structure of the assay [49].

4.14 Protease viability marker assay

The Protease Viability Marker Assay is a fluorescence-based method employed for assessing cell viability, cytotoxicity, and proliferation. This assay takes advantage of the presence of intracellular proteases, which are released from cells upon loss of membrane integrity, as a marker for cell viability [50]. These proteases specifically cleave nonfluorescent substrates, such as the commercially available Calcein-AM or CellEvent Caspase-3/7 Green Detection Reagent, to generate a highly fluorescent product, which can be detected using a fluorescence plate reader or a fluorescence microscope [32, 51].

One of the main advantages of the Protease Viability Marker Assay is its high sensitivity and specificity, as the fluorescent signal is only generated when the substrate is cleaved by the intracellular proteases, ensuring minimal background fluorescence. Additionally, the assay is nontoxic to the cells, allowing for real-time monitoring of cell viability over time and facilitating the assessment of cellular responses to various treatments or conditions [52]. However, there are some limitations to the Protease Viability Marker Assay. The assay may not be suitable for all cell types or conditions, as the presence and activity of intracellular proteases can vary depending on the cell type, culture conditions, or experimental treatments. Moreover, the fluorescent signal generated by the cleaved substrate may not be directly proportional to the number of viable cells, as the protease activity can be affected by various factors, such as cell density, cell size, or cell cycle stage. The Protease Viability Marker Assay can be combined with other assays to obtain more comprehensive information about cell viability, cytotoxicity, and the mechanisms underlying cellular responses to various stimuli. For example, the assay can be used alongside assays that measure apoptosis, necrosis, or autophagy to provide a more complete picture of the cellular response to a test substance. By integrating the Protease Viability Marker Assay with complementary methods, researchers can gain deeper insights into the complex biological processes that govern cell survival and death in response to various stimuli.

4.15 Clonogenic cell survival assay

The clonogenic cell survival assay, also known as colony formation assay, is a widely used method for evaluating the ability of cells to survive and proliferate following exposure to various stressors, such as radiation, chemotherapeutic agents, or other cytotoxic substances [53]. This assay is based on the principle that a single cell can give rise to a colony of cells, which can be counted and analyzed to determine the proportion of surviving cells with the ability to form colonies [53].

The assay involves seeding cells at a low density in culture dishes, followed by treatment with the agent of interest. After a suitable incubation period, typically 1–3 weeks, the cells are fixed, stained, and the number of colonies containing at least 50 cells is counted. The surviving fraction is calculated by comparing the colony formation efficiency of treated cells with that of untreated control cells. The clonogenic cell survival assay has several advantages. It provides a direct measure of the reproductive capacity of cells, allowing for the assessment of treatment-induced cytotoxicity at the level of individual cells. Additionally, the assay is highly sensitive and can detect changes in cell survival across a wide range of treatment doses [54]. There are also some limitations to the clonogenic cell survival assay. The assay can be time-consuming and labor-intensive, as it requires a long incubation period for colony formation and manual counting of the colonies. Additionally, the assay may not be suitable for all cell types, particularly non-adherent or slow-growing cells, which may not form distinct colonies under the experimental conditions [55].

4.16 DNA synthesis cell proliferation assays

DNA synthesis cell proliferation assays are a group of methods used to assess cell proliferation by measuring the incorporation of nucleotide analogs into newly synthesized DNA during the S phase of the cell cycle. These assays are valuable for studying the effects of various stimuli, such as growth factors and cytotoxic agents, on cell growth and division. One commonly used DNA synthesis cell proliferation assay is the 5-bromo-2′-deoxyuridine (BrdU) assay. BrdU is a thymidine analog that gets incorporated into newly synthesized DNA during the S phase of the cell cycle. After incorporation, the BrdU-containing DNA can be detected using specific antibodies, allowing for the quantification of proliferating cells. The BrdU assay has been used in various applications, including drug screening and evaluation of the cytotoxic effects of anticancer agents [56]. Another DNA synthesis cell proliferation assay is the 3H-thymidine incorporation assay. This assay involves the incorporation of radioactive tritiated thymidine (3H-thymidine) into newly synthesized DNA. The amount of 3H-thymidine incorporated into the DNA can be quantified using a scintillation counter, providing a measure of cell proliferation. The 3H-thymidine incorporation assay has been widely used in studies investigating cell proliferation in response to growth factors, cytokines, and other signaling molecules [57]. DNA synthesis cell proliferation assays offer several advantages. They provide a direct measurement of

DNA synthesis, reflecting cell proliferation rates, and they can be applied to a variety of cell types, including adherent and suspension cells. However, these assays also have some limitations, such as the potential for false-positive or false-negative results due to nonspecific incorporation of the nucleotide analogs and the need for proper controls to account for variations in DNA synthesis rates.

4.17 AGAR diffusion assay

The agar diffusion test is a cytotoxicity barrier testing method in which the test material is simply placed on an agar layer covering a monolayer cell culture that simulates the mucosal membrane [58]. In this technique, cells are incubated for 24 hours before evaluating cytotoxicity. They are stained with neutral red dye to identify viable cells, stressed, and lysed [59]. Toxicity is determined by the loss of viable cells around the test material, which appears as an unstained area under and around the material being tested. In cases of high concentration and cytotoxicity of the diffusing substance, a loss of dye within the cells may be observed as the leachable toxic substance causes cell lysis [60]. Agar diffusion assays can evaluate the nonspecific cytotoxicity of the tested material's leachable components.

This test has the advantage of being cost-effective and simple to perform as a cytotoxicity screening tool. However, one limitation of this test is that potentially cytotoxic leachates in a solid state may stick to the agar rather than spreading across the plate, resulting in cells being only partially exposed to the test substance. Additionally, this test can only be used for materials that diffuse through the agar covering the cell monolayer. If materials do not dissolve in or spread through the agar, they will not show toxicity by damaging cells. Nonetheless, such materials could still be cytotoxic in a clinical setting [58].

4.18 Raman micro-spectroscopy

Raman micro-spectroscopy is a nondestructive and label-free optical technique that combines Raman spectroscopy with microscopy to provide detailed information about the molecular composition, structure, and interactions within a sample. This technique relies on the inelastic scattering of light, also known as Raman scattering, which occurs when light interacts with molecular vibrations in a sample [61]. Raman micro-spectroscopy has been widely used in various fields, including materials science, biology, and medicine. In materials science, it is used to study the crystallographic structures, stress distributions, and phase transitions of materials [62]. In biology and medicine, Raman micro-spectroscopy has been applied to study cellular processes, molecular interactions, and disease diagnosis.

One of the significant advantages of Raman micro-spectroscopy is its ability to obtain high-resolution spatial information about the sample's molecular composition without the need for labeling or sample preparation. This allows for real-time, in situ analysis of living cells, tissues, and biomaterials. Additionally, the technique is sensitive to both chemical and structural information, enabling the identification and differentiation of various molecular species within the sample [63]. However, Raman micro-spectroscopy also has some limitations. The most significant challenge is its inherently weak signal, which often requires long acquisition times or high laser power, which can cause sample damage or photobleaching. Recent advancements in instrumentation, such as the development of near-infrared lasers and highly sensitive detectors, have significantly improved the signal-to-noise ratio and reduced the acquisition time.

5. Conclusions

Cytotoxicity screening assays provide a measure of cell death caused by biomaterials. The term cytotoxicity describes the cascade of molecular events that interfere with the macromolecular synthesis, causing specific cellular, functional, and structural damage. Different screening assays are available, and it is crucial to understand the advantages and limitations of the various available assays so that they can be selected for appropriateness and interpreted accurately.

Conflict of interest

The authors declare no conflict of interest.

Author details

Anil Sukumaran[1,2]*, Vishnupriya K. Sweety[3], Biba Vikas[4] and Betsy Joseph[3,5]

1 Department of Dentistry, Hamad Medical Corporation, Oral Health Institute, Doha, Qatar

2 College of Dental Medicine, Qatar University, Doha, Qatar

3 Pushpagiri Institute of Medical Sciences and Research Centre, Medicity, Tiruvalla, Kerala, India

4 Biomedical Research Centre, Tuskegee University, Tuskegee Alabama, USA

5 Saveetha Dental College and Hospital, Saveetha Institute of Medical and Technical Sciences, Chennai, India

*Address all correspondence to: drsanil@gmail.com and asukumaran1@hamad.qa

References

[1] Williams DF. On the mechanisms of biocompatibility. Biomaterials. 2008;**29**(20):2941-2953

[2] Schweikl H, Schmalz G. Toxicity parameters for cytotoxicity testing of dental materials in two different mammalian cell lines. European Journal of Oral Sciences. 1996;**104**(3):292-299

[3] Ciapetti G, Stea S, Cenni E, Sudanese A, Marraro D, Toni A, et al. Cytotoxicity testing of cyanoacrylates using direct contact assay on cell cultures. Biomaterials. 1994;**15**(1):63-67

[4] Groth T, Falck P, Miethke R-R. Cytotoxicity of biomaterials — Basic mechanisms and in vitro test methods: A review. Alternatives to Laboratory Animals. 2020;**23**(6):790-799

[5] Mosmann T. Rapid colorimetric assay for cellular growth and survival: Application to proliferation and cytotoxicity assays. Journal of Immunological Methods. 1983;**65**(1-2):55-63

[6] Adan A, Kiraz Y, Baran Y. Cell proliferation and cytotoxicity assays. Current Pharmaceutical Biotechnology. 2016;**17**(14):1213-1221

[7] Piccinini F, Tesei A, Arienti C, Bevilacqua A. Cell counting and viability assessment of 2D and 3D cell cultures: Expected reliability of the trypan blue assay. Biological Procedures Online. 2017;**19**:8

[8] Huzum B, Puha B, Necoara RM, Gheorghevici S, Puha G, Filip A, et al. Biocompatibility assessment of biomaterials used in orthopedic devices: An overview (review). Experimental and Therapeutic Medicine. 2021;**22**(5):1315

[9] Groth T, Altankov G. Fibroblast spreading and proliferation on hydrophilic and hydrophobic surfaces is related to tyrosine phosphorylation in focal contacts. Journal of Biomaterials Science. Polymer Edition. 1995;**7**(3):297-305

[10] Kirkpatrick CJ, Mittermayer C. Theoretical and practical aspects of testing potential biomaterials in vitro. Journal of Materials Science: Materials in Medicine. 1990;**1**(1):9-13

[11] Trabbic CJ, Dietsch HM, Alexander EM, Nagy PI, Robinson MW, Overmeyer JH, et al. Differential induction of cytoplasmic vacuolization and Methuosis by novel 2-Indolyl-substituted Pyridinylpropenones. ACS Medicinal Chemistry Letters. 2014;**5**(1):73-77

[12] Fawzi-Grancher S, Goebbels RM, Bigare E, Cornu O, Gianello P, Delloye C, et al. Human tissue allograft processing: Impact on in vitro and in vivo biocompatibility. Journal of Materials Science Materials in Medicine. 2009;**20**(8):1709-1720

[13] van Meerloo J, Kaspers GJ, Cloos J. Cell sensitivity assays: The MTT assay. Methods in Molecular Biology. 2011;**731**:237-245

[14] van Tonder A, Joubert AM, Cromarty AD. Limitations of the 3-(4,5-dimethylthiazol-2-yl)-2,5-diphenyl-2H-tetrazolium bromide (MTT) assay when compared to three commonly used cell enumeration assays. BMC Research Notes. 2015;**8**:47

[15] Sumantran VN. Cellular Chemosensitivity assays: An overview. In: Cree IA, editor. Cancer Cell Culture:

Methods and Protocols. Totowa, NJ: Humana Press; 2011. pp. 219-236

[16] Ghasemi M, Turnbull T, Sebastian S, Kempson I. The MTT assay: Utility, limitations, pitfalls, and interpretation in bulk and single-cell analysis. International Journal of Molecular Sciences. 2021;**22**(23):12827

[17] Berridge MV, Herst PM, Tan AS. Tetrazolium dyes as tools in cell biology: New insights into their cellular reduction. Biotechnology Annual Review. 2005;**11**:127-152

[18] Scudiero DA, Shoemaker RH, Paull KD, Monks A, Tierney S, Nofziger TH, et al. Evaluation of a soluble tetrazolium/formazan assay for cell growth and drug sensitivity in culture using human and other tumor cell lines. Cancer Research. 1988;**48**(17):4827-4833

[19] O'Brien J, Wilson I, Orton T, Pognan F. Investigation of the Alamar blue (resazurin) fluorescent dye for the assessment of mammalian cell cytotoxicity. European Journal of Biochemistry. 2000;**267**(17):5421-5426

[20] Al-Nasiry S, Geusens N, Hanssens M, Luyten C, Pijnenborg R. The use of Alamar blue assay for quantitative analysis of viability, migration and invasion of choriocarcinoma cells. Human Reproduction (Oxford, England). 2007;**22**(5):1304-1309

[21] Fields RD, Lancaster MV. Dual-attribute continuous monitoring of cell proliferation/cytotoxicity. American Biotechnology Laboratory. 1993;**11**(4):48-50

[22] Larson EM, Doughman DJ, Gregerson DS, Obritsch WF. A new, simple, nonradioactive, nontoxic in vitro assay to monitor corneal endothelial cell viability. Investigative Ophthalmology & Visual Science. 1997;**38**(10):1929-1933

[23] Zachari MA, Chondrou PS, Pouliliou SE, Mitrakas AG, Abatzoglou I, Zois CE, et al. Evaluation of the alamarblue assay for adherent cell irradiation experiments. Dose-response: A Publication of International Hormesis Society. 2014;**12**(2):246-258

[24] Allen M, Millett P, Dawes E, Rushton N. Lactate dehydrogenase activity as a rapid and sensitive test for the quantification of cell numbers in vitro. Clinical Materials. 1994;**16**(4):189-194

[25] Stevens JF, Tsang W, Newall RG. Measurement of the enzymes lactate dehydrogenase and creatine kinase using reflectance spectroscopy and reagent strips. Journal of Clinical Pathology. 1983;**36**(12):1371-1376

[26] Chan FK, Moriwaki K, De Rosa MJ. Detection of necrosis by release of lactate dehydrogenase activity. Methods in Molecular Biology. 2013;**979**:65-70

[27] Korzeniewski C, Callewaert DM. An enzyme-release assay for natural cytotoxicity. Journal of Immunological Methods. 1983;**64**(3):313-320

[28] O'Toole SA, Sheppard BL, McGuinness EP, Gleeson NC, Yoneda M, Bonnar J. The MTS assay as an indicator of chemosensitivity/resistance in malignant gynaecological tumours. Cancer Detection and Prevention. 2003;**27**(1):47-54

[29] Cory AH, Owen TC, Barltrop JA, Cory JG. Use of an aqueous soluble tetrazolium/formazan assay for cell growth assays in culture. Cancer Communications. 1991;**3**(7):207-212

[30] Ginouves M, Carme B, Couppie P, Prevot G. Comparison of tetrazolium

salt assays for evaluation of drug activity against Leishmania spp. Journal of Clinical Microbiology. 2014;**52**(6):2131-2138

[31] Roehm NW, Rodgers GH, Hatfield SM, Glasebrook AL. An improved colorimetric assay for cell proliferation and viability utilizing the tetrazolium salt XTT. Journal of Immunological Methods. 1991;**142**(2):257-265

[32] Riss TL, Moravec RA, Niles AL, Duellman S, Benink HA, Worzella TJ, et al. Cell viability assays. In: Markossian S, Grossman A, Brimacombe K, Arkin M, Auld D, Austin C, et al., editors. Assay Guidance Manual. Eli Lilly & Company and the National Center for Advancing Translational Sciences: Bethesda (MD); 2004

[33] Alpar B, Leyhausen G, Gunay H, Geurtsen W. Compatibility of resorbable and nonresorbable guided tissue regeneration membranes in cultures of primary human periodontal ligament fibroblasts and human osteoblast-like cells. Clinical Oral Investigations. 2000;**4**(4):219-225

[34] Ishiyama M, Shiga M, Sasamoto K, Mizoguchi M. HE P-g: A new sulfonated tetrazolium salt that produces a highly water-soluble formazan dye. Chemical Pharmaceutical Bulletin. 1993;**41**(6):1118-1122

[35] Scarcello E, Lambremont A, Vanbever R, Jacques PJ, Lison D. Mind your assays: Misleading cytotoxicity with the WST-1 assay in the presence of manganese. PLoS One. 2020;**15**(4):e0231634

[36] Castano A, Tarazona J. ATP assay on cell monolayers as an index of cytotoxicity. Bulletin of Environmental Contamination and Toxicology. 1994;**53**(2):309-316

[37] Skehan P, Storeng R, Scudiero D, Monks A, McMahon J, Vistica D, et al. New colorimetric cytotoxicity assay for anticancer-drug screening. Journal of the National Cancer Institute. 1990;**82**(13):1107-1112

[38] Vichai V, Kirtikara K. Sulforhodamine B colorimetric assay for cytotoxicity screening. Nature Protocols. 2006;**1**(3):1112-1116

[39] Orellana EA, Kasinski AL. Sulforhodamine B (SRB) assay in cell culture to investigate cell proliferation. Bio-Protocol. 2016;**6**(21):e1984

[40] Ates G, Vanhaecke T, Rogiers V, Rodrigues RM. Assaying cellular viability using the neutral red uptake assay. Methods in Molecular Biology. 2017;**1601**:19-26

[41] Repetto G, del Peso A, Zurita JL. Neutral red uptake assay for the estimation of cell viability/cytotoxicity. Nature Protocols. 2008;**3**(7):1125-1131

[42] Olivier P, Testard P, Marzin D, Abbott D. Effect of high polyol concentrations on the neutral red absorption assay and tetrazolium-MTT test of rat hepatocytes in primary culture. Toxicology In Vitro. 1995;**9**(2):133-138

[43] Shukla RK, Sharma V, Pandey AK, Singh S, Sultana S, Dhawan A. ROS-mediated genotoxicity induced by titanium dioxide nanoparticles in human epidermal cells. Toxicology In Vitro. 2011;**25**(1):231-241

[44] McGaw LJ, Elgorashi EE, Eloff JN. Cytotoxicity of African medicinal plants against Normal animal and human cells. In: Kuete V, editor. Toxicological Survey of African Medicinal Plants. Amsterdam, NL: Elsevier; 2014. pp. 181-233

[45] Avelar-Freitas BA, Almeida VG, Pinto MC, Mourao FA, Massensini AR, Martins-Filho OA, et al. Trypan blue exclusion assay by flow cytometry. Brazilian Journal of Medical and Biological Research = Revista brasileira de pesquisas medicas e biologicas. 2014;**47**(4):307-315

[46] Kim JS, Nam MH, An SS, Lim CS, Hur DS, Chung C, et al. Comparison of the automated fluorescence microscopic viability test with the conventional and flow cytometry methods. Journal of Clinical Laboratory Analysis. 2011;**25**(2):90-94

[47] Wu G, Fang YZ, Yang S, Lupton JR, Turner ND. Glutathione metabolism and its implications for health. The Journal of Nutrition. 2004;**134**(3):489-492

[48] Townsend DM, Tew KD, Tapiero H. The importance of glutathione in human disease. Biomedicine & Pharmacotherapy = Biomedecine & Pharmacotherapie. 2003;**57**(3-4):145-155

[49] Hamad A, Elshahawy M, Negm A, Mansour FR. Analytical methods for determination of glutathione and glutathione disulfide in pharmaceuticals and biological fluids. Reviews in Analytical Chemistry. 2019;**38**(4):0019

[50] Lundholt BK, Scudder KM, Pagliaro L. A simple technique for reducing edge effect in cell-based assays. Journal of Biomolecular Screening. 2003;**8**(5):566-570

[51] Freimoser FM, Screen S, Bagga S, Hu G, St. Leger RJ. Expressed sequence tag (EST) analysis of two subspecies of Metarhizium anisopliae reveals a plethora of secreted proteins with potential activity in insect hosts. Microbiology (Reading). 2003;**149**(Pt 1):239-247

[52] Riss TL, Moravec RA, Niles AL, Duellman S, Benink HA, Worzella TJ, et al. Cell viability assays. In: Assay Guidance Manual [Internet]. Bethesda, MD: Eli Lilly & Company and the National Center for Advancing Translational Sciences; 2016

[53] Franken NA, Rodermond HM, Stap J, Haveman J, van Bree C. Clonogenic assay of cells in vitro. Nature Protocols. 2006;**1**(5):2315-2319

[54] Puck TT, Marcus PI. Action of x-rays on mammalian cells. The Journal of Experimental Medicine. 1956;**103**(5):653-666

[55] Brix N, Samaga D, Hennel R, Gehr K, Zitzelsberger H, Lauber K. The clonogenic assay: Robustness of plating efficiency-based analysis is strongly compromised by cellular cooperation. Radiation Oncology. 2020;**15**(1):248

[56] Mead TJ, Lefebvre V. Proliferation assays (BrdU and EdU) on skeletal tissue sections. Methods in Molecular Biology. 2014;**1130**:233-243

[57] Ahmed SA, Gogal RM Jr, Walsh JE. A new rapid and simple non-radioactive assay to monitor and determine the proliferation of lymphocytes: An alternative to [3H]thymidine incorporation assay. Journal of Immunological Methods. 1994;**170**(2):211-224

[58] Liu X, Rodeheaver DP, White JC, Wright AM, Walker LM, Zhang F, et al. A comparison of in vitro cytotoxicity assays in medical device regulatory studies. Regulatory Toxicology and Pharmacology. 2018;**97**:24-32

[59] Murray PE, Garcia Godoy C, Garcia Godoy F. How is the biocompatibilty of dental biomaterials evaluated? Medicina Oral, Patología Oral y Cirugía Bucal. 2007;**12**(3):E258-E266

[60] Vulcani VAS, Bizarria MTM, d'Ávila MA, Mei LHI, Bernal C,

Perussi JR. Cytotoxicity tests for nanostructured chitosan/PEO membranes using the agar diffusion method. Materials Research. 2012;**15**(2):213-217

[61] Pence I, Mahadevan-Jansen A. Clinical instrumentation and applications of Raman spectroscopy. Chemical Society Reviews. 2016;**45**(7):1958-1979

[62] Cui S, Zhang S, Yue S. Raman spectroscopy and imaging for cancer diagnosis. Journal of Healthcare Engineering. 2018;**2018**:8619342

[63] Krafft C, Popp J. The many facets of Raman spectroscopy for biomedical analysis. Analytical and Bioanalytical Chemistry. 2015;**407**(3):699-717

Chapter 2

Toxicity Evaluation and Biocompatibility of Nanostructured Biomaterials

Nebu George Thomas, Nibu Varghese, Nandakumar Kalarikkal, Sabu Thomas, Mridula Sreedharan, Sherin Sara George, Saumya John, Mekha Grace Varghese and Valliaveettil Thomas George

Abstract

Biomaterials have occupied a prominent place in regenerative procedures to restore human health. Moreover, there is a greater need in understanding, analyzing and establishing their toxicity profile. These, when made into nano-sized constructions called nanostructured biomaterials, their regenerative potential is enhanced, which could influence their toxicity nature. This chapter intends to give comprehensive information on their nanotoxicology pathways at the cellular level, their entry pathways into the human body, and their potential consequences on human health. It clearly explains the cytocompatibility and biocompatibility of various nanostructured biomaterials for potential human health applications like drug delivery and tissue engineering. A detailed overview of various in vitro and in vivo evaluation methods of biocompatibility of nanomaterials are outlined in this chapter that researchers should address as they move forward in developing new systems for the field of regeneration.

Keywords: nanobiomaterials, nanotoxicity, cytotoxicity, biocompatibility, regeneration

1. Introduction

Organ and tissue transplantation has limitations like immune reactions and donor limited availability, leading to the necessity for bioactive tissue engineering biomaterials, which is a rapidly developing multidisciplinary field. Nanoparticles (NP) based biomaterials offer better control over their desirable properties, such as the controlled release of biological molecules at surgical sites and optimization of mechanical properties matching recipient sites [1, 2]. Nanotechnology includes the development of NP and their applications in wide areas of interest. The size of NP will be within the range of ~10–1000 nm which can be synthesized in colloid and solid states [3]. Due to the increased production and use of nanoparticles in various fields, the unintended toxicity of nano biomaterials is a growing concern in tissue engineering and regenerative medicine. Minimal data are available regarding their toxicity and end products in the human physiologic system. Exposure pathways for NP include dermal penetration, ingestion, and inhalation. Physical properties of NP that influence toxicity include surface chemistry, particle shape, and size [4].

This chapter presents toxicological profiles of NP used as biomaterials highlighting in vitro and in vivo analysis techniques. This will give an insight into overall patterns and further requirements for assessing the toxicity of these biomaterials.

2. Human exposure routes to nanomaterials

Nanomaterials or nano-sized particles can enter the body through various routes such as skin, olfactory tracts, weak or damaged skin, intestinal tracts, respiratory tract, and intravenous or intramuscular routes. Their entry can have adverse biological effects such as tumors, cardiac diseases, skin problems, allergic reactions, and respiratory diseases. As a result of human industrial activities, dust storms or volcanic eruptions, humans get exposed to nanoparticles continuously and their toxicity depends on the chemical composition of the nanoparticles. Based on the composition, nanoparticles are divided into inorganic, metallic and metal oxide, where their biological effects and toxicity level are explained. The body parts easily exposed to nanoparticles are skin, gastrointestinal tract and lungs.

2.1 Skin as the entry route

Human skin which has an average surface area of approximately 2 m^2 has a thick outer layer of keratinized dead cells and which does not absorb any essential elements other than solar radiation. Skin appendages and stratum corneum are the major routes of entry through the skin. The root of the hair or follicles gives space for nanoparticles to accumulate; during the follicle opening, they enter into deep layers of skin. Intercellular spaces along the lipid layers are the easiest route for the nanoparticles to penetrate the skin. The classical diffusion theory can understand the diffusion of aggregated particles through stratum corneum. Even though the stratum corneum is porous and allows facile entry of nanoparticles, the dose and exposure time affects the entry. Follicular penetration varies at different sites on the body. It is observed that the follicle morphology on the lateral forehead is the area of maximum penetration as it has maximum surface coverage. Follicle area is considered the best site for particle storage as it is not exposed to washing or fabric contact. This area is considered suitable for drug release and a shortcut to the systemic circulation [5, 6].

Polymeric nanoparticles or drug carriers cannot penetrate the skin without mechanical stress. Several studies have been conducted to study the drug release from polymeric particles through the skin. It has been found that the particles did not penetrate the skin without any mechanical stress. Studies on hairless animals showed that the drug could not penetrate and thus proved that hair follicles pave the way for entry. Most topical applicants or cosmetics include zinc oxide (ZnO) or titanium dioxide (TiO_2) nanoparticles, which are mostly studied for skin penetration. It is shown that the size of the particles and exposure time effect the penetration. Titanium dioxide nanoparticles (TiO_2 NP) of size 20 nm showed their presence in the first 3–5 layers of corneocytes. Very small sized TiO_2 NP 4 nm size showed presence in the deep epidermis layer of pig's ear after 30 days of exposure. ZnO nanomaterials of size 30–40 nm could not penetrate deep into epidermis layer. Liquid type liposomes can penetrate epidermis layer, whereas gel type cannot. Metallic nanoparticles show higher penetration rates. Another important aspect is the physical state of the skin. The damaged skin easily allows the nanoparticles whereas intact skin only paves the way for tiny nanoparticles below 10 nm to penetrate the epidermis layer [7, 8].

The shape of the nanoparticles also affects toxicity. Differently shaped silver nanoparticles penetrated differently into the various layers of skin. Tak et al. [9] showed that the intercellular penetration pathway plays a central role in the shape-dependent penetration of AgNPs through the lipid matrix between corneocytes. The possible skin penetration pathways are shown in **Figure 1**.

The rate and depth of penetration of nanoparticles affect the toxicity level. Metallic nanoparticles can enter systemic circulation easily and thus can add to the toxicity depending on their dose. Thus, safety evaluation must be done before using nanoparticles in cosmetic formulations and drug delivery. Rapid penetrating nanoparticles pose more toxicity than slow penetrating ones as the latter can have efficient drug delivery and will be attacked by the immunological system of the body. In contrast, rapid penetrating NP can circumvent the macrophages and cause systemic toxicity. Argyria is a diseased condition due to the toxicity of silver nanoparticles leading to permanent pigmentation of eyes and skin [9].

2.2 Inhalation as the entry route

Airborne nanoparticles enter the lungs and cause respiratory disease by accumulating in the alveolar cells. They not only affect the lungs but also affect the

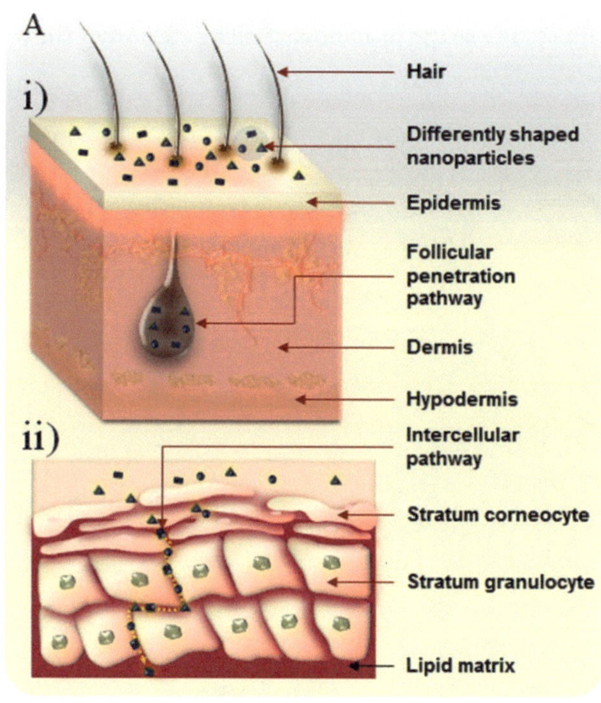

Figure 1.
The schematic diagram of possible skin penetration pathways of three differently shaped AgNPs. (A) Two main possible skin penetration pathways are illustrated: (i) enters via hair follicles (the follicular penetration pathway); and (ii) diffuses through the gaps between corneocytes (the intercellular penetration pathway). Reproduced with Creative Commons Attribution 4.0 International License (CC BY. 4.0) from Ref. [9].

extrapulmonary organs like the heart and liver. Metallic oxide and hydroxide NP cause pulmonary inflammation. The nanoparticles enter through olfactory tubes, pass the pharynx, and reach the alveoli. They have a good retention time, and the diameter of the particles affects the toxicity level. Some NP travel through the alveolar epithelium and capillary endothelial cells and reach the cardiovascular system and other internal organs [10]. Chronic obstructive pulmonary disease (COPD) is one of the major cause of death in the world as suggested by World Health Organization (WHO). Sarcoidosis and pulmonary fibrosis are other respiratory diseases caused by airborne pollutants. Wildfire, dust winds and volcanoes are sources of natural nanoparticles whereas engineered nanoparticles are found in gasoline exhausts, cosmetics and drug delivery carriers. These nanoparticles cause direct injury to lung tissues and cause inflammation. These nanoparticles then move to systemic circulation from the alveolar airspaces and thus cause toxicity to other internal organs. The high surface area of nanoparticles help them interact with enzymes or proteins of the immune system and thus causes inflammation and subsequent injury to the tissues, thus being detrimental to tissues [11].

The level of injury caused by nanoparticles is different in different regions of the lungs based on the clearance mechanism and cell types. In lung alveoli, there is only a single layer of cells; in bronchi and bronchioles, there is a layer of mucus for protection. Thus the nanoparticles which reach the alveoli can easily diffuse into blood capillaries and enter the systemic circulation, causing toxicity. Alveolar epithelial cells (AEC) are critical in protecting the respiratory tract and AEC in alveoli gets easily damaged by NP. AEC I gets damaged easily by NP and to replace it to maintain the alveolar structure, AEC II undergoes hyperplastic proliferation and repeated exposure to NP results in its aggregation in systemic circulation and thus leads to cardiovascular diseases. As cited above, the shape of nanoparticles determines the level of toxicity.

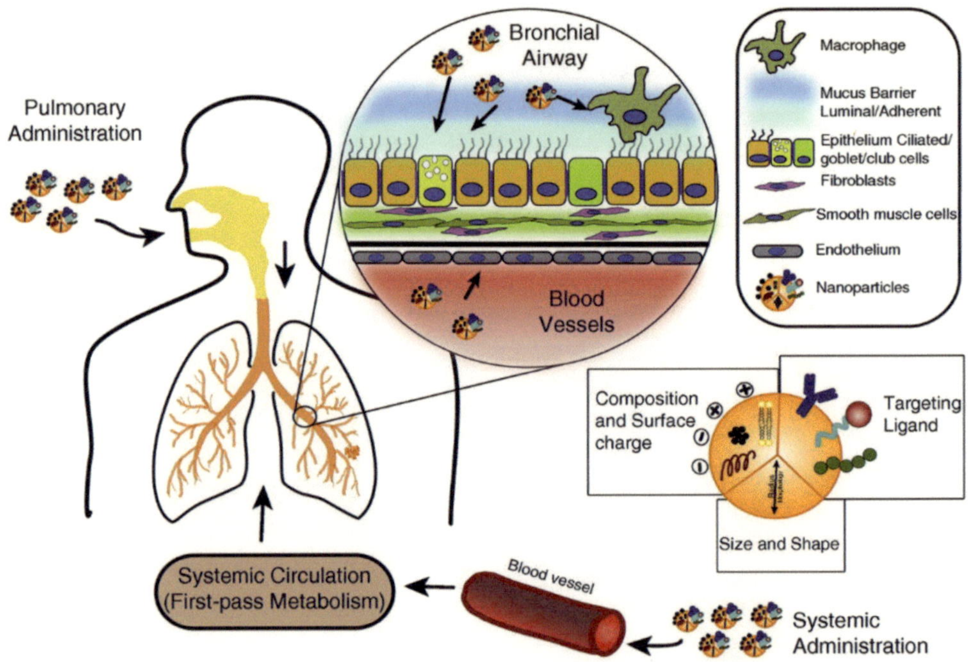

Figure 2.
Systemic and pulmonary administrations of nanoparticles to target respiratory cell types. Reproduced with Creative Commons Attribution 4.0 International License from Ref. [13].

Needle-like and rod-shaped NP cause more cellular destruction than spherical and flake-like structures. Other crucial factors to be considered are NP's degradation, solubility and chemical composition. The more they persist in the system, the more toxic they are to the cells. Metal NP like gold and silver persist in lungs for more than 7 days. This causes toxicity to all organs by creating oxidative stress, thus leading to cellular toxicity [12, 13] (**Figure 2**).

2.3 Nanoparticle ingestion

Metallic NP, carbon-based NP, ceramic NP, polymeric NP or dendrimers are the classes of nanoparticles that enter the gastrointestinal tract through ingestion. Implant materials have become very common, and orthopedic and dental implants use nanoparticle coatings to enhance their performance and bioactivity. However, intentionally or unintentionally, these implants release ions and NP over time, which cause adverse effects in the intestine and its related organs. The gastrointestinal tract (GIT) has a huge surface area of approximately 200 m^2, which is very much amenable to NP interaction. NP are absorbed by the GIT and thus enter the systemic circulation. Nanoparticles also damage the microbiome of the gut and thus affects digestion [14].

Targeted drug delivery through nanocarriers has been used in treating skeletal infections. But in many cases, these engineered nanomaterials create cytotoxicity. In order to achieve native bone tissue structure and more biocompatibility, nanomaterials have been used in orthopedic implants. Dental implants use a wide range of engineered materials, including nanomaterials. Titanium is used as dental implants, which can release titanium NP due to many activities like wear and tear caused by chewing or bacterial activity and chemical or physical deterioration. Titanium oxide nanoparticles from dental implants can dissolve in saliva reaching the intestine and other organs like the liver, spleen, kidney, and heart. Risk assessment of the engineered nanomaterials should be very focused on criteria like physicochemical characterization and thorough biological evaluation as per the regulatory agencies' guidelines [15]. Most implant failures occur due to wear and the release of particles which cause a chronic inflammatory response. A cobalt-chromium hip implant faced rejection due to the release of cobalt and chromium nanoparticles triggering an inflammatory response. A study conducted by Posada, 2015 showed the involvement of lymphocytes in such implant allergies, and within 48 hours of treatment, the metal nanoparticles caused apoptosis in the cells [16].

The most common nanoparticles that enter GIT are silver, iron oxide, titanium dioxide, zinc oxide etc. Organic NP, liposomes, engineered protein NP etc. pose little concern about their toxicity as many of them have been used by humans for centuries. Not many studies have been done in this area of nanotoxicology, and very little information is known about the toxicity of such nanoparticles. But the most important fact to be noted is that the properties of each type of nanoparticle differs on many factors and that these factors determine their toxicity. Dimension, morphology, composition, surface charge, aggregation state affects the fate of the nanoparticles in vivo. Most of the nanoparticles, after passing through various parts of the GIT change their properties due to interaction between the proteins and varying pH conditions. Thus, their fate is altered when compared with that in vitro. The properties of pristine nanoparticles will be very much different from the nanoparticles in vivo and thus, the studies related to their toxicity should be taken into consideration only for in vivo results. Most of the studies which show cytotoxicity in cell lines do not show any toxic effects when performed in animal models unless it is at a very high dose. Thus the in vivo studies should be concentrated more on the physiological aspects of the cells rather than on the organs as a whole [17, 18].

3. Pathways of cellular uptake

In drug delivery applications, the nanocarriers should go inside the cells to efficiently deliver drugs to target organelles or cells. Thus, the uptake route or mechanism of NP into cell should be considered in designing a nanocarrier system. Cell membrane is the protective coating of a cell and is not permeable to all particles as it maintains homeostasis. The lipid molecules of the cell membrane have hydrophilic heads and hydrophobic tails. Lot of small molecules enter the cell by passive diffusion. The cell uptake of molecules with the help of energy from ATP (Adenosine triphosphate) is called active diffusion [19].

There are a lot of factors which influence the internalization of the NP and their interaction with intracellular components such as size, surface modifications, net charge, hydrodynamic volume and stiffness. The persistence of nanoparticles in the cell and their release of free radicals to induce oxidative stress bring toxicity and cell death. Endocytosis, an active transport, is the most important pathway by which molecules enter the cell and trigger responses. The main types of endocytosis are phagocytosis (cell eating) and pinocytosis. In the process of endocytosis, the particle

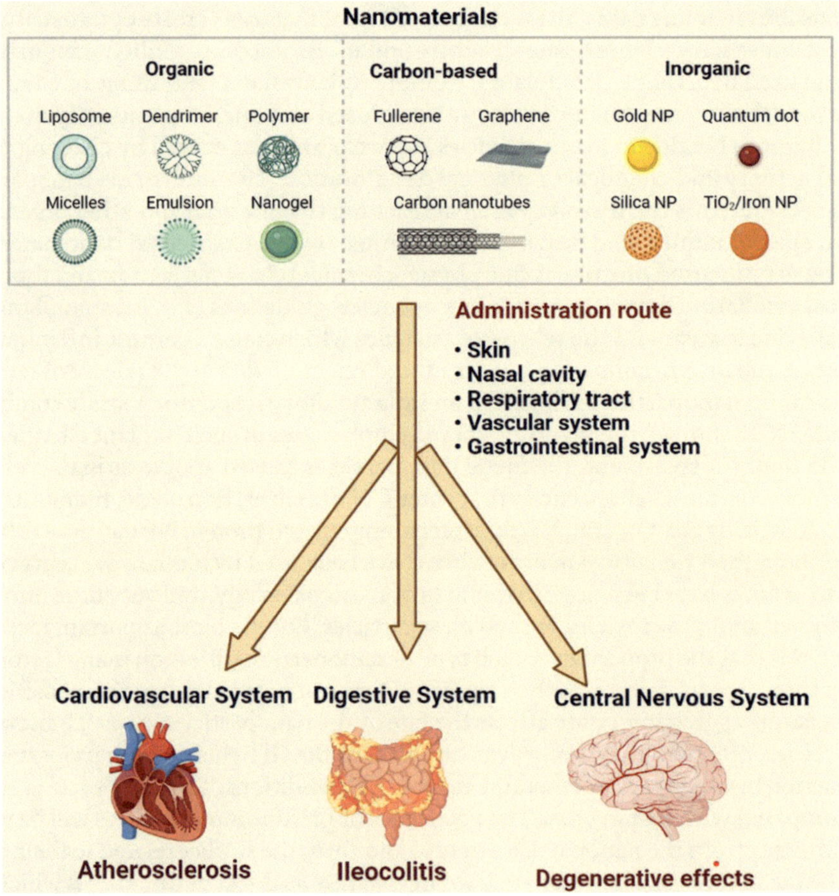

Figure 3.
Schematic diagram showing the types of nanomaterials and their impact on various organ systems. Created with the help of Biorender.com [18].

to be internalized with be covered by a part of cell membrane which the buds off inside the cell to form vesicle containing the particle [20].

Phagocytes are immune cells that protect from infections and pollution by engulfing foreign bodies. The 3 main phagocytes are macrophages, neutrophils and monocytes. In phagocytosis opsonins (group of proteins found in serum) play a major role by adsorbing onto NP surface and form protein corona. Phagocytes recognize these and a signaling cascade is triggered which internalizes the particles forming phagosome [18] (**Figures 3** and **4**).

Different cell types use different uptake mechanisms for the same NP. It can occur through phagocytosis, macropinocytosis, clathrin and caveolin mediated endocytosis, non-clathrin and non-caveolin mediated endocytosis are some among them. Nanoparticles less than 100 nm in size can enter the cells and size less than 40 nm can enter the nucleus. The surface charge of the NP are a crucial factor for cell internalization where the negatively charged cell membrane is attracted towards positively charged NP. Small NPs are engulfed by F-actin mechanism, dynamin and lipid rafts which are energy-dependent. NP of size more than 500 nm are engulfed by phagocytosis and macropinocytosis. Aggregated NPs, TiO_2 or carbon black undergo opsonization in the biological fluid and undergo phagocytosis or macropinocytosis [19].

3.1 Adverse effects of the cellular uptake

The bio resistance of certain nanomaterials to degradation is the root cause of toxicity. These nanoparticles retain in the endosomal compartment of cell. Certain other nanoparticles like ZnO, when taken into the acidic part of lysosome will get dissolved into Zn^{2+} ions which in excess amounts will cause cytokine production leading to cytotoxicity. Non-soluble nanoparticles like TiO_2 and carbon black NP were found to be free in cytoplasm and even in the D.N.A. The reasons for this toxicity could be that these NP diffuse through the cell membrane through transient holes or they may

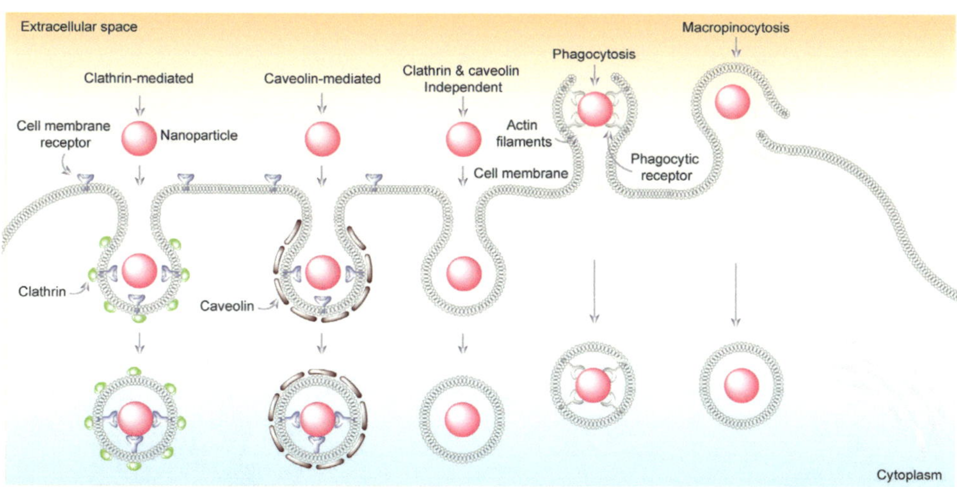

Figure 4.
Schematic representation showing the mechanisms of nanoparticle cellular internalization such as clathrin-mediated; caveolin-mediated; clathrin- and caveolin-independent; phagocytosis and macropinocytosis pathways. Reproduced with creative commons attribution 4.0 (CC-BY-4.0) license from Augustine et al. [19].

be accumulated in the lysosome where later on lead to membrane rupture and release in the cytoplasm. These cytoplasmic NP during mitosis can enter the nucleus through microtubules and are found in the DNA [21].

The engineered NP persisting in the cell result in oxidative stress leading to apoptosis and inflammation. Oxidative stress is a result of imbalance between reactive oxygen species [ROS] and antioxidant capacity of cell. NP trigger the ROS production and thus leads to the imbalance and oxidative stress. ROS has various cellular roles by acting as secondary messengers. If the stress caused by ROS is at a higher level, it can result in cell membrane and organelle injury leading to necrosis or apoptosis. The smaller the size of the nanoparticles, the higher the surface area and thus produce more ROS. Thus, the oxidative stress caused by the small NP affect lipid membrane and cause their disorganization in structure and function. Inactivation of certain sensitive proteins which have methionine or cysteine in their active site are also the toxic effects of the NP which entered the cells. Genotoxicity is also a serious effect of NP which damages DNA because of intrachain adducts and strand breakage [22].

4. Effect of physicochemical properties of nanomaterials

The cellular uptake and dispersion of nanomaterials depend mainly on the physicochemical properties of the particles. Evaluating the role of physicochemical properties in toxicity issues is very important (**Figure 5**).

4.1 Size

The size of the nanoparticle is proportional to its surface area. The lesser the size, more is the surface area and more its reactivity. The size of the particle and its effect in vivo is difficult to observe as its structure changes in vivo. The ability of the nanoparticles to enter the cells and modify the macromolecules accounts for its level of toxicity. The nanomaterials below 100 nm can enter cells easily as their size is comparable with that of protein globules, DNA., and cell membrane thickness. Very small NP of size less than 20 nm which can enter the nucleus are more toxic than particles

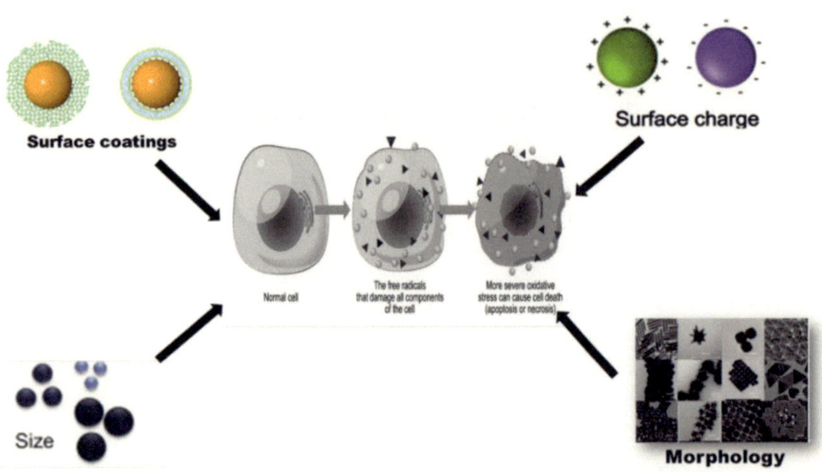

Figure 5.
Schematic diagram showing the various physicochemical properties of nanoparticles and their effect on cell [23].

which cannot enter the nucleus. Studies have done on gold nanoparticles in relation to their size and toxicity in vivo. It was found that particles of size 1.4 nm whose size is comparable with that of a major groove of DNA will easily block transcription by interacting with the sugar-phosphate in DNA. It is shown by Zhang et al. that those particles of size less than 5 nm enter cells through translocation while those above 5 nm enter through specific pathways like macropinocytosis and phagocytosis [24]. Small sized particles circulate rapidly all around the body while particle above 50 nm were found in organs like liver and spleen. Star shaped nanoparticles have shown accumulation in lungs while larger nanoparticles are recognized by the immune system and eliminated from the body. Size and shape influence the kinetics of excretion and accumulation of the particles in vivo. The larger surface area of the particles help in adsorption onto the cell surface where particles above 600 nm destructed the cell membrane and caused haemolysis. In contrast, particles of size 100 nm did not cause membrane destruction [23–25].

Larger particles show negligible toxicity when compared with smaller particles. The shape of the particles also makes them behave differently. Spherical particles are easier to be engulfed by cells and they are the ones that cause less toxicity. Whereas rod shaped and fiber-shaped nanomaterials are difficult to eliminate and their toxicity is higher when compared to that of their spherical counterparts. Nanoparticles that enter the body can agglomerate or agglomerates can disperse into primary particles. When primary particles come in contact with lung fluid, they agglomerate. Agglomerates or particles of size >1 μm are up taken by phagocytosis and their clearance mechanism is easy. But the smaller the particles or the agglomerates, the more difficult it becomes to eliminate them and toxicity like genotoxicity happens. The more time they retain in the body organs like lungs, the chances of depositing in secondary organs is more and they damage the lungs. Clearance rate by translocation is less than 0.5% compared to their exposure and these can result in their persistence, causing toxicity and organ damage [26].

4.2 Morphology

The shape of particles is very important in determining their fate inside the body. In nature we can see different shapes for microorganisms like spherical shaped HIV virus, star shaped bacteria and rod shaped TMV which confer different effects in vivo. Thus the function of a nanoparticle with a specific shape determines their effect in nature. The nanoparticle uptake, toxicity, biodistribution and inflammatory response is highly dependent on the structure of the nanoparticle. It is studied that rod shaped particles enter inside the cell easily even compared to spherical particles. But the shape alone does not add to this effect, we need to consider the aspect ratio where higher aspect ratio particles expose more to the cell surface thus easily engulfed into the cells. Nanowires and nano worms due to their length have higher aspect ratio and they enter cells with ease. Particles with highly curved and sharp edges are not very easily up taken by cells. High aspect ratio particles have shown proinflammatory effects leading to cytotoxicity when conducted studies in in vitro models. The persistence of nanoparticles in vivo depends on the geometry where it has been found that filament-shaped micelles prolong the persistent time compared to spherical-shaped micelles. In a study conducted on gold nanoparticles in ovarian cancer xenograft by Arnida et al., it was shown that rod-shaped gold particles were more in circulation and accumulated in tumor cells than spherical-shaped gold NP [27].

The behavior of nanoparticles in vivo cannot be related to just one physicochemical factor alone, whereas the toxicity or fate of a nanoparticle depends on many factors like size, shape, ligands, material, charge, route of exposure etc. Among these factors, the surface chemistry of a nanoparticle has more influence than the particle's geometry since receptor ligand binding strength is the limiting factor which determines the uptake and thus leads to toxicity of these NP. But when you consider the same surface chemistry of a rod shaped and sphere-shaped NP., the rod-shaped NP will be favored by the cells. Apart from this, the fluid motion of rod-shaped particles is favorable. Spheres and short rod NP accumulate in liver, higher aspect ratio NP are found in spleen and lungs. Nanofibers that enter through lungs have more cytotoxic effects where the asbestos particles inhalation is an example. They can disrupt the cell membrane's lipid bilayer and cause a pro-inflammatory response [28].

Nanoparticle toxicity strongly depends on their morphology. In studies conducted on gold nanostructures, it was found that gold nanostars bind to serum proteins better than gold rods. Spherical gold nanoparticles showed higher binding affinity when compared with branched particles. The radius of curvature of the nanoparticles is a factor for protein corona formation. A planar surface is provided by a large radius of curvature, thus resulting in more effective protein binding [29].

4.3 Surface charge

The cytotoxic effects of nanoparticles depend on their charge density and charge polarity. Positively charged nanoparticles of gold, zinc oxide and silica have more profound effects than their counter parts. For certain nanoparticles of polymers, the charge does not have much effect on their properties in vivo. Some nanoparticles which are porous does not depend on charge such as mesoporous silica nanoparticles. But phagocytic cells interact more with negatively charged particles. In case of nonphagocytic cells where they interact with cationic charges on NP surface, its cellular uptake is more and thus cause cytotoxicity by plasma membrane disruption. But the effect of serum reduces their uptake. Whereas phagocytic cells interact more with anionic NP and ingest them effectively and serum has a positive effect on their uptake [30].

Cell type is an important factor which determines the uptake of NP. The same NP behave differently in various cell types. Effect of surface charge on Mesoporous silica NP in human mesenchymal stem cells revealed that a strong positive charge on the NP showed a good uptake in mesenchymal stem cells whereas their uptake was inhibited in 3 T3 L1 cell lines. When the positive charge density is less, the uptake efficiency is low [31].

Ionic Interactions of NP and cell membrane is important in understanding the fate and effect of NP in the living system. Even though the cell membrane is negatively charged, it behaves differently to positively charged molecules depending on the charge density and other factors. Gold nanoparticles of different surface charges (positive, negative, neutral, zwitterionic) were applied to various cell types. It was shown that positively charged AuNPs depolarized the membrane more than their counterparts. The authors also suggested that changing or varying the surface charge density or applying both positive and negative charges on the nanoparticle may result in an organized cellular uptake and thus target various organelles [32].

The toxicity of silver nanoparticles in the environment and to biological systems is greatly influenced by the surface charge of silver nanoparticles. The electrostatic barrier between silver NP and cell membrane highly affects the fate of NP and influence of other factors such as shape and size become negligible. The authors suggest that surface charge measurement of NP can be used as an analyzing tool to find the toxicity of the NP [33].

Curcumin which is considered as a boon for many diseases can become toxic to cells, especially alveolar macrophages when the surface charge is altered. Curcumin nanoparticles when given positive charge by coating with the polymer polyvinylpyrrolidone and was given negative charge by coating with polyvinyl alcohol and neutral with dextran. The surface charge varied from −20 mV to +5.5 mV. The positively charged curcumin NP resulted in lysosomal and mitochondrial destabilization leading to ROS generation and apoptosis. They entered the cells through clathrin-coated endocytosis and damaged the lysosomes. Thus the effect of surface charge on materials, especially nanomaterials are a very important aspect to be looked upon to determine cytotoxicity and during the designing of nanoparticles [34]. Apart from these facts, another important factor is the density of cationic charges. When the density of the cationic charge is less, the toxicity effect will also be less. The zeta potential of the Nanoparticles does not have any effect if the surface charge density is very high [35].

4.4 Coating

Nanoparticles are given coating with various materials depending on the result to be achieved. When they are used for cell targeting, the NP must be coated with protein ligands identified by the cell receptors or if targeted to cancer cells then attached with RGD sequence. These coatings can affect the properties of nanoparticles when used in pristine form. The coating materials determine the toxicity of the NP, where they can increase or decrease the toxicity levels. There are a plethora of coating mechanisms and some methods prefer coating the NP with more than one material. Silver nanoparticles tend to aggregate, thus, coating helps improve the NP's stability in suspension. Certain coatings make the silver NP more toxic such as citrate compared with PVP, or protein coated when compared to bare silver NP [36].

Type of coating material, whether they are hydrophilic or hydrophobic, influences the toxicity of metal-based nanoparticles. Hydrophobic coatings have proved to be safer than hydrophilic coatings and this also depends on the density of the coating. Zinc oxide used in various cosmetic formulations when coated with hydrophobic material showed very little toxicity and is recommended for commercial use. This can be explained by the reduced bioactivity of the hydrophobic particles. Theoretical models such as QASR quantitative structure-activity relationships should be applied to evaluate the effect of various coating on nanoparticles [37].

5. Physiological impact due to nanomaterials

Nanostructured biomaterials are being analyzed in detail currently for regenerative applications. However, their physiological impact due to the prolonged presence of these foreign agents within the body or their degradation byproducts can be broadly divided as the impact of ROS generation and oxidative stress, inflammation and cellular injury due to nanoparticle dissolution.

5.1 Reactive oxygen species (ROS) generation and oxidative stress

Nanoparticles, including transitional metals, have been reported to be oxidized and release reactive oxygen species of which hydroxyl radicals are considered to cause maximum damage, including affecting cell signaling pathways and affecting the cellular lipids, protein production, alteration of DNA and gene transcription [38]. Also inflammatory cells including neutrophils and macrophages are attracted by phagocytosis of these nanoparticles which in turn leads to the production of proinflammatory cytokines and oxidative stress [38]. Researchers have reported the extent and severity of inflammation is dependent on characteristics of the nanoparticles such as size and shape.

In response to low levels of oxidative stress generated by nanoparticles, the cells release antioxidants such as ferritin, N-acetylcysteine (NAC), which nullify the oxidative stress; however cellular damage occurs due to excessive reactive species [38].

5.2 Inflammation

Deng et al. reported that the decreased size of poly aryl acid-coated gold nanomaterials (<20 nm) have been associated with the activation of the Mac-1 receptor of the monocytes [39]. This in turn, leads to upregulation of the proinflammatory cytokines via the NF-κB pathway. Poly(d,L-lactide-co-glycolic acid) (PLGA) nanoparticles 75 nm or lesser reported lesser PMN in bronchial lavage fluid than in 200 nm size particles. Also, the varying shapes of these nanoparticles have been found to elicit inflammatory responses. Albanese et al. reported that the size and shape of these nanoparticles dictate the ligand and receptor interactions that in turn determine the cellular uptake and the further downstream reactions [40]. However, nanoparticles can be used as drug-delivery agents to suppress these inflammatory reactions.

Nanoporous scaffolds similar in architecture to the native tissues have been found to have lesser associated inflammatory reactions [41]. Silica and hydroxyapatite based nanoporous scaffolds have integrated better than cancellous bone substitutes in association with implant-based in vivo studies. Nanoporous scaffolds made of native matrix proteins or with a coating of bioinert polymers (such as poly ethyl glycol PEG) along with sustained release of anti-inflammatory agents and antioxidants would have better integration with host tissues with minimal inflammation.

Certain nanotopographical alterations in these nanomaterials are capable of altering the host inflammatory response attenuating or at other times exacerbating the same [41]. It needs to be understood to ensure successful incorporation of nanomaterials in tissue engineering applications.

5.3 Cellular injury due to nanoparticle dissolution

Nanoparticle dissolution is a crucial property which determines the extent of its availability, its toxicity and also its impact on the host environment. Their greater surface area leads to increased physical and chemical interactions which leads to dissolution.

It may lead to cell death or in case of non-degradable NP can accumulate within the cells leading to damage. Quantum dots localize in varying cellular locations, Silica (40–80 nm) is deposited in the nucleoplasm, and gold nanoparticles have been found in the major groove of DNA (leading to the formation of human cancer cells) [38]. Nanomaterials have been found to induce autophagy which occurs as a response to the cellular changes in the aftermath of oxidative stress.

In vitro studies have shown that cationic Polyamidoamine (PAMAM) dendrimers are more likely to lead to autophagy than the anionic dendrimers [41]. Nanomaterials are thought to alter the autophagic degradation activity which might lead to toxicity [42]. Gold nanoparticles as compared to Silicon dioxide nanoparticles have been found to lower the lysosomal degradation which in turn affects the normal functioning of the cells. Atomic Force Microscopy studies on osteoblast cells reveal alterations in osteoblast cell membranes leading to changes in cell adhesion in response to nanoparticles [43]. Also, Bhabra et al. reported that nanoparticles (Cobalt-chromium) also indirectly affect the cells, including DNA damage without even permeating through the cellular barrier [44].

6. Protein corona formation

When nanomaterials are exposed to biological fluids a dynamic covering of various serum proteins and biomolecules attach to the nanomaterial and this structure is called protein corona (PC). The PC formation on NP can influence its toxicity and targeting behavior due to immunological effects [45].

PC has two different layers one is a hard corona which is the first tightly bound layer and another is soft corona which is the second layer which consists of proteins that are easily exchanged. The composition of corona can change any time by replacement of proteins, but the quantity will remain constant. Initial proteins that bind to the NP will be the most abundant in the blood or the biological fluid exposed to the NP later on the proteins are replaced according to the affinity. This is the 'Vroman Effect' which explains the change in the composition of the protein corona on NP surface in response to time. The proteins compete to get adsorbed onto NP surface. Different types of corona are formed based on the binding affinity and surface ligands on the engineered NP.

Engineered nanoparticles have high free energy and thus adsorb protein to a low-energy state by increasing their dispersion. In less than 0.5 min of exposure to blood plasma, silicon NP adsorbed almost 300 proteins onto its surface [46] (**Figure 6**).

Toxicity of the NP are greatly influenced by the protein corona adsorbed. The fate of an engineered nanoparticle is different in cell lines and in vivo. According to the type of proteins adsorbed, the targeting of the NP and their cellular uptake also changes. There

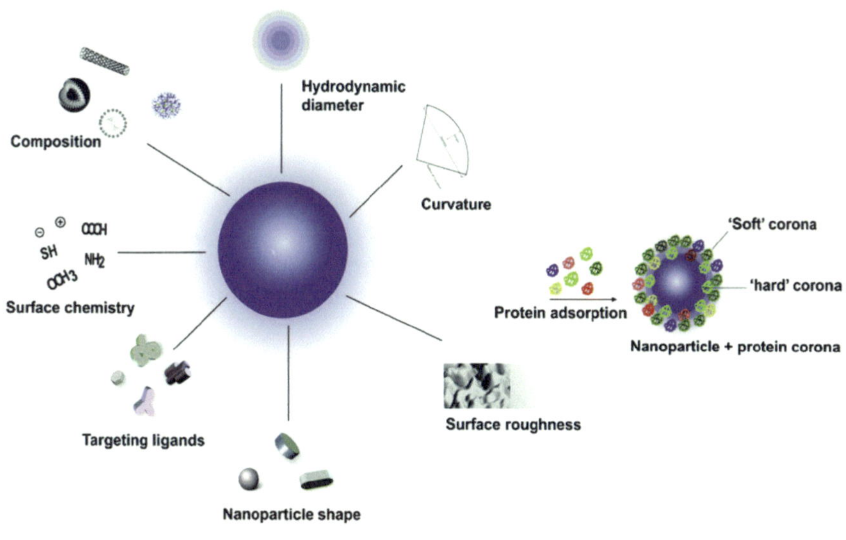

Figure 6.
Factors influencing protein corona formation on nanoparticles. Reproduced with permission from RSC with license id 1257299-1 [47].

are chances of non-specific uptake by receptor-mediated endocytosis. Opsonization plays a major role in determining the fate of ENPs. Proteins act as opsonin and these NP are engulfed by macrophages, it can also lead to uncontrolled aggregation. The protein corona, when made by cell penetrating peptides and antibodies will result in active targeting. It is observed that certain protein corona like blood proteins reduce the toxicity of carbon nanotubes when compared to the fate in cell lines. Overall, the protein corona affects the cellular uptake, target organ and thus toxicity [47].

7. In vitro nano biomaterial cytotoxicity assessment

The minimal ethical concerns associated with in vitro assays make them in demand for toxicity assays. It is much faster, and the cost is also reasonable. It is further subdivided into a few.

7.1 Experimental models used

7.1.1 Cell lines

Various types of cell lines are used in the study of the cytotoxicity of nanoparticles. Using various cell lines for toxicity assay has added risks, which have been documented by Donaldson et al. [48]. This is because of the different toxic responses that the cells experience in vitro when compared to in vivo. Moreover, when carcinoma cell lines are used for in vitro toxicity testing of nanoparticles, the results may conflict with that of normal cells, as the carcinoma cell lines have different pathophysiology than normal cells [49]. The various types of cell lines that are used include murine cell lines (mouse fibroblast cells, mouse macrophage cells, rat mesenchymal cells, etc.), mammalian cells, human osteoblast cell lines, human alveolar and bronchial epithelial cells, human hepatocytes, human macrophages, cancerous cell lines (lung cancer cells, hepatocellular carcinoma cells, colon and cervix carcinoma cell lines, human epidermoid like carcinoma cells—HELA, etc.) and various hematic cells from murine, and mammalian species and humans [50].

7.2 In vitro cytotoxicity assessment methods

7.2.1 Proliferation assay

The cell viability assay is the most crucial investigation to understand a foreign agent's toxicity. Cell viability confirms the number of healthy cells in a sample indicated by their proliferation potential. It also gives information on cell death in the given sample and monitors cytotoxicity [51].

The most prevalent cell viability assays are based on the estimation of the metabolic activity of the cells. Examples are the MTT assay, protease activity assay, and the reduction of resazurin salts [52]. The mitochondrial and the cytoplasmic enzymes in the viable cells can react with these substrates to bring out colored products or fluorescence, which corresponds to the number of live cells.

The most commonly involved method is the MTT assay where a tetrazolium dye MTT (3-(4,5-dimethyl-2-thiazolyl)-2,5-diphenyl-2H-tetrazolium bromide) penetrates the cells and due to the activity of mitochondrial enzymes get converted to formazan, a colored product. They are insoluble crystals and accumulate within the cells, which are later solubilized using reagents like sodium dodecyl sulfate or isopropanol. The absorbance is then recorded, and maximum absorbance indicates higher survival [53, 54]. With an increasing number of viable cells, the intensity of color deepens. The advantages are quick results and the requirement of limited manipulations [55]. The positively charged MTT dye can readily penetrate the cell membrane. There are negatively charged tetrazolium dyes too, like the MTS (3-(4,5-dimethyl-2-thiazolyl)-5-(3-carboxymethoxyphenyl)-2(4-sulfophenyl)-2H-tetrazolium), XTT, and WST-1(2-(4-iodophenyl)-3-(4-nitrophenyl)-5(2,4-disulfophenyl)-2H-tetrazolium, monosodium salt) that can penetrate the cells only in the presence of an electron acceptor. Since the end products are soluble, they are also preferred.

In protease viability assay, a marker known as glycylphenylalanyl-aminofluoro-coumarin (GF-AFC). The aminopeptidase enzyme in the cytoplasm of viable cells acts upon this substrate breaking down the compound into glycine, phenylalanine, and aminoflourocoumarin (AFC) that exhibits fluorescence corresponding to the number of viable cells [56].

Alamar blue/Resazurin dye (7-hydroxy-10-oxidophenoxazin-10-ium3-one) method is more sensitive, offers simpler sample preparation protocols and is inexpensive compared to MTT. The deep blue-colored resazurin gets converted to pink-colored resorufin by the activity of inner mitochondrial enzymes. It is less successful due to difficulty in the biochemical assaying and unwanted reactions [53].

ATP cell assay is also a preferred method for assessing cell viability. The advantages include faster results, higher sensitivity, and lesser artifacts. The principle behind this is that cells with damaged membranes cannot synthesize ATP and cellular ATP concentration gets depleted faster by the endogenous ATPase.

The clonogenic cells assay is a qualitative assessment that investigates the proliferation capacity of cells. The cell sample is exposed to a known quantity of nanomaterials and allowed to proliferate. After 1–3 weeks, they are stained and quantified per growth in number or size. A survival curve is plotted based on percentage survival and increasing dose of nanomaterials. Increasing dosage gradually results in a lower number or size of colonies [57].

DNA synthesis cell proliferation assay is where a radioactive tracker like 3H-thymidine is incubated with a cell sample. This gets incorporated into the DNA of proliferating cells. The radioactivity of the DNA of the daughter cells is measured using a scintillation beta counter, thereby quantifying the number of viable cells. Another method is to use a non-radioactive compound which is 5 Bromo-2′-deoxyuridine (BrdU). BrdU-specific antibodies are then used to obtain a colorimetric estimation [58]. Developed thymidine analogues like 5-ethynyl-2′-deoxyuridine (EdU), 5-iodo-2′-deoxyuridine (IdU) and 5-chloro-2′deoxyuridine (CldU) are also sensitive methods that to have additional advantages over BrdU assay [59].

A precise cell viability assay available is flow cytometry. A cell suspension is introduced into the flow cytometer which allows particles of less than 150 μm to pass through. During the passage, a laser light beam of a known wavelength interrogates the solution, interacts with every single cell, and gets scattered. These scattered lights are converted to voltage signals, the intensity of which gives information on cell viability and cellular kinetics.

7.2.2 Apoptosis assay

Another important parameter observed in nanoparticle toxicity assays is the occurrence of apoptosis. Studies have confirmed that exposure to silver nanoparticles has resulted in apoptosis and DNA damage with the release of apoptosis markers viz. caspase-3 and caspase-9 [60]. Various tests for assessing apoptosis in the cell culture systems are as follows.

Annexin-V and propidium iodide (PI.) are two cell death markers that give information on apoptosis. The activation of the caspase-dependent pathway during apoptosis causes externalization of the plasma membrane, which is indicated by increased fluorescence. This is because of the binding of Annexin-V to phosphatidylserine. PI is usually impermeable, but it can stain the nucleus when the integrity of the cell membrane is lost. This also relates to the later stages of apoptosis [61].

Comet assay, or the single cell gel electrophoresis assay (SCGE) is a sensitive test can be used both in vitro and in vivo. It detects the breaks in the DNA strands of individual cells [62]. The basic principle is that when an electric current is applied to the cell system, the damaged DNA fragments migrate out of the cell, whereas the undamaged DNA will remain within the nucleus. A comet shape can be seen with undamaged DNA forming the head and damaged DNA forming the tail. The size and shape of the comet give information about the extent of DNA damage. DNA-specific fluorescent dyes will stain the samples, and later the amount of fluorescence in the head and the tail portions, as well as the tail length, are analyzed [63, 64].

The morphological changes specific to apoptosis are another parameter of importance. Irregular reduction in the cell size and fragmentation of DNA confirms the presence of apoptosis. Typical ladder-like patterned DNA fragmentation and irregular cell sizes can be easily identified by Agarose gel electrophoresis [65].

TUNEL assay or the terminal deoxynucleotidyl transferase dUTP nick end labeling assay—it is a method of staining initially described for identifying cells that have undergone programmed cell death and DNA fragmentation [66]. Later it was confirmed that the test could also detect DNA damage due to non-apoptotic events like necrosis [67]. Hence it cannot distinguish between apoptosis and necrosis. It is based on the activity of enzyme terminal deoxynucleotidyl transferase on fluorescent-labeled dUTP. The cells with fragmented DNA will bind to the fluorescent labeled assay molecule and later be estimated by fluorescent microscopy or by immunohistochemical staining. This method gives a quantitative estimation of the viable cells [68].

7.2.3 Necrosis assay

Necrosis assay determines the viability of cells by identifying the loss of membrane integrity. The dye exclusion method is a preliminary method to identify dead or the dying cells. Dying cells exhibit a loss of membrane integrity, so there will be permeation of the dye within the cell. The most used dyes are trypan blue, eosin, or propidium iodide. The cell suspensions are added with the dyes and manual cell counting using a Neubauer hemocytometer conventionally to determine the number of live cells [69].

PI dye can stain the nucleus and binds to the DNA of nonviable cells forming a fluorescent complex. The amount of fluorescent light emitted is proportional to the voltage signal output. The higher the output, the higher the number of nonviable cells will be. Even though the method is faster and more convenient, it is time-consuming and requires expensive instrumentation. A recently designed microchip and microcell counter (Adam, Nanoentek, Seoul, Republic of Korea) is established on which PI stain can be used to distinguish viable and nonviable cells by direct cell counting technique [70].

Neutral Red is another dye that is weakly cationic and slightly acidic in nature. They can easily diffuse through the plasma membrane and get concentrated in the lysosomes binding to the anionic sites of the lysosomal matrix [71]. Exposure of cells to nanoparticles leads to cell surface alterations and increased lysosomal fragility that favors binding of neutral red thus helping in identifying viable and dead cells [72].

7.2.4 Oxidative stress assay

Exposure to nanoparticles leads to the production of reactive oxygen species (ROS) and reactive nitrogen species (RNS) [73]. These compounds can be detected by X-band electron paramagnetic resonance (EPR) [74], fluorescent probes, and non-fluorescent probes. EPR use is limited due to its high cost. Fluorescent probes are cost-effective but inefficient due to high reactivity, which gives misleading results [75].

Oxidative stress can also be assessed by measuring the lipid peroxidation BODIPY-C_{11} assay. ROS can induce membrane lipid peroxidation. BODIPY-C_{11} is a fluorescent dye that inserts into the lipid bilayers and helps in identifying the oxidized and unoxidized lipids by their respective green and red colors. Later these are quantified fluorimetrically [76].

ROS-induced membrane lipid peroxidation TBA assay for malondialdehyde [77] also gives information on oxidative stress. Various other assays like the measurement of lipid hydroperoxide using Amplex Red assay, antioxidant depletion by 5,5′-dithio-bis-(2-nitrobenzoic acid) (DTNB), and superoxide dismutase activity by Nitro blue tetrazolium assay is also convenient procedures [78].

7.2.5 Endotoxin assays

Nanoparticles can absorb contaminants onto their surface, resulting in an exaggerated inflammatory response as seen in the case of endotoxins. This acute inflammatory response results in the activating the signaling cascades and releasing inflammatory mediators like cytokines.

The main assays in this regard are the gel clot assay, the coagulogen-based (turbidity) assay, and the chromogenic assay. In the gel clot method, the endotoxin sample solution is combined with LAL solution (Limulus amebocyte lysate) leading to cleavage of coagulogen, which is then checked for subsequent clotting. Chromogenic assay substitute chromogens for coagulogen and releases chromophores when cleaved. Commercial kits like Endosafe®-PTS (Portable Test System) based on chromogenic LAL assays are also available [79, 80].

8. In vivo nano biomaterial cytotoxicity assessment

8.1 Experimental models used

8.1.1 Animal models

Animal models are particularly useful in studying those aspects of nanomaterials that cannot be mimicked by in vitro systems, such as particokinetics (absorption, distribution, metabolism, elimination: ADME), which are caused by cellular responses of the nanomaterials in vivo, and nanomaterial-caused intercellular communication, multicellular interactions, and immune regulation, all of which may be quite difficult to study in vitro [81]. Therefore animal experiments are indispensable in nanotoxicology.

The selection of the animal models is the first and critical step, as a predictive animal model is still underdetermined, although many models have been built. Mice, rats, zebrafish, rabbits, and *Caenorhabditis elegans* are used in nanotoxicology studies, while mice and rats are more appropriate models due to their explicit genome for toxicity tests of organs. Laboratory mice are the most used mammalian research model, popular because of their availability, size, relatively low cost, and ease of handling. Hamsters, as they have wide cheek pouches and mouths that can be widely opened, are used to study pulmonary clearance and toxicity of nanomaterials. Rabbits are also frequently used as they are mildly tempered, easy to confine, and breed. Their ease of handling benefits intravenous injections, blood sample collection, and dermal tests. Zebrafish (*Danio rerio*), a tropical freshwater fish, have a small size, short life cycle, high gene homology with superior animals (also human), and an availability of a wealth of genetic databases, which makes them effective in vivo models to study the toxicity of nanomaterials. Adult zebrafish may be exploited to study nanomaterial transport, biodistribution, bioaccumulation, and chronic toxicity [82].

For the selection of suitable animal models for the assessment of nanomaterial toxicity, Christop et al. [83], suggested the following: the animal model chosen for the toxicity assessment should be suitable in such a way that the pathways of NP penetration into the organism and excretion from the organism conform from the biochemical and anatomical perspective. For the assessment of nanosafety, using a healthy animal may not be suitable when the results are to be extrapolated to the general population, as interorgan compensatory mechanisms may lower the toxic effects of the nanoparticles. Therefore, they suggested that the suitable test animal model choice should be based on the intended application of engineered nanomaterial. Healthy animal models are suitable only for nanoparticle-based therapeutics, and special animal models with unique features should be used when nanoparticles for therapeutic approaches.

8.1.2 Organs

In vivo assessments include analysis of tissue structure changes, apoptosis, and inflammation infiltration in main organs (kidney, spleen, lung, brain, and heart). Target organ systems include those that may concentrate nanoparticles due to their structural specificities, such as hepatic sinusoid and Kupffer cells and the renal filtration membrane [84].

Hepatic assessment: Immunohistochemistry to detect liver fibrosis and inflammation, serum enzymology for hepatic function analysis, hematology and chemistry analyzer for comprehensive analysis.

Renal system: Histopathology and immunohistochemistry, kidney indices for assessment of renal function, and renal variables assessed from blood and urine.

Gastrointestinal system: Histological assay of the GIT, absorption function measured indirectly by the evaluation of metal content and electrolyte.

Pulmonary system: MRI to visualize the pulmonary accumulation of nanoparticles, intratracheal instillation for long-term and short-term toxicity assessments, LDH index, and assessment of oxidative stress.

Cardiovascular system: Assessment for signs and symptoms of phlebitis, hemolysis, thrombosis and signs of cardiac injury.

Nervous system: Assessment of drug delivery of solid lipid nanoparticles across the blood-brain barrier, by using various radiographic techniques such as PET and PET/CT.

Immune and reproductive system: Any modifications of the nanoparticles are assessed to understand their immunogenicity, the generation and release of inflammatory mediators are examined upon the use of nanoparticles, histological changes of major immune organs are assessed by H&E staining, long term studies on the reproductive organs, reproductive index and offspring survival, growth and development are assessed [84].

8.2 In vivo toxicity evaluation methods

In vivo methods of assessment of nanomaterial toxicity have certain advantages over in vitro testing methods. The advantages lie in the fact that in most of the in-vitro testing methods, the growth medium that is used for culturing the cells is supplemented with the lowered concentration of serum proteins as the source of nutrition, and usually, serum of bovine origin is used. The concentration of the serum proteins in the medium influences the nanoparticle-cell interaction, which may range from mild effects on the cells to cell death. A study to assess the role of serum concentration on nanoparticle-cell interactions was conducted by Kim et al. [85]. They suggested that while comparing the results of various in-vitro studies, the serum concentration in the medium used to culture the cell lines should be considered. Moreover, when serum of bovine origin is being used, it may not represent a condition relevant to real in vivo exposure situations.

In vivo toxicity assessment methods assess the toxicity and reactions of the nanomaterials in the organism from the point of administration through the subcutaneous vein, inhalation, skin, oral administration, and intraperitoneal routes; interaction with biological components (such as cells and proteins); spread to different organs, metabolism, penetration into the cells of organs and their excretion [86]. The nanomaterial is introduced into the body of the test animal, and the biodistribution, clearance, hematology, serum chemistry, and histopathology are monitored. This provides more practical data on the interaction of the nanoparticles with the immune system, proteins, and dynamic body fluids at the systemic level.

In order to evaluate the acute *in vivo* toxicity of nanomaterials, the Organization for Economic Cooperation and Development (OECD) guidelines recommend oral toxicity test, eye irritation, corrosion and dermal toxicity, and lethal Dose 50 (LD50).

OECD (Organization for Economic Cooperation and Development) guideline 423 provide data on the methods to be used to determine the LD_{50} (lethal dose 50%) of the nanoparticles being studied. The vehicle used to carry the nanoparticle dose must be non-reactive, and particles must disperse appropriately in it.

8.2.1 Acute oral toxicity assessment

Different concentrations of the nanoparticles may be administered to the test animals orally. Changes in body weight, behaviors, and other toxic signs and symptoms may be recorded regularly. The animals may be observed daily for 14 days for skin symptoms like edema, erythema, ulcers, body scabs, discoloration, and scars. Toxic signs like weight loss, water and food consumption, and the animals' behavior are also assessed. Skin biopsies and blood may be taken periodically for histopathological evaluation and biochemical and hematological investigations, respectively. After 14 days, the animal may be sacrificed, and all organs collected. The following formula may be used for the calculation of organosomatic index: [weight (g) of the organ/total body weight (g)] × 100 [87, 88].

8.2.2 Acute dermal toxicity assessment

Tests may be performed on animals using the test guideline 404 published by the Organization for Economic Co-operation and Development [89, 91]. The substance to be tested may be applied to a small area of the skin of the experimental animal in a single dose for an exposure period of 4 h. The animals will to be examined for signs of erythema edema during the next 14 days. Skin biopsies are taken periodically, and after 14 days, the animal is sacrificed, and the skin is collected for histopathological examination.

8.2.3 Acute eye irritation and corrosion assessment

The standard eye irritation test established by the Organization for Economic Co-operation and Development (OECD) for the testing of chemicals is used as the standard to measure nanomaterials eye toxicity. Five minutes prior to the application of the test substance, 2 drops of a topical ocular anesthetic is to be applied to minimize pain or distress and then the test. One eye may serve as a as a reference control. A small amount of the test substance (0.1 g or 0.1 ml of its colloidal suspension) may be applied to the conjunctival sac of the test eye. The animals are to be observed for toxic symptoms periodically at 1, 24, 48 and 72 h, and grading for ocular lesions of cornea, iris, conjunctivae, and chemosis may be done over 14 days according to OECD test guideline TG 405 [90, 91].

8.2.4 Biodistribution studies

Biodistribution of the nanoparticles may be detected in live or killed animals after conjugating with a radioactive label or organic dye and tracking in blood and tissue at different periods. Some metallic nanoparticles intrinsic properties may be probed by using specific instruments. These tests are especially important for nanomaterials used for drug delivery. Suspension of drug loaded nanoparticles may be radio-labeled and administered intra-venously into the test animal. Blood is collected at regular periods for 24 h, plasma is separated, and radioactivity levels of the residues may be measured. Main organs and tissues (lung, liver, kidney, heart, spleen, pancreas, brain, fat, and muscle) may also be collected, weighted and radioactivity may be measured [87, 92].

8.2.5 Changes in serum chemistry and cell type

Blood from the test animals may be collected for biochemical (triglyceride, cholesterol, glucose, glutamic oxaloacetic transaminase (GOT), and glutamic pyruvic transaminase (GPT)) and hematological investigations [88].

Chrishtop et al. [83], in their review on nanosafety versus nanotoxicology, in 2021, referred to studies that have demonstrated animal models with chronic diseases being sufficiently susceptible to nanomaterials even with low toxicity. Considering that chronic diseases, like bronchial asthma, are widely prevalent in various populations, they suggested that nanosafety should also be considered along with toxicity assessment of nanomaterials [83].

9. Conclusion

Techniques for evaluating the biocompatibility of nanostructured is not fully comprehensive since it is a continuously developing field where biomaterials are being constantly developed for tissue engineering applications in regenerative medicine. Extensive use of nano biomaterials in regenerative medicine necessitates efficient techniques for evaluating toxicity at an affordable cost. Literature analysis clearly states that the cytocompatibility of biomaterials influences the output of in vitro experiments. The determination of parameters, such as cell type and nanoparticle concentration, required a detailed understanding of anticipated nanoparticle exposure and its metabolic activity within the human body. Hence selecting suitable methods for the analysis of in vitro toxicity will provide a proper idea about nanomaterials toxicity mechanism and can be effectively used in regenerative medicine.

Acknowledgements

Pushpagiri Research Centre supported this work, and all the authors especially acknowledge the contributions and support of Dr. Nibu Varghese, Director and Head, Pushpagiri Research Centre, and Dr. Anil Sukumaran, Professor, Pushpagiri Research Centre.

Conflict of interest

The authors declare no conflict of interest.

Author details

Nebu George Thomas[1*], Nibu Varghese[2], Nandakumar Kalarikkal[3], Sabu Thomas[4], Mridula Sreedharan[3], Sherin Sara George[5], Saumya John[1], Mekha Grace Varghese[6] and Valliaveettil Thomas George[1]

1 Department of Periodontology, Pushpagiri College of Dental Sciences, Tiruvalla, Kerala, India

2 MACFAST, Tiruvalla, Kerala, India

3 International and Inter University Centre for Nanoscience and Nanotechnology, Mahatma Gandhi University, Kerala, India

4 Mahatma Gandhi University, Kottayam, Kerala, India

5 Department of Paediatric and Preventive Dentistry, Pushpagiri College of Dental Sciences, Tiruvalla, Kerala, India

6 Pushpagiri Research Centre, Tiruvalla, Kerala, India

*Address all correspondence to: nebugt@gmail.com

References

[1] Park J, Wrzesinski SH, Stern E, Look M, Criscione J, Ragheb R, et al. Combination delivery of TGF-beta inhibitor and IL-2 by nanoscale liposomal polymeric gels enhances tumour immunotherapy. Nature Materials. 2012;**11**:895-905. DOI: 10.1038/nmat3355

[2] Bahal R, Ali McNeer N, Quijano E, Liu Y, Sulkowski P, Turchick A, et al. In vivo correction of anaemia in β-thalassemic mice by γPNA mediated gene editing with nanoparticle delivery. Nature Communications. 2016;**7**:13304. DOI: 10.1038/ncomms13304

[3] Colson YL, Grinstaff MW. Biologically responsive polymeric nanoparticles for drug delivery. Advanced Materials. 2012;**24**:3878-3886. DOI: 10.1002/adma.201200420

[4] Lisunova MO, Lebovka NI, Melezhyk OV, Boiko YP. Stability of the aqueous suspensions of nanotubes in the presence of nonionic surfactant. Journal of Colloid and Interface Science. 2006;**299**:740-746

[5] Zaiter T, Cornu R, El Basset W, et al. Toxicity assessment of nanoparticles in contact with the skin. Journal of Nanoparticle Research. 2022;**24**:149. DOI: 10.1007/s11051-022-05523-2

[6] Simkó M, Nentwich M, Gazsó A, Fiedeler U. How nanoparticles enter the human body and their effects there. Nano trust dossiers No.0003en 2010. Available from: https://epub.oeaw.ac.at/0xc1aa5576%200x0024c7a5.pdf

[7] Schneider M, Stracke F, Hansen S, Schaefer UF. Nanoparticles and their interactions with the dermal barrier. Dermatoendocrinology. 2009;**1**(4):197-206. DOI: 10.4161/derm.1.4.9501

[8] De Matteis V. Exposure to inorganic nanoparticles: Routes of entry, immune response, biodistribution and in vitro/in vivo toxicity evaluation. Toxics. 2017;**5**(4):29. DOI: 10.3390/toxics5040029

[9] Tak Y, Pal S, Naoghare P, et al. Shape-dependent skin penetration of silver nanoparticles: Does it really matter? Scientific Reports. 2015;**5**:16908. DOI: 10.1038/srep16908

[10] Sonwani S, Madaan S, Arora J, Suryanarayan S, Rangra D, Mongia N, et al. Inhalation exposure to atmospheric nanoparticles and its associated impacts on human health: A review. Frontiers in Sustainable Cities. 2021;**3**:690444. DOI: 10.3389/frsc.2021.690444

[11] Nho R. Pathological effects of nano-sized particles on the respiratory system. Nanomedicine: Nanotechnology, Biology and Medicine. 2020;**29**

[12] Lu X, Zhu T, Chen C, Liu Y. Right or left: The role of nanoparticles in pulmonary diseases. International Journal of Molecular Sciences. 2014;**15**(10):17577-17600. DOI: 10.3390/ijms151017577

[13] Deng Z, Kalin GT, Shi D, Kalinichenko VV. Nanoparticle delivery systems with cell-specific targeting for pulmonary diseases. American Journal of Respiratory Cell and Molecular Biology. 2021;**64**(3):292-307. DOI: 10.1165/rcmb.2020-0306TR

[14] Bergin IL, Witzmann FA. Nanoparticle toxicity by the gastrointestinal route: Evidence

and knowledge gaps. International Journal of Biomedical Nanoscience and Nanotechnology. 2013;**3**(1-2). DOI: 10.1504/IJBNN.2013.054515

[15] Scimeca JC, Verron E. Nano-engineered biomaterials: Safety matters and toxicity evaluation. Materials Today Advances. 2022;**15**. DOI: 10.1016/j.mtadv.2022.100260

[16] Ajdary M, Moosavi MA, Rahmati M, et al. Health concerns of various nanoparticles: A review of their in vitro and in vivo toxicity. Nanomaterials. 2018;**8**:634. DOI: 10.3390/nano8090634

[17] Asmatulu E, Andalib MN, Subeshan B, et al. Impact of nanomaterials on human health: A review. Environmental Chemistry Letters. 2022;**20**(4):2509-2529. DOI: 10.1007/s10311-022-01430-z

[18] Barhoum A, García-Betancourt ML, Jeevanandam J, et al. Review on natural, incidental, bioinspired, and engineered nanomaterials: History, definitions, classifications, synthesis, properties, market, toxicities, risks, and regulations. Nanomaterials. 2022;**12**:177. DOI: 10.3390/nano12020177

[19] Augustine R, Hasan A, Primavera R, Wilson RJ, Thakor AS, Kevadiya BD. Cellular uptake and retention of nanoparticles: Insights on particle properties and interaction with cellular components, materials today. Communications. 2020;**25**

[20] Déciga-Alcaraz A, Medina-Reyes EI, Delgado-Buenrostro NL, Rodríguez-Ibarra C, Ganem-Rondero A, Vázquez-Zapién GJ, et al. Toxicity of engineered nanomaterials with different physicochemical properties and the role of protein corona on cellular uptake and intrinsic ROS production. Toxicology. Sep 2020;**442**:152545. DOI: 10.1016/j.tox.2020.152545. Epub 2020 Aug 2. PMID: 32755642

[21] Marano F, Rodrigues-Lima F, Dupret J-M, Baeza-Squiban A, Boland S. Cellular mechanisms of nanoparticle toxicity. Encyclopedia of Nanotechnology. 2016:498-505. DOI: 10.1007/978-94-017-9780-1_175

[22] Nel A, Xia T, Madler L, Li N. Toxic potential of materials at the nanolevel. Science. 2006;**311**:622-627

[23] Sukhanova A, Bozrova S, Sokolov P, et al. Dependence of nanoparticle toxicity on their physical and chemical properties. Nanoscale Research Letters. 2018;**13**:44. DOI: 10.1186/s11671-018-2457-x

[24] Zhang S, Gao H, Bao G. Physical principles of nanoparticle cellular endocytosis. ACS Nano. 2015;**9**(9):8655-8671

[25] De Jong WH, Hagens WI, Krystek P, Burger MC, Sips AJAM, Geertsma RE. Particle size-dependent organ distribution of gold nanoparticles after intravenous administration. Biomaterials. 2008;**29**(12):1912-1919

[26] Braakhuis HM, Park MV, Gosens I, et al. Physicochemical characteristics of nanomaterials that affect pulmonary inflammation. Particle and Fibre Toxicology. 2014;**11**:18. DOI: 10.1186/1743-8977-11-18

[27] Arnida, Janát-Amsbury MM, Ray A, Peterson CM, Ghandehari H. Geometry and surface characteristics of gold nanoparticles influence their biodistribution and uptake by macrophages. European Journal of Pharmaceutics and Biopharmaceutics. 2011;**77**:417-423. DOI: 10.1016/j.ejpb.2010.11.010

[28] Kinnear C, Moore TL, Rodriguez-Lorenzo L, Rothen-Rutishauser B, Petri-Fink A. Form follows function: Nanoparticle shape and its implications for nanomedicine. Chemical Reviews. 2017;**117**(17):11476-11521

[29] Seo S, Lee JE, Lee K, Kim HN. Effects of microenvironmental factors on assessing nanoparticle toxicity. Environmental Science: Nano. 2022;**9**(2):454-476. DOI: 10.1039/D1EN00925G

[30] Fröhlich E. The role of surface charge in cellular uptake and cytotoxicity of medical nanoparticles. International Journal of Nanomedicine. 2012;**7**:5577-5591. DOI: 10.2147/IJN.S36111

[31] Chung TH, Wu SH, Yao M, Lu CW, Lin YS, Hung Y, et al. The effect of surface charge on the uptake and biological function of mesoporous silica nanoparticles in 3T3-L1 cells and human mesenchymal stem cells. Biomaterials. Jul 2007;**28**(19):2959-2966. DOI: 10.1016/j.biomaterials.2007.03.006. Epub 2007 Mar 19. PMID: 17397919

[32] Arvizo RR, Miranda OR, Thompson MA, Pabelick CM, Resham Bhattacharya J, Robertson D, et al. Effect of nanoparticle surface charge at the plasma membrane and beyond. Nano Letters. 2010;**10**(7):2543-2548. DOI: 10.1021/nl101140t

[33] Amro M, Badawy E, Silva RG, Morris B, Scheckel KG, Suidan MT, et al. Surface charge dependent toxicity of silver nanoparticles. Tolaymat Environmental Science & Technology. 2011;**45**(1):283-287. DOI: 10.1021/es1034188

[34] Loo C-Y, Siew EL, Young PM, Traini D, Lee W-H. Toxicity of curcumin nanoparticles towards alveolar macrophage: Effects of surface charges. Food and Chemical Toxicology. 2022;**163**

[35] Weiss M, Fan J, Claudel M, et al. Density of surface charge is a more predictive factor of the toxicity of cationic carbon nanoparticles than zeta potential. Journal of Nanbiotechnology. 2021;**19**:5. DOI: 10.1186/s12951-020-00747-7

[36] Yang X, Gondikas AP, Marinakos SM, Auffan M, Liu J, Hsu-Kim H, et al. Mechanism of silver nanoparticle toxicity is dependent on dissolved silver and surface coating in Caenorhabditis elegans. Meyer Environmental Science & Technology. 2012;**46**(2):1119-1127. DOI: 10.1021/es202417t]

[37] Lai RWS, Kang H-M, Zhou G-J, Yung MMN, He YL, Ng AMC, et al. Hydrophobic surface coating can reduce toxicity of zinc oxide nanoparticles to the marine copepod *Tigriopus japonicus*. Environmental Science & Technology. 2021;**55**(10):6917-6925. DOI: 10.1021/acs.est.1c01300

[38] Ali A, Suhail M, Mathew S, Shah MA, Harakeh SM, Ahmad S, et al. Nanomaterial induced immune responses and cytotoxicity. Journal of Nanoscience and Nanotechnology. 2016;**16**(1):40-57

[39] Deng ZJ, Liang M, Monteiro M, Toth I, Minchin RF. Nanoparticle-induced unfolding of fibrinogen promotes mac-1 receptor activation and inflammation. Nature Nanotechnology. 2011;**6**(1):39-44

[40] Albanese A, Tang PS, Chan WCW. The effect of nanoparticle size, shape, and surface chemistry on biological systems. Annual Review of Biomedical Engineering. 2012;**14**:1-16

[41] Padmanabhan J, Kyriakides TR. Nanomaterials, inflammation and tissue engineering. Wiley Interdisciplinary Reviews: Nanomedicine and Nanobiotechnology;**20157**(3):355-370

[42] Manshian BB, Pokhrel S, Mädler L, Soenen SJ. The impact of nanoparticle-driven lysosomal alkalinization on cellular functionality. Journal of Nanobiotechnology. 2018;**16**(1):85

[43] Rasel MAI, Singh S, Nguyen TD, Afara IO, Gu Y. Impact of nanoparticle uptake on the biophysical properties of cell for biomedical engineering applications. Scientific Reports. 2019;**9**(1):5859

[44] Bhabra G, Sood A, Fisher B, Cartwright L, Saunders M, Evans WH, et al. Nanoparticles can cause DNA damage across a cellular barrier. Nature Nanotechnology. 2009;**4**(12):876-883

[45] Riccardo R, Sara C, Paolo C, Salvatore P, Marco A. Recent advances in understanding the protein corona of nanoparticles and in the formulation of "stealthy" nanomaterials. Frontiers in Bioengineering and Biotechnology. 2020;**8**. DOI: 10.3389/fbioe.2020.00166

[46] Tenzer S, Docter D, Kuharev J, et al. Rapid formation of plasma protein corona critically affects nanoparticle pathophysiology. Nature Nanotechnology. 2013;**8**:772-781

[47] Kamaly N, Farokhzad OC, Corbo C. Nanoparticle protein corona evolution: From biological impact to biomarker discovery. Nanoscale. 2022;**14**:1606-1620

[48] Donaldson K, Borm P, Castranova V, Gulumian M. The limits of testing particle-mediated oxidative stress in vitro in predicting diverse pathologies; relevance for testing of nanoparticles. Particle and Fibre Toxicology. 2009;**6**(13):99

[49] Shetab-Boushehri SV, Abdollahi M. Current concerns on the validity of in vitro models that use transformed neoplastic cells in pharmacology and toxicology. International Journal of Pharmacology. 2012;**8**:594-595

[50] Bahadar H, Maqbool F, Niaz K, Abdollahi M. Toxicity of nanoparticles and an overview of current experimental models. Iranian Biomedical Journal. 2016;**20**(1):1-11. DOI: 10.7508/ibj.2016.01.001. Epub 2015 Aug 19

[51] Hansen MB, Nielsen SE, Berg K. Re-examination and further development of a precise and rapid dye method for measuring cell growth/cell kill. Journal of Immunological Methods. 1989;**119**(2):203-210

[52] Denizot F, Lang R. Rapid colorimetric assay for cell growth and survival. Modifications to the tetrazolium dye procedure giving improved sensitivity and reliability. Journal of Immunological Methods. 1986;**89**(2):271-277

[53] Martin A, Sarkar A. In vitro toxicity testing of nanomaterials. In: Mehanna RA, editor. Cell Culture [Internet]. London: IntechOpen; 2018

[54] Sayes CM, Reed KL, Warheit DB. Assessing toxicity of fine and nanoparticles: Comparing in vitro measurements to in vivo pulmonary toxicity profiles. Toxicological Sciences. 2007;**97**(1):163-180

[55] Ernst C, Christie BR. Isolectin-IB 4 as a vascular stain for the study of adult neurogenesis. Journal of Neuroscience Methods. 2006;**150**(1):138-142

[56] Niles AL, Moravec RA, Eric Hesselberth P, Scurria MA, Daily WJ, Riss TL. A homogeneous assay to measure live and dead cells in the same sample by detecting different protease markers. Analytical Biochemistry. 2007;**366**(2):197-206

[57] Franken NAR, Rodermond HM, Stap J, Haveman J, Van Bree C.

Clonogenic assay of cells in vitro. Nature Protocols. 2006;**1**(5):2315-2319

[58] Priyanka Ganguly AB, Pillai SC. Toxicity of nanomaterials: Exposure, pathways, assessment, and recent advances. ACS Biomaterials Science & Engineering. 2018;**4**(7):2237-2275

[59] Viggiani CJ, Knott SR, Aparicio OM. Genome-wide analysis of DNA synthesis by BrdU immunoprecipitation on tiling microarrays (BrdU-IP-chip) in Saccharomyces cerevisiae. Cold Spring Harbor Protocols. Feb 2010;**2010**(2):pdb.prot5385. DOI: 10.1101/pdb.prot5385. PMID: 20150148

[60] Ahamed M et al. DNA damage response to different surface chemistry of silver nanoparticles in mammalian cells. Toxicology and Applied Pharmacology. 2008;**233**(3):404-410

[61] Berghe TV et al. Determination of apoptotic and necrotic cell death in vitro and in vivo. Methods. 2013;**61**(2):117-129

[62] Bajpayee MKA, Dhawan A. The comet assay: Assessment of in vitro and in vivo DNA damage. In: Dhawan A, Bajpayee M, editors. Genotoxicity Assessment, Methods in Molecular Biology (Methods and Protocols). Vol. 1044. Totowa, NJ: Humana Press; 2013. pp. 325-345

[63] Fairbairn DW, Olive PL, O'Neill KL. The comet assay: A comprehensive review. Mutation Research/Reviews in Genetic Toxicology. 1995;**339**(1):37-59

[64] Lemay M, Wood KAA. Detection of DNA damage and identification of UV-induced photoproducts using the CometAssay [TM] Kit. BioTechniques. 1999;**27**:846-848

[65] Suman S, Pandey A, Chandna S. An improved non-enzymatic "DNA ladder assay" for more sensitive and early detection of apoptosis. Cytotechnology. 2012;**64**(1):9-14

[66] Loo DT, Taaoot ID, Vladimir V, editors. Situ Detection of DNA Damage: Methods and Protocols. New York, United States: Humana Press; 2002. pp. 21-30

[67] Ansari B et al. In situ end-labelling detects DNA strand breaks in apoptosis and other physiological and pathological states. The Journal of Pathology. 1993;**170**(1):1-8

[68] Charriaut-Marlangue C, Ben-Ari Y. A cautionary note on the use of the TUNEL stain to determine apoptosis. Neuroreport. 1995;**7**(1):61-64

[69] Strober W. Trypan blue exclusion test of cell viability. Current Protocols in Immunology. 2001;**Appendix 3**: Appendix 3B

[70] Nicoletti IMG, Pagliacci M, Grignani F, Riccardi C. A rapid and simple method for measuring thymocyte apoptosis by propidium iodide staining and flow cytometry. Journal of Immunological Methods. 1991;**139**(2):271-279

[71] Kim JSN, Nam MH, An SSA, Lim CS, Hur DS, Chung C, et al. Comparison of the automated fluorescence microscopic viability test with the conventional and flow cytometry methods. Journal of Clinical Laboratory Analysis. 2011;**25**(2):90-94

[72] Nemes Z et al. The pharmacological relevance of vital staining with neutral red. Experientia. 1979;**35**(11):1475-1476

[73] Borenfreund E, Puerner JA. Toxicity determined in vitro by morphological alterations and neutral red absorption. Toxicology Letters. 1985;**24**(2-3):119-124

[74] Magder S. Reactive oxygen species: Toxic molecules or spark of life? Critical Care. 2006;**10**(1):208

[75] Zang LYMB, van Kuijk FJ, Misra HP. EPR studies on the kinetics of quenching singlet oxygen. Biochemistry and Molecular Biology International. 1995;**37**(6):1187-1195

[76] Gomes A, Fernandes E, Lima JL. Fluorescence probes used for detection of reactive oxygen species. Journal of Biochemical and Biophysical Methods. 2005;**65**(2):45-80

[77] Schins RP et al. Surface modification of quartz inhibits toxicity, particle uptake, and oxidative DNA damage in human lung epithelial cells. Chemical Research in Toxicology. 2002;**15**(9):1166-1173

[78] Fantel AG. Reactive oxygen species in developmental toxicity: Review and hypothesis. Teratology. 1996;**53**(3):196-217

[79] Hussain SM et al. The interaction of manganese nanoparticles with PC-12 cells induces dopamine depletion. Toxicological Sciences. 2006;**92**(2):456-463

[80] Savage DT, Hilt JZ, Dziubla TD. In vitro methods for assessing nanoparticle toxicity. Methods in Molecular Biology. 2019;**1894**:1-29

[81] He X. In vivo nanotoxicity assays in animal models. In: Zhao Y, Zhang Z, Feng W, edtiors. Toxicology of Nanomaterials. Wiley-VCH Verlag GmbH & Co. KGaA; 2016. DOI: 10.1002/9783527689125.ch7

[82] George S, Xia T, Rallo R, Zhao Y, Ji Z, Lin S, et al. Use of a high-throughput screening approach coupled with in vivo zebrafish embryo screening to develop hazard ranking for engineered nanomaterials. ACS Nano. 2011;**5**(3):1805-1817

[83] Chrishtop VV, Prilepskii AY, Nikonorova VG, Mironov VA. Nanosafety vs. nanotoxicology: Adequate animal models for testing in vivo toxicity of nanoparticles. Toxicology. Oct 2021;**462**:152952. DOI: 10.1016/j.tox.2021.152952. Epub 2021 Sep 17. PMID: 34543703

[84] Yang Y, Qin Z, Zeng W, Yang T, Cao Y, Mei C, et al. Toxicity assessment of nanoparticles in various systems and organs. Nanotechnology Reviews. 2017;**6**(3):279-289. DOI: 10.1515/ntrev-2016-0047

[85] Kim JA, Salvati A, Åberg C, Dawson KA. Suppression of nanoparticle cytotoxicity approaching in vivo serum concentrations: Limitations of in vitro testing for nanosafety. Nanoscale. 2014;**6**(23):14180-14184. DOI: 10.1039/c4nr04970e

[86] Iosub CS, Olaret E, Grumezescu AM, Holban AM, Andronescu E. Nanostructures for Novel Therapy. Romania: Elsevier; 2017. pp. 793-809

[87] Kim SC et al. In vivo evaluation of polymeric micellar paclitaxel formulation: Toxicity and efficacy. Journal of Controlled Release. 2001;**72**(1):191-202

[88] Clichici S, Filip A. In vivo assessment of nanomaterials toxicity. In: Soloneski S, Larramendy ML, editors. Nanomaterials—Toxicity and Risk Assessment [Internet]. London: IntechOpen; 2015. Available from: https://www.intechopen.com/chapters/48547. DOI: 10.5772/60707

[89] OECD. Test No. 404: Acute dermal irritation/corrosion, OECD guidelines for the testing of chemicals, section 4.

2002. Available from: http://dx.doi.org/10.1787/9789264070622-en

[90] OECD. Test guideline 405. Acute eye irritation and corrosion. In: OECD guidelines for the testing of chemicals. Paris, France: Organization for Economic Cooperation and Development (OECD); 2002

[91] Kim SH, Park MK, Seol JK, Im JM, Park HS, Seo HS, et al. Evaluation of potential eye or skin irritation/corrosion in rabbit exposed to TiO2 photocatalyst (GST). Environmental Analysis Health and Toxicology. 2021;**36**(3):e2021022-0. DOI: 10.5620/eaht.2021022. [Epub 2021 Sep 28]

[92] Li Y, Pei Y, Zhang X, Gu Z, Zhou Z, Yuan W, et al. PEGylated PLGA nanoparticles as protein carriers: Synthesis, preparation and biodistribution in rats. Journal of Controlled Release. 2001;**71**(2):203-211. DOI: 10.1016/s0168-3659(01)00218-8

Chapter 3

Colorimetric Cytotoxicity Assays

Eslin Ustun Karatop, Cagla Eren Cimenci and Ayla Melisa Aksu

Abstract

Cytotoxicity experiments are carried out to evaluate whether a chemical has cytotoxic potential. Because of its ease of use and compatibility with data collected from *in vivo* investigations, cell-based cytotoxicity studies have emerged as a viable alternative to animal trials in research. Cell-damaging events such as apoptosis, autophagy, and necrosis may occur after exposure to cytotoxic substances. Thanks to the cell-based cytotoxicity studies, basic information is obtained about the cytotoxic effects of the tested substance. To measure cell viability, a variety of techniques are used. Regardless of the sort of cytotoxicity investigation that was carried out, the crucial thing is to figure out how much metabolic activity there is in the cells at the end of the experiment. Cytotoxicity detection methods are generally colorimetric, luminescent, and enzymatic methods. In colorimetric methods, measurement is based on color change using tetrazolium salts, such as MTT, MTS, XTT, WST. Three main steps are followed in tetrazolium compound toxicity tests. Toxic compounds are introduced to cells in the initial stage. The poisonous chemical is eliminated in the second phase and followed by the addition of the tetrazolium compound. The metabolically active cells are determined in the last stage by using a spectrophotometric approach to measure color change.

Keywords: cytotoxicity, colorimetric assay, formazan, *in vitro*, metabolic activity

1. Introduction

Cytotoxicity assays measure the destructive capacity of substances on living cells or tissue systems. These assays play a key role in the biomedical field as they provide information on the therapeutic potential of the biochemical molecules. Cytotoxicity assays can be applied to both *in vivo* and *in vitro* systems where each method has some advantages and disadvantages. In this chapter, we present and discuss conventional *in vitro* colorimetric cytotoxicity assays that were optimized for spectrophotometric analysis.

There are many cytotoxicity tests with different mechanisms and sensitivities. However, cytotoxicity determination methods are generally examined in three groups. These are colorimetric, luminescence, and enzymatic methods. Luminometric methods are divided into fluorescence and bioluminescence. Fluorometric assays utilize fluorescent substances such as resazurin and are performed with a fluorometer or fluorescence microplate reader. On the other hand, cytotoxicity detection in bioluminescent methods is made by an enzyme called luciferase. In addition, in the real-time bioluminescence method, the exposure of cells to cytotoxic substances has become possible to be examined during the exposure. Enzymes leaking into the medium following cell damage or death have also been considered markers of dead cell counts. Among these enzymes, lactate dehydrogenase (LDH), which stands out with its stability, has taken its place among viability tests as a marker of cell death.

Colorimetric methods are techniques based on color change using tetrazolium salts or specific staining of cells using crystal violet and neutral red dyes. The first colorimetric method was described by Mossman et al., in 1983, which was developed as an alternative to the labor-intensive, time-consuming, and costly radioactive methods of measuring surviving and/or proliferating mammalian cells [1]. Furthermore, they can measure large numbers of samples with a high degree of precision directly in the plate by using a spectrophotometer or plate reader [2, 3]. There are also methods used especially in routine cell culture processes, such as microscopy (staining with trypan blue-Thoma slide) or an automatic cell counter that can be used in cytotoxicity studies. However, these methods are very time-consuming and are not suitable for studies with a large number of samples.

The basic principle of colorimetric assays lies in the color development as a result of metabolic activity rate of the cells. When a reagent meets with the viable cells, it is converted to a colored product by the metabolic activity of the cells. Such color change can easily be detected, measured, and quantified via spectrophotometric measurements as absorbance at a specific wavelength (**Figure 1**). Collected signal reflects the viable cell population or metabolic activity of the cells. As dying cells lose their function to convert substrate into a detectable colored product, the observed signal would be less than the active population.

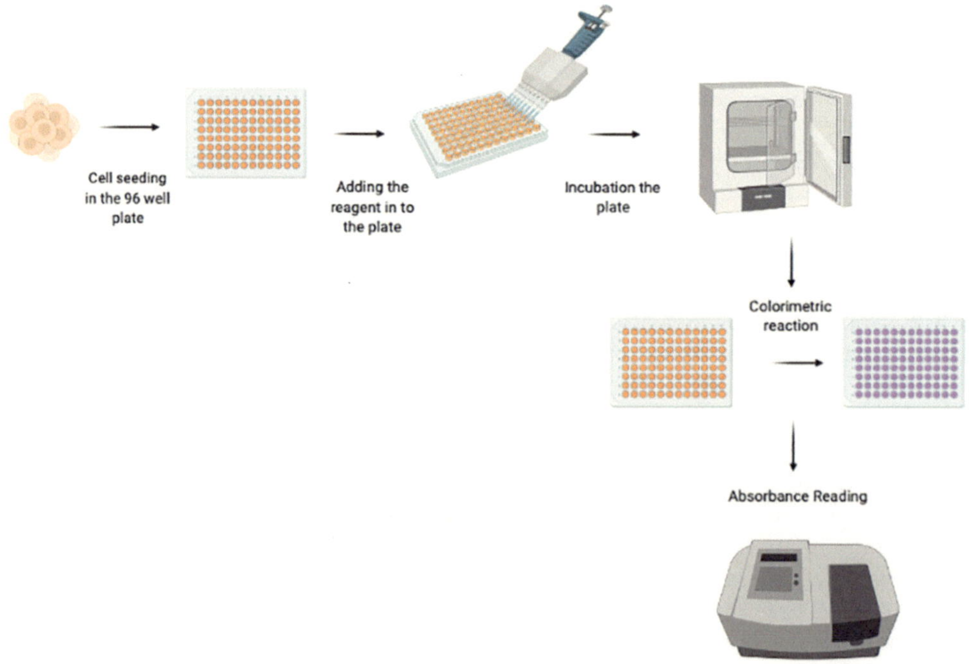

Figure 1.
A simple representation of colorimetric detection assays.

2. Colorimetric assays with tetrazolium salts

In colorimetric cytotoxicity methods, measurements are made based on color change upon addition of tetrazolium salts. Tetrazolium salts are heterocyclic organic compounds. The reduction of tetrazolium salts by gaining electrons enables them to transform into a formazan structure and leads to a color change. The tetrazolium ring can be broken by active mitochondria, and therefore, color change can only occur in metabolically active cells. Taking advantage of these properties, many tetrazolium compounds have been developed [4]. However, only a few of the tetrazolium salts have been accepted and adapted to biological systems. The most widely used tetrazolium salts are listed below:

- MTT: 3-(4,5 dimethylthiazol-2-yl)-2,5 diphenyl-2tl tetrazolium bromide
- XTT: 2,3-bis(2-methyloxy-4-nitro-5-sulfophenyl)-5-[(phenylamino)carbonyl]-2H-tetrazolium hydroxide
- MTS: 3-(4,5-dimethylthiazol-2-yl)-5-(3-carboxymethoxyphenyl) 2-(4-sulfophenyl) -2H-tetrazolium
- WST-1: 2-(4-iodophenyl)-3-(4-nitrophenyl)-5-(2,4-disulphophenyl)-2H tetrazolium
- WST-8: 2-(2-methoxy-4-nitrophenyl)-3-(4-nitrophenyl)-5-(2,4-disulphophenyl)-2H tetrazolium

These compounds can be divided into two groups based on their ability to enter the cell and dissolve in the medium. MTT sits in one of these groups that the tetrazolium salt that has the ability to enter the cell. This salt is a positively charged compound and can be reduced within the cell by conveniently passing the membrane of eukaryotic cells. However, the formazan from reduction is insoluble in water and therefore precipitates in the medium as crystals. Unlike MTT compounds, MTS, XTT, and WST compounds are naturally negatively charged and cannot fully penetrate the cell membrane. Therefore, they use an electron-accepting molecule with themselves. The electron acceptor molecule enters the cell, takes electrons from the cytoplasm or plasma membrane, and returns to the medium to reduce the tetrazolium compound. Formazan, which is formed as a result of the reduction of these compounds, is soluble in water and the medium. Phenazine methyl sulfate and phenazine ethyl sulfate are generally used as electron acceptors in these reactions [5].

3. Spectrophotometry-based colorimetric assays

Tetrazolium salts and their formazan products have been popular candidates for spectrophotometric methods. In this section, the most commonly used salts and their usage methods will be explained.

3.1 MTT assay

Tetrazolium salts are a large group of heterocyclic organic compounds that form highly colored and generally insoluble formazan upon reduction. These compounds, which were first prepared and used in 1894, have been widely used in tests for both biological redox systems and indicators of vitality. MTT is one of the first tetrazolium salts introduced by Mossman in 1983 [1, 6].

MTT molecule is a mono tetrazolium salt and its reagent is 3-(4,5 dimethylthiazol-2-yl)-2,5 diphenyl-2tl tetrazolium bromide, consisting of a positively charged quaternary tetraazide ring nucleus containing four nitrogen atoms surrounded by three aromatic rings containing two phenyl moieties and a thiazyl ring (**Figure 2**) [7, 8].

MTT reduction is one of the most commonly used methods to measure cell proliferation and cytotoxicity. It is an important substance in ongoing tests to determine cytotoxic responses to mitogens, antigenic stimuli, and growth factors. The MTT test is used not only for cell samples, but also for testing tissue cultures. It is used in skin irritation, skin corrosion, and eye irritation tests [9].

The decrease, that is, reduction, of MTT, which passes through the eukaryotic cell membrane very easily, causes the nuclear tetrazole ring to deteriorate and the formation of a violet-colored water-insoluble formazan (**Figure 3**).

As MTT is a positively charged molecule, it can easily pass through the cell and the mitochondrial inner membrane of metabolically active cells. Such viable cells reduce MTT into formazan in their mitochondria as they keep their metabolism in regular activity. The intensity of intracellular formazan produced by this redox reaction is measured via colorimetric-based system in spectrophotometry *in vitro*.

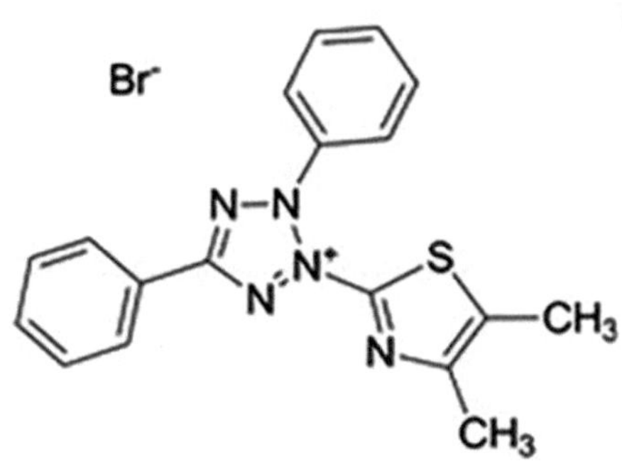

Figure 2.
The molecular structure of MTT tetrazolium salt.

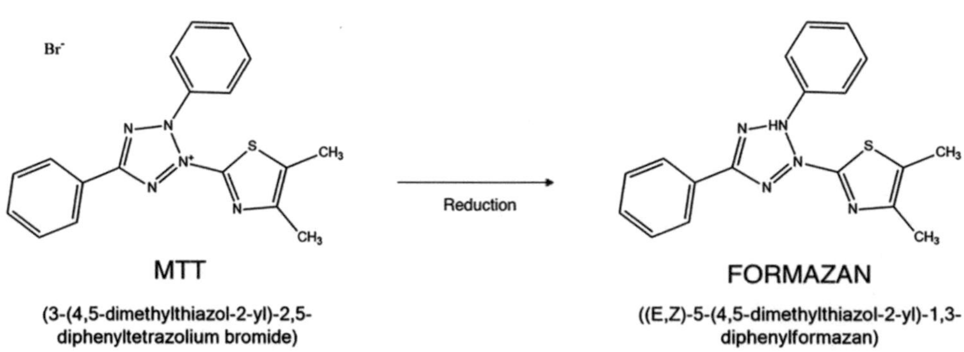

Figure 3.
Redox reaction principle of MTT assay.

The MTT test takes place as a three-stage process. In the first step, the cells or tissue samples to be tested are exposed to the toxic substance. Tetrazolium salt is added to the sample obtained after the toxic substance is removed. In this process, MTT is used as the tetrazolium salt [8].

MTT is originally a yellow-colored substance. During cell proliferation, a reduction leads to an increase in the mitochondrial dehydrogenase enzyme activity and forms a purple crystal structure, formazan (**Figure 4**). The MTT assay is typically performed in several hours (1–4 hours) after incubation of cells with MTT. The formazan crystals that are formed after reaction are insoluble in water. In order to measure the absorbance, they must be dissolved in a suitable solvent, which solubilizes the product. Therefore, the formazan is usually dissolved with a solvent such as dimethyl sulfoxide (DSMO) or isopropanol before the measurement recording [10, 11].

The concentration of formazan dissolved in the appropriate solvent is determined by optical density at 570 nm. As a final step, the color change in the sample should be measured by the spectrophotometric method (**Figure 5**). The reason for this step is to measure cell viability. The measured OD values are considered to be a representation of the intracellular reduction of formazan concentration and thus of MTT. The cellular viability rate of the untreated control group is taken as 100% and proportional calculations are made according to this control group.

One of the reasons why the MTT assay is preferred is that it is a reliable and fast method. The fact that the MTT assay has fewer steps compared with other tests increases the reproducibility. Among the positive aspects of the MTT assays are that more than one sample can be examined with a single test mechanism and the result is sensitive [12].

As with every assay, MTT assays also have disadvantages. If these disadvantages are well understood and necessary precautions are taken, it does not show a negative effect during the application of the test.

As stated earlier, MTT formazan is insoluble in water. After the reaction, it forms needle-shaped crystal structures in the cells. Therefore, these formed crystals must be dissolved before proceeding to the measurement phase. If this dissolution step is not performed well, there may be a difference in absorbance between the wells [13].

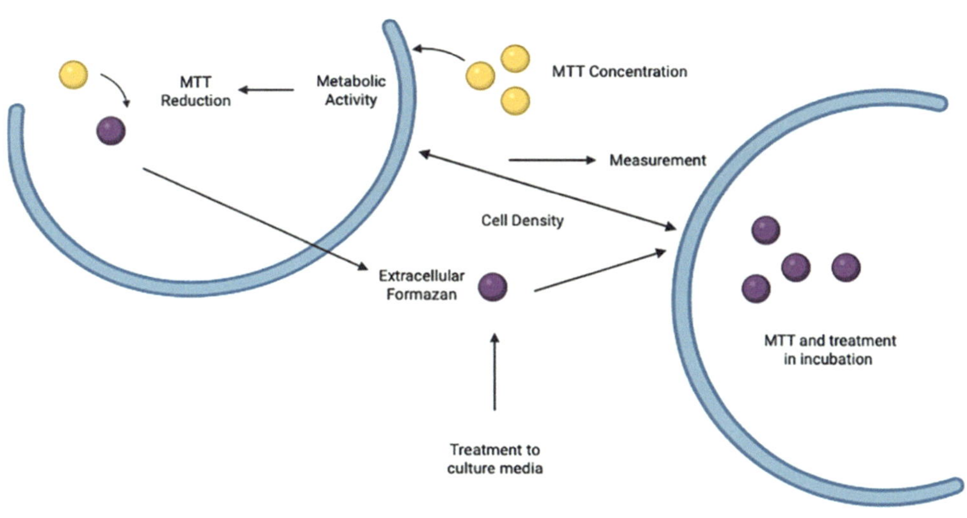

Figure 4.
The illustration of the mechanism of MTT.

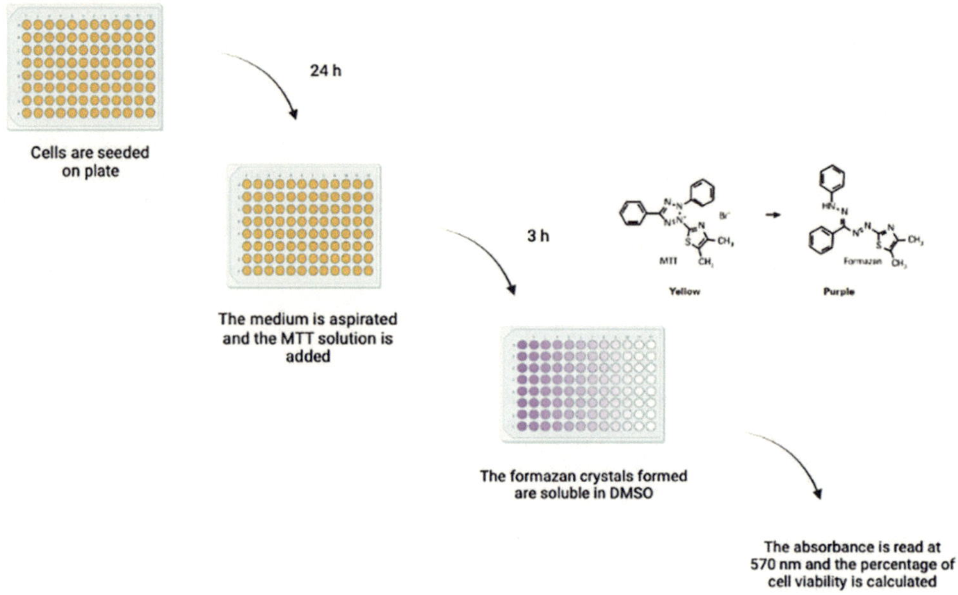

Figure 5.
General steps of MTT assay.

The DSMO substance used during the dissolution process of MTT formazans can have a toxic effect. Care should also be taken when adding DSMO to the wells. The pipette used during the application may damage the formazan crystals. This can create undesirable deviations in the results.

MTT formazan can be a toxic substance due to its structure. Therefore, a control group should be established for cell death observed due to MTT formazan toxicity. In this way, false-negative or -positive results can be avoided.

3.2 XTT assay

After the MTT test developed by Mossman (1983), other tetrazolium compounds such as XTT (2,3-bis(2-methyloxy-4-nitro-5-sulfophenyl)-5-[(phenylamino) carbonyl]-2H-tetrazolium hydroxide) were described by Scudiero et al. in 1988 for the colorimetric method [14]. The XTT procedure is easy to measure proliferation, so it is an acceptable solution for quantifying cells and determining their viability. XTT is a method that is used to determine how cells respond to different growth factors, foreign chemicals, drugs, etc.

XTT testing is a fast, responsive, simple, and safe strategy for using the determination of cytotoxicity. It also offers a high level of sensitivity and precision yet the XTT assay's performance is highly dependent on the mitochondrial dehydrogenase activity of living cells. Therefore, some factors can affect the final absorbance reading. These factors are changes in the reducing capacity of living cells resulting from enzymatic regulation, pH, cellular ion concentration, cell cycle variation, and other environmental factors [15].

Having negatively charged ions on the structure, XTT cannot penetrate the cellular and mitochondrial membrane. Therefore, it is required to incorporate electron acceptor molecules such as phenazine methyl sulfate, phenazine ethyl sulfate, etc.

Phenazine methosulfate (PMS) is an electron acceptor often used in XTT tests. The oxidized (cationic) form of PMS has a yellow color, whereas the reduced derivative shows no color. The reduced PMS is utilized in this assay as an electron carrier since it is rapidly oxidized by oxygen. PMS can also be reduced non-enzymatically by NADH and NADPH. The reduction at the cellular level is gained by PMS at the plasma membrane level and reduces XTT outside the cells, thereby increasing the water solubility of the dye and formazan [5].

As illustrated in **Figure 6**, the tetrazolium salt XTT is reduced to orange-colored formazan by metabolically active cells. The electron acceptor molecules receive electrons from the cell and initiate the redox reaction to reduce the tetrazolium compound. The orange-colored formazan is dissolvable in water, thus the color intensity can be measured with a spectrophotometer. The number of metabolically active cells is proportional to the intensity of the color detected [3].

XTT testing is a fast, responsive, easy to use, and safe method for using the determination of cytotoxicity. It also has high sensitivity and accuracy. In contrast to some other salts, the formazan dye is soluble in aqueous solutions and can directly be quantified using a scanning multiplate spectrophotometer (ELISA-based, **Figure 7**). This enables a high degree of accuracy, allows online data processing by computers, and thus allows a high number of samples to be handled quickly and conveniently (4).

3.3 MTS assay

The MTS assay, which is produced as an alternative to the MTT assay, is an MTT analog that is generally formed by adding sulfonate, methyl, or similar groups to the MTT tetrazolium salt. The structure of the MTS tetrazolium salt is (3- (4,5-dimethylthiazol -2-yl) -5-(3-carboxymethoxyphenyl) 2-(4-sulfophenyl) -2H-tetrazolium) (**Figure 8**). It is also called one-step MTT assay because it is an analog developed to facilitate the MTT assay [16].

MTS tetrazolium salt is negatively charged by nature. As this situation does not allow passage through the cell membrane, the help of intermediate electron acceptor (IEA) molecules is needed. Electron acceptor molecules such as phenazine methyl sulfate (PMS) or phenazine ethyl sulfate (PES) enter the cell, and electron acquisition occurs from the cytoplasm or plasma membrane, and the reduction reaction of the tetrazolium salt takes place. The formazan (dark pink/red color), which is produced as a product after the reduction reaction, is easily soluble in water. Unlike the MTT assay, since the formazan formed after the reduction reaction is water-soluble, there is no need for a second procedure during the test (**Figure 9**) [17–19].

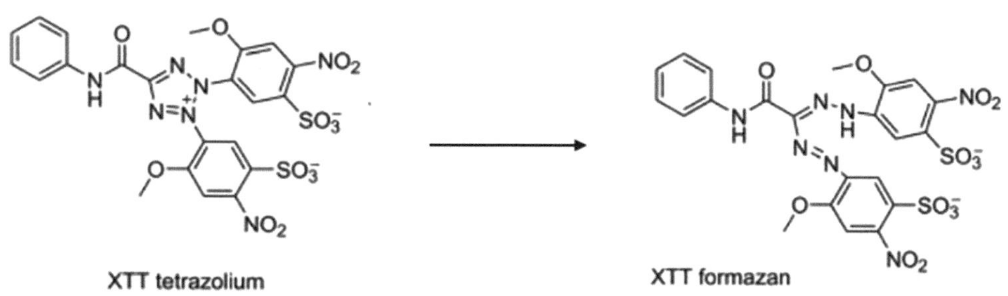

Figure 6.
Conversion of XTT to formazan by mitochondrial dehydrogenase.

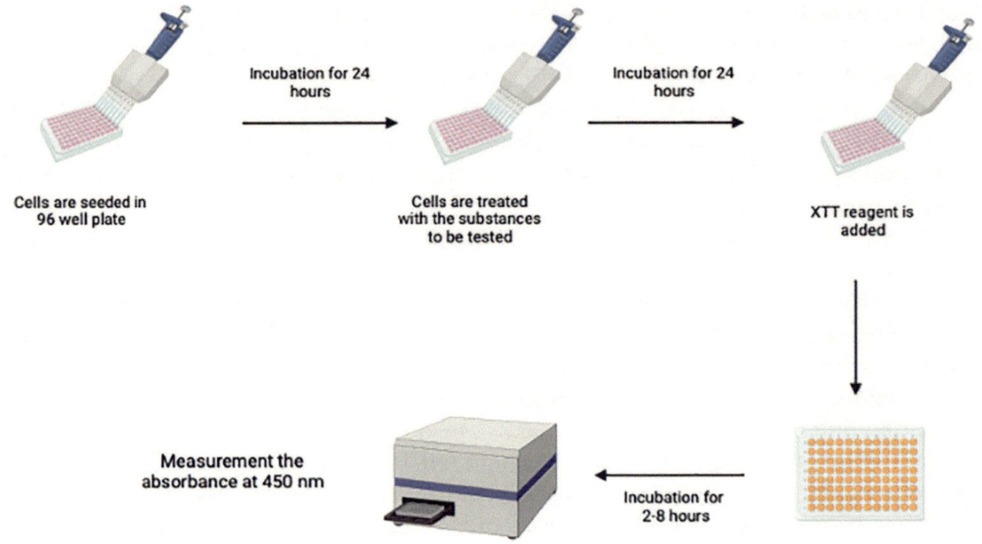

Figure 7.
Illustration of XTT assay principle.

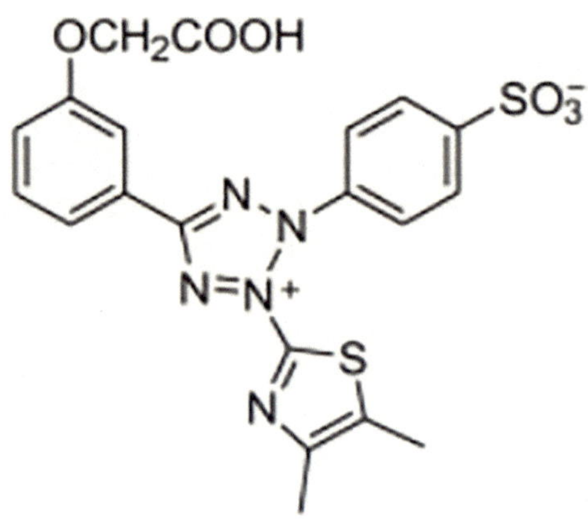

Figure 8.
Chemical structure of MTS tetrazolium salt.

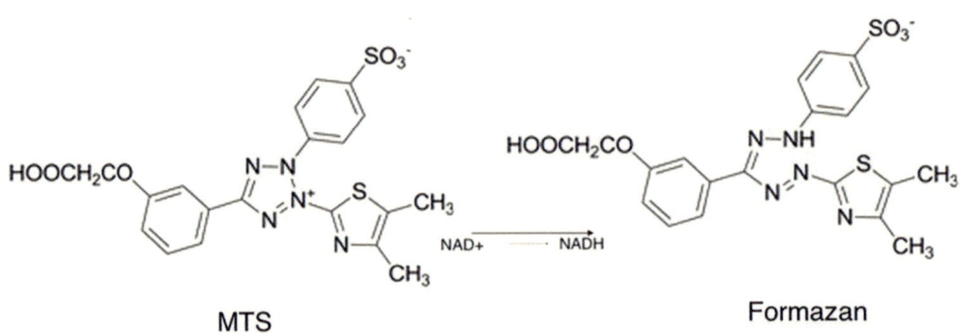

Figure 9.
Reduction of MTS to an aqueous soluble formazan through the transfer of electrons from NADH in the cytoplasm.

The test procedure performed is the same as the MTT assay. MTS salt added to the cells in the culture medium is measured by spectrophotometer at 492 nm after the determined incubation period (30 min–4 hours). MTS assay, which has taken its place among *in vitro* cytotoxicity tests due to its advantages, is sensitive, fast, and easy to apply. Although the MTS assay provides ideal properties for cytotoxicity measurements, the level of absorbance at 492 nm depends on the incubation time applied, the cell type, and the number of cells tested. Considering all these substances, it is a suitable and prone test for use in toxicological test evaluations in the right places [10, 19].

3.4 WST assay

The WST method is another colorimetric method based on the principle of tetrazolium salts that produces a water-soluble formazan product. Among the water-soluble tetrazolium salts, the most frequently used one today is WST-1 in the form of 2-(4-iodophenyl)-3-(4-nitrophenyl)-5-(2,4-disulphophenyl)-2H tetrazolium and another frequently used WST compound is WST-8 (2-(2-methoxy-4-nitrophenyl)-3-(4-nitrophenyl)-5-(2,4-disulphophenyl)-2H tetrazolium) (**Figure 10**) [20].

After the formed formazan crystals dissolve, they can be quantified quickly and easily at an absorbance value of 450 nm in a conventional microplate reader. The absorbance value measured spectrophotometrically in the WST-1 method is related to the number of viable cells (**Figure 11**). As the proliferation increases, the absorbance value increases due to the formation of formazan salt. Since proliferating cells

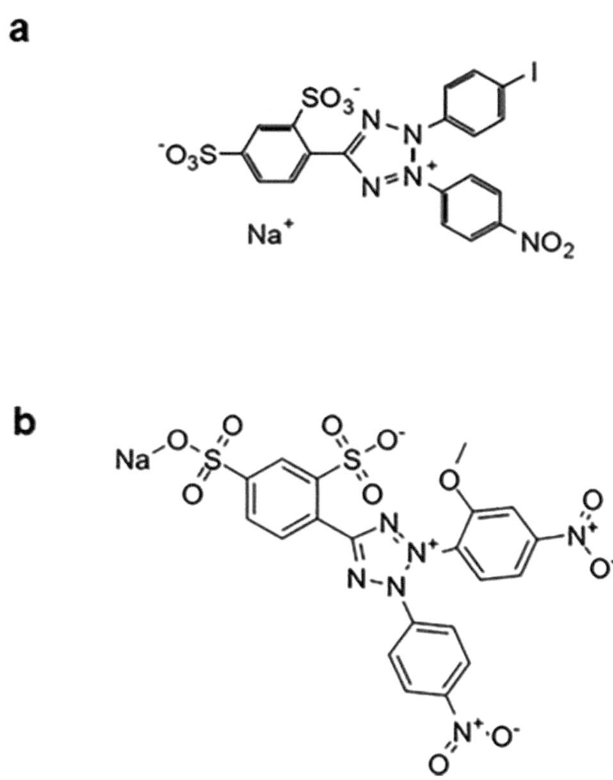

Figure 10.
Chemical structure of a) WST-1 tetrazolium salt and b) WST-8 tetrazolium salt.

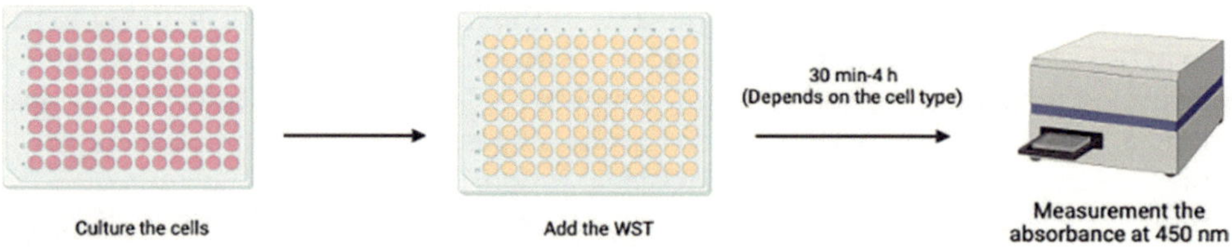

Figure 11.
Principle of WST assay.

show more metabolic activity than non-proliferating cells, this method determines not only cell viability and cytotoxicity, but also determines cell activation and proliferation. In this method, measurements can be made even at low cell concentrations without the use of additional agents or cell washing processes. However, it is a perfect solution for the quantification of cells and the determination of their viability without using radioactive isotopes. This method is based on the principle of measuring cell proliferation concerning various growth factors and nutritional components [21].

4. Discussion and conclusion

Evaluating cellular cytotoxicity is one of the most essential parts of studying cellular functions in biology. These assays are used to test a substance's effects on *in vitro* systems in the fields of including but not limited to oncology, biotechnology, drug discovery, pharmacology, product development, and medical device biocompatibility testing [17, 22]. When exposed to cytotoxic compounds, cells can undergo necrosis, apoptosis, and autophagy, or they could stop proliferating. Detecting this dynamic event is crucial for evaluating the mechanisms in action of cellular actions and pathways involved in cell death after exposure to toxic agents.

There are a variety of assays that can be used in general, such as dye exclusion assays, colorimetric, fluorometric, and luminometric assays [23]. Here, in this chapter, we reported the uses of colorimetric cytotoxicity assays where data are recorded using a multiplate reader. The idea for all is to use a compound to treat the cell and addition of a dye that changes its absorption spectra upon cellular reduction, which is directly proportional to the number of metabolically active cells. Thus, the principles of cytotoxicity assays are different from cellular viability assays, which typically measure viable cells rather than the metabolic activity. As cytotoxicity and viability assays can be utilized separately, additional tests may also be required based on the research aim. As such, they could be used together as complementary methods to get a better understanding of a cell's metabolic reaction.

It should be noted that most of the cytotoxic measurement assays actively affect cellular integrity, protein production, cellular trafficking and alter the cell fate by activating programmed cell death [24]. This makes them irreversible assays where the cells could not be used after the assay. Nonetheless, they provide rapid, robust, sensitive, and cost-effective means to determine whether a material contains potentially biologically harmful activity or substances.

One of the key factors in selecting the assay/dye type in cytotoxicity assays is the biological endpoint. As some experiments require certain types of cells, cellular sources should be carefully selected depending on the endpoint used in the cytotoxicity test.

Colorimetric assay	Mechanism	Advantages	Disadvantages
MTT	Determination of metabolic activity from mitochondrial dehydrogenase enzyme	Cost effective, robust	Needs additional step to solubilize. Less sensitive. Toxic to cells
MTS		Produced formazan is soluble. More efficient, more accurate. Less toxic to cells.	More expensive technique. Absorbance at 492 nm could be easily affected by incubation time.
XTT		Highly sensitive, accurate, safe to cells	Performance may be affected by environmental factors
WST		Easy to use, safe. Can be used in phenol red media.	More time consuming. Expensive. Performance may be affected by environmental factors

Table 1.
Summary, advantages, and disadvantages of different colorimetric biological assays used in vitro.

For example, in cytokine release studies, monocytes and fibroblasts should be the cell of choice [25, 26]. Similarly, cardiac-regeneration-related studies could be conducted with cardiac fibroblasts or cardiomyocytes for better results [27, 28].

Overall, colorimetric assays are simple, inexpensive, accurate, rapid, and sensitive methods in determining cellular toxicity. They are also applicable to both cell suspensions and adherent cells that make them attractive molecules in optimization studies. Although their main principle is based on metabolic activity determination, they all show some unique properties. We summarized their advantages and disadvantages in **Table 1**.

Main difference between the assays is that the MTT is not soluble in water, which requires an additional formazan dissolution step. In contrast, MTS, XTT, and WST assays use a different kind of tetrazolium salt, which produces a soluble formazan, reducing one step in the MTT assay procedure. Therefore, such assays are more efficient and less time-consuming when compared with the MTT assay.

It should be noted that environmental factors such as enzymatic regulation, pH, incubation time, temperature, cellular ion concentration, and variations in cell cycle could affect the performance of the colorimetric assays [29]. As such, an ideal cytotoxicity assay could differ from study to study depending on the aim, action mechanism, and environmental factors.

Conflict of interest

The authors declare no conflict of interest.

Author details

Eslin Ustun Karatop[1*], Cagla Eren Cimenci[2] and Ayla Melisa Aksu[3]

1 Department of Electrical and Computer Engineering, University of Ottawa, Ottawa, Ontario, Canada

2 Department of Cellular and Molecular Medicine, University of Ottawa, Ottawa, Ontario, Canada

3 Department of Bioengineering, Yildiz Technical University, Istanbul, Turkey

*Address all correspondence to: eustu100@uottawa.ca

References

[1] Mosmann T. Rapid colorimetric assay for cellular growth and survival: Application to proliferation and cytotoxicity assays. Journal of Immunological Methods. 1983;**65**:55-63. DOI: 10.1016/0022-1759(83)90303-4

[2] Aslantürk ÖS. In vitro cytotoxicity and cell viability assays: Principles, advantages, and disadvantages. Genotoxicity-A Predictable Risk to Our Actual World. 2018;**2**:64-80. DOI: 10.5772/ intechopen.71923

[3] Riss TL, Moravec RA, Niles AL, Duellman S, Benink HA, Worzella TJ, et al. Assay Guidance Manual [Internet]. 2016

[4] Riss TL, Moravec RA. In: J.E.B.T.-C.B, editor. Chapter 4 - Cell Proliferation Assays: Improved Homogeneous Methods Used to Measure the Number of Cells in Culture. Burlington: Academic Press; 2006. pp. 25-31

[5] Tokur O, Aksoy A. In Vitro Sitotoksisite Testleri,Harran Üniversitesi Vet. Fakültesi Derg. 2017;**6**:112-118

[6] Ghasemi M, Turnbull T, Sebastian S, Kempson I. The MTT assay: Utility, limitations, pitfalls, and interpretation in bulk and single-cell analysis. International Journal of Molecular Sciences. 2021;**22**:12827. DOI: 10.3390/ ijms222312827

[7] Stockert JC, Horobin RW, Colombo LL, Blázquez-Castro A. Tetrazolium salts and formazan products in Cell Biology: Viability assessment, fluorescence imaging, and labeling perspectives. Acta Histochemica. 2018;**120**:159-167

[8] Liu Y, Peterson DA, Kimura H, Schubert D. Mechanism of cellular 3-(4,5-dimethylthiazol-2-yl)-2,5-diphenyltetrazolium bromide (MTT) reduction. Journal of Neurochemistry. 1997;**69**:581-593. DOI: 10.1046/j.1471-4159.1997.69020581.x

[9] Kim H, Choi J, Lee H, Park J, Yoon B-I, Jin SM, et al. Skin corrosion and irritation test of nanoparticles using reconstructed three-dimensional human skin model, EpiDerm(TM). Toxicology Research. 2016;**32**:311-316. DOI: 10.5487/ TR.2016.32.4.311

[10] Erkekoğlu P, Baydar T. Güncel in vitro Sitotoksisite Testleri, Hacettepe Univ. Journal of Faculty Pharmacy. 2021;**41**:45-63

[11] Pascua-Maestro R, Corraliza-Gomez M, Diez-Hermano S, Perez-Segurado C, Ganfornina MD, Sanchez D. The MTT-formazan assay: Complementary technical approaches and in vivo validation in Drosophila larvae. Acta Histochemica. 2018;**120**:179-186. DOI: 10.1016/j. acthis.2018.01.006

[12] Sumantran VN. Cellular chemosensitivity assays: An overview. Methods in Molecular Biology. 2011;**731**:219-236. DOI: 10.1007/ 978-1-61779-080-5_19

[13] Lim S-W, Loh H-S, Ting K-N, Bradshaw TD, Allaudin ZN. Reduction of MTT to Purple Formazan by Vitamin E Isomers in the Absence of Cells. Tropical Life Science Research. 2015;**26**:111-120

[14] Scudiero DA, Shoemaker RH, Paull KD, Monks A, Tierney S, Nofziger TH, et al. Evaluation of a soluble tetrazolium/formazan assay for cell growth and drug sensitivity in culture using human and other tumor cell lines. Cancer Research. 1988;**48**:4827-4833

[15] Page B, Page M, Noel C. A new fluorometric assay for cytotoxicity measurements in-vitro. International Journal of Oncology. 1993;**3**:473-476

[16] Cory AH, Owen TC, Barltrop JA, Cory JG. Use of an aqueous soluble tetrazolium/formazan assay for cell growth assays in culture. Cancer Communications. 1991;**3**:207-212. DOI: 10.3727/095535491820873191

[17] Riss TL, Moravec RA, Niles AL, Duellman S, Benink HA, Worzella TJ, et al. Assay Guidance Manual [Internet]. 2019

[18] Berridge MV, Herst PM, Tan AS. Tetrazolium dyes as tools in cell biology: New insights into their cellular reduction. Biotechnology Annual Review. 2005;**11**:127-152. DOI: 10.1016/S1387-2656(05)11004-7

[19] Arab-Bafrani Z, Shahbazi-Gahrouei D, Abbasian M, Fesharaki M. Multiple MTS assay as the alternative method to determine survival fraction of the irradiated HT-29 colon cancer cells. Journal of Medical Signals Sensing. 2016;**6**:112-116

[20] Präbst K, Engelhardt H, Ringgeler S, Hübner H. Basic colorimetric proliferation assays: MTT, WST, and Resazurin. Methods in Molecular Biology. 2017;**1601**:1-17. DOI: 10.1007/978-1-4939-6960-9_1

[21] Protocol Guide: WST-1 Assay for Cell Proliferation and Viability. n.d.. https://www.sigmaaldrich.com/CA/en/technical-documents/protocol/cell-culture-and-cell-culture-analysis/cell-counting-and-health-analysis/cell-proliferation-reagent-wst-1. [Accessed May 29, 2022]

[22] Adan A, Kiraz Y, Baran Y. Cell proliferation and cytotoxicity assays. Current Pharmaceutical Biotechnology. 2016;**17**:1213-1221. DOI: 10.2174/1389201017666160808160513

[23] Wahab NFAC, Kannan TP, Mahmood Z, Rahman IA, Ismail H. Methods in cytotoxicity testing: A review. Recent Patents Material Science. 2017;**10**. DOI: 10.2174/1874464810666170411105114

[24] Groth T, Falck P, Miethke RR. Cytotoxicity of biomaterials –Basic mechanisms and in vitro test methods: A review. Alternatives to Laboratory Animals. 2020;**23**:790-799

[25] Schildberger A, Rossmanith E, Eichhorn T, Strassl K, Weber V. Monocytes, peripheral blood mononuclear cells, and THP-1 cells exhibit different cytokine expression patterns following stimulation with lipopolysaccharide. Mediators of Inflammation. 2013;**2013**:697972

[26] Zhang Y. Cell toxicity mechanism and biomarker. Clinical and Translational Medicine. 2018;**7**:34. DOI: 10.1186/s40169-018-0212-7

[27] Mishra PK, Adameova A, Hill JA, Baines CP, Kang PM, Downey JM, et al. Guidelines for evaluating myocardial cell death. American Journal of Physiology and Circulation Physiology. 2019;**317**:H891-H922. DOI: 10.1152/ajpheart.00259.2019

[28] Bolt HM. Highlight report: Cell type selection for toxicity testing. EXCLI Journal. 2018;**17**:1180-1181

[29] Kamiloglu S, Sari G, Ozdal T, Capanoglu E. Guidelines for cell viability assays. Food Frontiers. 2020;**1**:332-349. DOI: 10.1002/FFT2.44

Chapter 4

Cellular Cytotoxicity and Multiple Sclerosis

Annie M.L. Willson and Margaret A. Jordan

Abstract

Multiple sclerosis (MS) is an autoimmune disease in which discrete central nervous system lesions result from perivascular immune cell infiltration associated with damage to myelin (demyelination), oligodendrocytes and neurons. This culminates in debilitating neurological symptoms, primarily affecting women in their child-bearing years. Both the innate and adaptive branches of the immune system have been implicated in disease initiation and progression, and although the underlying cause remains elusive, there is compelling evidence for a complex interaction between genetic and environmental factors, leading to inflammation and neurodegeneration. Both direct cellular toxicity and antibody-dependent cellular cytotoxicity (ADCC) involving several cell types have been identified in playing major roles. These cells and their interactions in the pathogenesis of MS will be discussed.

Keywords: multiple sclerosis, cellular cytotoxicity, cell subsets, human, MS mouse models

1. Introduction

Multiple sclerosis (MS) is the most prevalent neurological disease among young adults in developed countries, with approximately 2.8 million people being affected worldwide [1]. It principally affects women in their prime, with diagnosis typically occurring between the ages of 20 and 40. The disease is debilitating due to central nervous system (CNS) damage resulting from activated lymphocytes migrating across the blood brain barrier (BBB) and engaging in a proinflammatory response. This causes cells to attack and destroy the myelin sheaths that coat the axons of neurons of the brain, spinal cord and optic nerve, as well as the myelinating cells or oligodendrocytes, and the axons themselves [2]. As neurons receive sensory input from external sources and send motor commands to the muscles by relaying interneuron electrical impulses, breakdown causes interruption to the signals being sent around the body, and dependent on where the damage occurs, results in different signs and symptoms. These can include vision impairment, muscle spasms and numbness, bladder and bowel issues, fatigue and difficulty walking [3]. Most people with MS have progressive neurological disability which, though not usual, can culminate in death [4]. The area of damage or scarring caused by the immune system attack is called a lesion or plaque, and can be visualised by magnetic resonance imaging (MRI). A definitive diagnosis of MS is made when these plaques are shown to be reoccurring and when there is the clear presence of clinical symptoms [3].

Two major types of MS have been recognised, primary progressive multiple sclerosis (PPMS), diagnosed in approximately 15% of patients and which results in steady progression of disease from onset, and relapsing remitting multiple sclerosis (RRMS), which affects approximately 80% of patients and is characterised by periods of relapse separated by periods of remit without worsening of symptoms [5–7]. Most patients with an initial diagnosis of RRMS will, within 20 years of diagnosis, progress to secondary progressive multiple sclerosis (SPMS) where the stages between relapse and remit shorten and there is a steady decline with an increase in symptoms and disease progression [8]. Up to approximately 5% of MS patients have progressive relapsing multiple sclerosis (PRMS) and this characterised by steady disease progression with occasional relapses [9].

The exact cause of MS is still unknown, however research has determined that it is an autoimmune disease, arising from complex interactions between environmental and genetic influences. There is a latitude incidence variance, with prevalence of MS increased the further one is from the equator; sunlight and vitamin D are therefore being investigated as disease triggers [1, 10–12]. Childhood exposure to bacteria and viruses have also been investigated, due to a person's disease risk being set as the incidence of the region they moved to prior to puberty [13, 14]. Of note, every patient with MS have previously been exposed to Epstein–Barr virus (EBV) [15, 16]. Smoking also increases a person's risk and worsens symptoms following diagnosis [17].

Although the disease is not inherited, it has a genetic component, with those having an affected first degree relative exhibiting an increased incidence of disease [18], and twin studies indicate that there is a 30% chance of developing disease in the second twin if the first has been diagnosed with MS [19]. Genome wide association studies (GWAS) have identified more than 230 genes associated with a person's MS risk, several being immune genes, particularly those of T cells, B cells, natural killer (NK) cells, monocytes and microglia, implicating involvement of both major branches of the immune system, the innate and adaptive immune responses, in initiation and progression of disease [20–23]. These studies are supported by several human and animal model functional studies [24–26].

Cellular toxicity, or the ability to kill other cells, is an important effector mechanisms of the immune system to protect us from infections, cancer or autoimmune diseases. There is a close association between inflammation and neurodegeneration, and cellular toxicity has been implicated as a having a major role in MS [27]. The main players are CD8, or cytotoxic, T cells and NK cells. Cellular toxicity can operate by many mechanisms including NK cell release of lytic granules containing perforin or granzymes to kill directly, or by inducing death receptor-mediated apoptosis via tumour necrosis factor (ligand) superfamily member 10 (TRAIL) or Fas Ligand (FasL) expression on CD8 T cells [28]. There are also antibody-dependent cell-mediated cytotoxic mechanisms (ADCC), where B cells produce antigen specific antibodies or immunoglobulins, that will coat a pathogen or foreign body, marking them for killing or destruction through cell to cell cytoloysis by effector immune cells expressing FcγRIIIA (CD16A), including classical NK cells, monocytes/macrophages, neutrophils, eosinophils, NKT cells, or γδT cells (reviewed in [29]).

2. Innate and adaptive immunity

The immune system of vertebrates is commonly divided into two main complementary parts, innate and adaptive immunity, the bridge between which is critical for an efficient and effective immune response.

The innate immune system is evolutionary the most primitive, where there is non-specific response to a broad class of antigens. The haematopoietic cells involved include macrophages, dendritic cells, mast cells, neutrophils, eosinophils, NK cells and NKT cells. Although 1908 Nobel Prize winner, Elie Metchnikoff, first described an important role for the innate immune system [30], it is only now being recognised as a critical regulator of human inflammatory disease. Innate immunity involves the recognition of infected cells through surface recognition receptors. These are termed pattern recognition receptors (PRRs) which recognise pathogen associated molecular patterns (PAMPs) unique to non-vertebrate cells, including bacteria and fungi. They are also on internal vesicle membranes for recognition of viral ssRNA and dsRNA and for distinguishing lysed bacterial components [31]. Cytotoxic innate lymphocytes can lyse abnormal or infected cells through the release of cytotoxic granules containing perforin or granzymes, and antigen presenting cells (APCs) can be activated by the innate immune system to present pathogen antigens on their surface. Once activated they will migrate to secondary lymph organs to present their antigen to T cells, and in so doing also activate the adaptive immune system response [32, 33]. The innate immune system therefore functions through a combination of cellular defences and humoral components to defend against nonspecific antigens before activating B and T cells, triggering an adaptive immune response. Speed is the main advantage of innate immunity, with a protective inflammatory response being generated within minutes of pathogen exposure.

Another part of innate immunity is the complement system, which is made up of several small proteins that have been synthesised in the liver and circulate in the blood as active precursors that when stimulated are proteolytically cleaved to release cytokines, leading to a cascade of reactions, ultimately resulting in complement activation or fixation [34]. As the name suggests, they complement or enhance the ability of antibodies and phagocytic cells to clear damaged or diseased cells by promoting inflammation and attack of the cell membrane of the pathogen. Antibodies, generated by the adaptive immune system, can activate the complement system.

Adaptive immunity, sometimes referred to as acquired immunity, is highly specialised and helps to protect the body by recognising antigens, whether they are foreign to the host's immune system (exogenous), produced by intracellular bacteria or viruses (intracellular) or produced by the host (autoantigen). The adaptive immune system also remembers previously encountered antigens, leading to quicker response times [35]. T and B lymphocytes are the main cells mediating adaptive immunity, with T cells being further divided into the cytotoxic CD8 T cells and CD4 T cells that constitute several classes of what are commonly referred to as "helper T cells". These cell have produced highly specific receptors for recognition of hundreds or even thousands of antigens through genetic recombination, and this facilitates pathogen specific immunologic effectors pathways, the generation of immunological memory and the regulation of host immune homoeostasis [36].

CD8 T cells recognise infected cells through interaction of T cell receptors with antigens presented by major histocompatibility complex (MHC) class I on the infected cell. The target cell is then killed by the release of cytotoxins, such as perforin and granzymes, from the CD8 T cell [28]. CD4 T cells, on the other hand, recognise antigens presented in the context of MHC II on an APC. Binding to MHC II molecules activates CD4 T cells to release cytokines, which can stimulate CD8 T cells, macrophages and B cells to form an immune response (reviewed in [37]). They can, for example, release cytokines as instructors to CD8 T cells to release cytotoxins, or to B cells to produce pathogen specific antibodies. They therefore instigate and shape adaptive immune responses dependent on the cytokines they release. These can be mainly Th1, or inflammatory, in nature, such as IFN-γ and IL-12, responsible for the control of intracellular pathogens, or polarised to a more anti-inflammatory Th2 response, where cytokines such as IL-4, IL-5 or IL-13 are produced [38, 39]. A disturbance in this Th1/Th2 response can have severe consequences, be they more Th2 in nature, driving asthma and allergy, or Th1 driven, resulting in autoimmune diseases, including MS (reviewed in [40]). A couple of the more recently identified CD4 T cells subsets include Th17 cells that are characterised by production of IL-17 and IL-23, and have been linked to inflammatory diseases, and T regulatory (Treg) cells, which are important in maintaining homeostasis and tolerance of the immune system [41–43]. Tregs express the transcription factor FoxP3 which is essential for their development and function [44–46]. In humans, mutations in FOXP3 have been found to result in immune dysregulation, polyendocrinopathy, enteropathy, X-linked (IPEX) syndrome, providing evidence that anomalies of Tregs can cause autoimmune disease and allergy [47].

During production of the T cell receptor (TCR) on T cells and B cell receptor (BCR) on B cells, random genetic recombination events can lead to receptors being produced that are specific to autoantigens [48, 49]. To prevent reaction to self, cells undergo central and peripheral tolerance events through which autoreactive cells are apoptotically removed, first in the primary lymphoid organs of the thymus (T cell) and bone marrow (B cell), and if this fails, in the secondary lymphoid organs after cells migrate to the periphery [49]. Self-reactive antibodies account for 55–75% of all antibodies expressed by early immature B cells, including polyreactive and anti-nuclear specificities [49]. However, it is estimated that the majority of newly produced B cells do not reach maturity, and during central and peripheral tolerance most of the self-reactive B cells are removed. If both selection processes fail in T or B cells, this will result in T and B cells able to react with the body's own cells and tissues. These events lead to autoimmune disease.

3. Autoimmune disease

Inflammation as a response of the body to infection or cell injury is a well-known concept that dates back to the beginning of medicine. However, Metchnikoff pointed out that although normally a method of protection, inflammation that exceeds normal bounds can cause disease [27]. Even with this knowledge, it was not until the 1950s that inflammation was recognised as inducing an autoimmune reaction responsible for disease. Autoimmune disease is characterised by an excessive immune response against self, often resulting in inflammation and tissue destruction, in the absence of a threat to the organism [50, 51]. Aberrant immune responses have been associated with over 80 disorders, including multiple sclerosis, and affects 5–7% of the population [52]. Clinical observations over the past decade have suggested that the prevalence of all autoimmune disease, not just MS, is increasing, bringing the issue to the forefront of scientific interest [53, 54]. Successful treatment of autoimmune disease is also of great societal interest, as they are commonly characterised by chronic natures, ongoing health care costs, and debilitating issues resulting in loss of productivity.

Immunological self-tolerance is maintained in part by Tregs. Tregs are CD4 T cells that actively and dominantly supress lymphocytes, particularly self-reactive T cells in the normal periphery that exist despite the deletion mechanisms in the thymus [43]. Natural CD25+ CD4 Tregs utilise several modes of suppression, including cell contact dependent mechanisms, such as the killing of APCs or responder T cells by granzyme and perforin, and by mediation of soluble factors, such as the secretion of immunosuppressive cytokines like IL-10, TGF-β or IL-35, or deprivation of cytokines necessary for expansion and survival of responder T cells (reviewed in [55, 56]).

Optimal T cell function relies on a carefully maintained state of equilibrium. When one subpopulation of T helper cells is activated, others are modulated or inhibited to promote the most specific effector response to the threat [57]. The cellular development of Tregs shares a common cytokine with Th17 cells, TGF-β [41, 42]. Th17 cells are the opposing force to Tregs, serving as an effector lymphocyte population that plays a key role in autoimmunity [41, 42]. At homeostasis, Th17 cells promote gut barrier defence, granulopoiesis, granulocyte chemotaxis and immunity against extracellular pathogen [58]. IL-17 induces granulopoesis indirectly through the stimulation of fibroblasts, epithelial and endothelial cells to secrete GM-CSF, IL-6, IL-8 and MIP-2, with IL-8 and MIP-2 enhancing chemotaxis of neutrophils [59, 60]. While Th17 cell mediated immunity is crucial for maintaining mucosal and haematopoietic homeostasis, too strong a response can induce autoimmunity. The relationship between Tregs and effector Th17 must remain balanced to provide the optimal functional immunity and health of an organism.

Another theory of immune regulation is the hypothesis of homeostasis between Th1 and Th2 cells. The subpopulations can be distinguished by the cytokines they produce and the expression of difference cell surface molecules. Th1 cells are responsible for cell mediated immunity, phagocyte dependent protective responses, B cell activation and production of opsonising antibodies such as IgG1, whereas Th2 cells produce cytokines that are responsible for strong antibody production, eosinophil activation and inhibition of several macrophage functions, thus providing phagocyte independent protective responses [61]. Th2 cells are also responsible for the general activation of B cells. When the Th1/Th2 paradigm is thrown out of balance by failure of central or peripheral tolerance, immunological disorders can occur due to uncontrolled responses [61].

4. MS and the immune system

MS arises when there is an imbalance in the body's immune response, shifting it from a beneficial immune process that fights infection and disease towards a self-aggressive immune attack on the cells within the CNS (**Figure 1**). Genetic and environmental factor interaction may facilitate movement of autoreactive T cells, macrophages and NK cells and demyelinating antibodies from the periphery to the CNS. In the periphery self-antigens can be presented on MHC II molecules by APCs to TCRs on T cells, thereby activating proinflammatory T cells [48]. The activated T cells can then migrate through the blood brain barrier to the brain and spinal cord [2]. Once in the CNS the T cells can be reactivated by CNS antigens presented on MHC II by other APCs, primarily microglial cells [62]. Secretion of proinflammatory Th1 cytokines by the reactivated T cells can induce CNS inflammation by activating

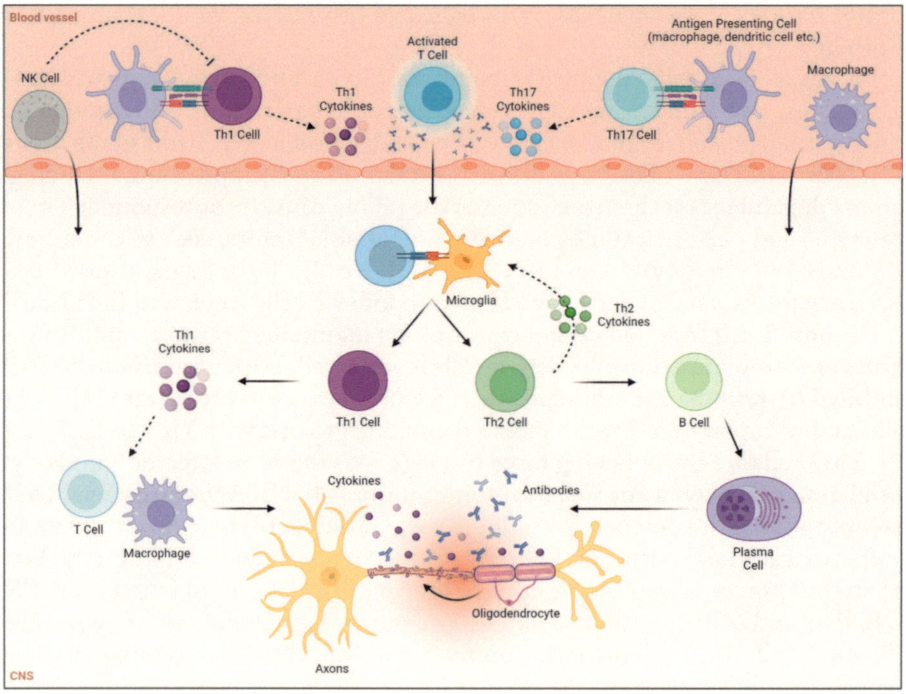

Figure 1.
Multiple sclerosis pathogenesis. Autoreactive T cells, macrophages and NK cells, and demyelinating antibodies, may migrate across a compromised blood brain barrier. T cells are reactivated in the central nervous system by antigen presenting cells (APC). Anti-inflammatory cytokines released by Th2 cells can stimulate B cells to differentiate into plasma cells that secrete demyelinating antibodies. Alternatively, the release of proinflammatory cytokines by Th1 cells can enhance immune response, via activation of other immune cells such as CD8 T cells and macrophages to attack the myelin sheath and oligodendrocytes causing demyelination and the development of clinical symptoms of MS (created with BioRender.com by A Willson).

macrophages, B cells and other T cells [63]. The antibodies can also initiate a complement cascade resulting in assembly of the membrane attack complex, forming pores in the myelin membranes.

5. Cell types

While inflammation and neurodegeneration are correlated in active lesions, research suggests that neurodegeneration may become independent from inflammation in progressive disease [64]. There are many MS therapeutics that suppress proinflammatory cytokines or their effector functions, but not all treatments show equal efficacy and can cause unintended effects. Currently, there is no cure. It is thus becoming clear that there is a need to elucidate the different populations important in initiating and progressing disease, and by studying their interactive networks, identify possible areas for targeted intervention.

5.1 T cells

While there is overwhelming evidence of a role for T cells in the pathogenesis of MS, further studies in humans and in the mouse model of disease, experimental autoimmune encephalomyelitis (EAE), provides compelling evidence that other cell types play major roles. Linkage to the Human Leukocyte antigen (HLA) locus, including MHC I and II genes, was the first genetic locus identified, and still provides today the strongest linkage to MS. Further studies have identified an extended HLA haplotype, HLA-DRB1*15:01, DQA1*0102, DQB1*0602, within the MHC class II region [65]. As MHC II molecules specifically present peptide antigens to activate CD4 T cells, this suggests that CD4 T cells are important in initiation and progression of MS.

Th1 cells are a lineage of CD4 effector T cells that promote cell mediated immune responses and are necessary for defence against intracellular viral and bacterial pathogens. They were originally believed to be the main pathogenic T cells in MS, not only because susceptibility genes were linked to MHC II molecules, but also because immune surveillance of a healthy brain to scan for infection, showed favouring towards infiltration by Th1 cells, and therapeutic strategies designed to induce a shift from Th1 to Th2 immune response resulted in beneficial outcomes in MS patients [66–68].

The development of Th1 cells is coupled to the involvement of cell-extrinsic and cell-intrinsic factors, including signal transducer activator 1 (STAT1), the transcription factor Tbx21, IL-21 and STAT4 [69]. The CD4-Th1 model for MS was further supported by a trial performed in 1987, which found that administering IFN-γ to RRMS patients exacerbated disease. An accompanying increase in circulating monocytes bearing class II (HLA-DR) surface antigens suggested that the attacks induced by the treatment were immunologically mediated [70].

Th1 cells are also known to drive EAE. However, it was found that transgenic mice that lacked Th1 cells developed more severe EAE, thereby contradicting the Th1 cell theory for MS [71]. This conundrum was partially resolved following further investigation involving IL-23, a heterodimer cytokine composed of a unique p19 subunit and a common p40 subunit shared with IL-12. IL-23 promotes development of Th17 cells as opposed to Th1 cells [72]. Early studies on Th17 cells therefore dismissed a role for the previously favoured Th1 cells, but more recent research suggests that both cell types may play distinct roles in pathology [73]. It was suggested that Th1 cells accessed the CNS initially and subsequently facilitated the recruitment of Th17 cells [73].

Analysis of CNS tissue revealed distinct histopathological features and immune profiles depending on cytokine modulated T cells. IL-12p70 driven disease was characterised by macrophage-rich infiltrates, however in IL-23 driven lesions it was found that neutrophils and the growth factor, granulocyte colony stimulating factor (CSF), were the most prominent [74]. Research has shown that while IL-23 is commonly associated with the expansion of Th17 cells or the stabilisation of the Th17 phenotype, a similar course of EAE has been reported following the transfer of MOG-specific T cells into either wild type or IL-23 knockout mice [75]. This suggests that once encephalitogenic cells have been generated, EAE can develop in the absence of IL-23. IL-23 may therefore only be necessary for disease induction and not the effector phase of disease.

While MHC II molecules were found to be the strongest associated with MS in genetic studies, the MHC I HLA-A*0301 allele, independent of the HLA II haplotype DRB1*15,DQB1*06, was found to be increased in MS patients [65]. There was also a negative association with the MHC I HLA-A*0201 and disease [76]. As MHC I molecules are recognised by CD8 T cells, this suggests that CD8 T cells play a role in MS.

In one of the first studies that shifted from a CD4 T cell focus, CD8 T cells outnumbered the CD4 T cell subset in all parenchyma samples from MS patients, regardless of the MS type, duration or speed of disease progression [77]. Research has also shown that APCs, including dendritic cells (DCs), interact with T cells and proliferating lymphocytes, predominately CD8 T cells, at the margins of chronic active MS lesions [78]. CD8 T cells have also been found within active lesions of RRMS patients [77]. These T cells, and to a lesser extent, compartmentally differentially distributed B cells, have been shown to correlate with disease progression and damage.

CD8 T cells are an important subpopulation of MHC I restricted T cells, and are mediators of adaptive immunity. Cytotoxic T cells specialise in direct killing of cells that are infected, particularly with viruses, or are cancerous or damaged in other ways. Cytotoxic cells rely on two mechanisms for lytic activity: granule-dependent cytotoxicity (reviewed in [79]) and death receptor dependent cytotoxicity (reviewed in [80]). The principle mechanism used is granule-dependent cytotoxicity. In lesion prone areas of the CNS, T lymphocytes, including CD8 cytotoxic T lymphocytes (CTLs), are recruited to the affected tissue and brain cells are stimulated to present antigens to the T lymphocytes via de novo expression of MHC molecules. Although levels of MHC I and MHC II are very low in normal CNS parenchyma, neural injury leads to a massive increase in activated and phagocytotic microglial, which can serve as competent APCs [81]. To develop into functioning CD8 T cells, the TCR must recognise the MHC-peptide combination along with the costimulatory signal from APCs. While classical MHC I molecules necessary for CD8 T cell activation are not usually expressed on neural cells, they are induced in most inflammatory and degenerative CNS diseases [82].

Oligodendrocytes lack expression of costimulatory molecules and are thus unable to trigger the full effector of T cells, however they have been known to express MHC I in vitro [83]. Therefore, despite the lack of complete activation of the T cells, oligodendrocytes may still be targets of primed CTLs. MHC I expressing oligodendrocytes are susceptible to lysis by blood donor derived CD8 CTLs [83]. IFN-γ treated human oligodendrocytes also express Fas/CD95, and are therefore susceptible to death receptor dependent cytotoxicity [84]. Another component of the CNS, the neurons, were found to be capable of expressing MHC I when treated with IFN-γ [85, 86]. Medana and colleagues in 2000 discovered that hippocampal neurons were highly susceptible to direct application of cytotoxic granules, but showed no signs of perforin mediated lysis or membrane damage following attack by CTLs [87]. This effect was not observed in any other cell type.

Research to date indicates that all cellular elements of the CNS may act as targets to CTLs but that susceptibility and cytotoxic pathways involved vary dependent on the cell type and the immune activations during the course of the inflammatory process.

5.2 B cells

Historically, B cells have not been recognised as major players in regulatory function in the development of autoimmune diseases, although the identification of autoantibodies produced by autoreactive plasma cells and their pathogenic consequences are widely accepted [88]. B cells are considered effector cells as well as cells with immunoregulatory potential. B cells in MS patients express increased levels of costimulatory molecules, increasing the stimulation of antigen-reactive T cells [89]. It has been reported that MS patients have increased levels of IL-6 and GM-CSF, correlating with increased Th17 cells [90, 91]. B cell targeted therapies utilise B cell depleting monoclonal antibodies against the B cell marker CD20. These antibodies trigger B cell lysis through antibody dependent cellular cytotoxicity, complement dependent cytotoxicity or apoptosis induction [92].

5.3 NK cells

Administration of daclizumab, an alpha subunit of IL-2 receptor blocking monoclonal antibody, to MS patients was found to strongly reduce brain inflammation. This therapy, while being associated with a decline in circulating CD4 and CD8 T cells, also correlated with a significant expansion of $CD56^{bright}$ NK cells in vivo. This provided supporting evidence of NK cell-mediated negative immunoregulation of T cells during daclizumab treatment [93], and the identification of NK cells in association with MS, where positive outcome was possibly due to the treatment's effect of increasing the NK cell numbers [94, 95].

For decades, NK cells have been classified as a component of the innate immune system. However, evidence suggests that, like B and T cells, NK cells are educated during development, possess antigen-specific receptors, undergo clonal expansion and generate memory cells (reviewed in [96]). Research originally suggested that NK cells developed and underwent differentiation within the bone marrow, however more recent extensive ex vivo characterisation of haematopoietic precursor cells (HPCs) and downstream NK cell development intermediates (NKDIs) reveals that they are enriched in secondary lymphoid tissues (STLs), including the tonsils, spleen and lymph nodes [97–100]. This suggests that NK cells in humans can differentiate in the SLTs, and may do so preferentially.

Human NK cells are phenotypically defined by expression of CD56 and the lack of CD3 expression [101]. CD56 is the 140-kDa isoform of neural cell adhesion molecule (NCAM) found on NK cells and a minority of T cells [102]. NK cells are categorised into two distinct populations depending on the cell surface density of CD56. The majority of human NK cells, approximately 90%, express low levels of CD56 ($CD56^{dim}$) and high levels of FCyRIII (CD16), while the minority express higher levels of CD56 ($CD56^{bright}$) [103]. $CD56^{bright}$ NK cells have long being associated with an immunoregulatory role, due to increased production of NK-derived immunoregulatory cytokines, including IFN-γ, TNF-β, IL-10, IL-13 and GM-CSF, and reduced cytotoxicity compared to $CD56^{dim}$ NK cells [104]. $CD56^{bright}$ NK cells express receptors for cytokines such as IL-12, IL-15 and IL-18, produced by APCs, which can trigger proliferation of $CD56^{bright}$ NK cells and their production of molecules, including IFN-γ, IL-10 and IL-13 [104]. It has been demonstrated that DCs are a key source of cytokines for the activation of $CD56^{bright}$ NK cells [105]. Modulation and proliferation of $CD56^{bright}$ NK cells can also occur due to DC-derived IL-27 [105]. Activated NK cells can modulate the function of APCs by stimulating monocytes to produce TNF-α and kill immature DCs by a perforin-dependent process referred to as DC editing [106, 107]. However, more recent research has challenged this commonly accepted concept of $CD56^{bright}$ as the primary source of immunoregulatory cytokines. Studies have shown that $CD56^{dim}$ NK cells are also a major source of proinflammatory cytokines and chemokines that are induced rapidly after target cell recognition [108, 109].

The absence of MHC class I molecules, as indicated by virally infected cells or cancerous cells with MHC I downregulated, is not always sufficient to induce NK cell mediated death, suggesting that there must be activating receptors on NK cells whose affinity for target cell ligands dominates over the inhibitory signals of the NK cell. Some activating receptors identified include NKG2D, the NCR, and NKp80 [110–112]. NKG2D is the best characterised of these activating NK cell receptors. It is a C-type lectin-like receptor expressed on the surface of all human NK cells and recognises at least six ligands, each with a MHC class I homology [113]. Following receptor-ligand interaction, NKG2D phosphorylates an adaptor protein that recruits and activates phosphatidylinositol-3 (PI-3) kinase, which results in perforin-dependent cytotoxicity [114, 115]. Gunesh et al. found that the deletion of CD56 on the NK92 cell line lead to impaired cytotoxic function. The knockout CD56 cells failed to polarise during immunological synapse formation and had severely impaired exocytosis of lytic granules [116].

Treatment of MS patients with IFN-β caused an expansion of $CD56^{bright}$ NK cells, and resulted in the population of $CD56^{dim}$ cells being diminished [117]. The study also found that the proportion of $CD56^{bright}$ NK cells was significantly higher in the secondary lymphoid tissues compared to the peripheral blood for the control group [117]. This suggested that $CD56^{bright}$ NK cells may preferably locate within secondary lymphoid tissues, where they are able to interact with T cells and contribute to control of disease activity in MS [117].

There is an ongoing debate as to whether NK cells have a predominately beneficial or detrimental role in EAE, made even more complex by the lack of CD56 expression on murine NK cells. Studies have shown that enhancing the regulatory features of NK cells ameliorates the disease course of EAE. When the interaction between NKG2A and its ligand Qa-1 (the murine equivalent to the human HLA-E) expressed on target cells were blocked by antibodies specific for either antigen, it was found that NKG2A-expressing NK cells in particular decreased CNS inflammation by killing microglial and T cells [118, 119].

Enrichment of NK cells through treatment with IL-2 coupled with a monoclonal antibody specific for IL-2 (IL-2 mAb) was also found to ameliorate EAE [120]. The IL-2 mAb supplements the proliferation of NK and CD8 T cells in mice by increasing the biological activity of the pre-existing IL-2 by formation of immune complexes [121]. Increased levels of IL-2 was also found to expand Tregs while preventing the induction of Th17 during EAE development [122]. However, NK cells have different effects during the early stages of EAE, and possibly MS, compared to the late stages. In the early stages NK cells were found to protect the CNS whereas NK cells were found to kill neural stem cells (NSCs) during the late stages of EAE, as a result of reduced expression of Qa-1 on NSCs [120, 123].

5.4 NKT cells

NKT cells are unique T lymphocytes that express NK cell lineage markers, and act as a bridge between the innate and adaptive immune system. NKT cells account for a small percentage of lymphocytes, but have profound immunomodulatory roles in a variety of diseases [124]. There are two categories of NKT cells, type I and type II. Type I NKT cells, also known as invariant NKT cells (iNKT cells), express a semi-invariant Vα24-Jα18 (Vα14-Jα18 in mice), paired with a restricted range of β chains, that recognises α-galactosylceramide (α-GalCer) presented by CD1d [125, 126]. Type II NKT cells use TCRα and β chains that are reactive to a broad range of antigens, but do not recognise α-Galcer [127].

Nonobese diabetic (NOD) mice are susceptible to MOG-induced EAE. However, if NKT cells are increased either by transgenesis or adoptive transfer, the mice show protection from disease [128]. EAE protection has been correlated with inhibition of Ag-specific IFN-γ production in the spleen, modulating the encephalitogenic Th1 response [128]. There is conflicting evidence as to the effects of deletion of NKT cells on EAE. Some studies resulted in no effect on disease [129], with other studies showing disease exacerbation in CD1d-deficient and Jα18-deficient mice [130, 131]. Activation of type I NKT cells by α-GalCer has been shown to improve EAE outcome. These improvements arise by indirectly enhancing Th2 response and reducing the Th1 response, or potentiating the differential of immunosuppressive myeloid cells [131–134]. However conflicting studies showed that high doses of α-GalCer could worsen EAE by directly enhancing Th17 and Th1 differentiation through phosphorylation of STAT3 and activation of NK-κB [135].

NKT cells from MS patients have been reported to have an increased production of cytokines. IL-4 production was increased by CD4 NKT cell clones in RRMS compared to other MS progression types, causing significant Th2 bias [136]. However, NKT cells in progressive MS patients displayed proinflammatory profiles [137]. It has also been suggested that the current available drugs for MS treatment may function through NKT cell targeting. A large reduction of type I NKT cells in peripheral blood was associated with remission of MS [136]. Type 1 interferon-β (T1IFN-β), a popular disease modifying therapy (DMT) for RRMS treatment, has been noted to promote expansion and functionality of type I NKT cells in vitro and to prevent disease in in vivo models of MS [138]. Research indicates a diverse role for NKT cells in MS pathology due to cytokine production.

5.5 Monocytes and macrophages

Besides imbalances in cytokine levels in the CNS and cerebrospinal fluid (CSF), immune imbalances also occur in the blood of MS patients, as reflected by altered levels of cytokines and cytokine producing cells (reviewed in [139]). The cause of these imbalances are thought to be due to circulating monocytes, with monocytes and macrophages influencing early MS, mediating both pro and anti-inflammatory responses [140, 141].

Surface expression of CD14 and CD16 are used to distinguish three distinct monocyte subsets: classical ($CD14^{++}CD16^{-}$), intermediate ($CD14^{++}CD16^{+}$) and nonclassical ($CD14^{+}$ $CD16^{++}$) [142]. Monocytes and macrophages perform the key functions of antigen presentation and co-stimulation vital to the body's immune response, with important roles in T and B cell activation and differentiation via the CD40-CD154 interaction (reviewed in [143]). Macrophages are primarily derived from blood borne monocytes, are present at sites of active demyelination in MS, and are assumed to be a part of the demyelinating process [144]. These inflammatory cells produce a range of toxic oxygen metabolites which mediate host tissue destruction. During MS progression, there is a significant expansion of the $CD16^{+}$ monocyte population, which can primarily be attributed to nonclassical monocytes [145]. Depletion of these nonclassical monocytes may be an alternative to T and B cell depletion with the advantage of leaving the major classical monocyte population untouched. Selective subset depletion of monocytes may also supplement existing therapies to increase efficacy [145].

6. Conclusions

Multiple sclerosis is a complex autoimmune disease. Due to the many cell types involved in pathogenesis of the disease, therapeutics and treatments are often broad ranged and relatively inefficient. Further studies are necessary to uncover the genetic and environmental triggers leading to aberrant cellular toxicity and its role in MS pathogenesis. Discovering these and the related pathways will potentially lead to more targeted therapeutics and the elimination of not only MS but other autoimmune and neurological diseases in the future.

Acknowledgements

AMLW is supported by an Australian Government Research Training Program Stipend and MAJ was supported by an MS Research Australia/NHMRC Research Betty Cuthbert Fellowship. Project funds were obtained from Multiple Sclerosis Research Australia (MSRA), Lions' club, Australia, and Australian Health Research Alliance-Women's Health Research Translation Network (WHRTN).

Conflict of interest

The authors declare no conflict of interest.

Author details

Annie M.L. Willson and Margaret A. Jordan*
Molecular and Cell Biology, College of Public Health, Medical and Veterinary Sciences, The Science Place, James Cook University, Townsville, QLD, Australia

*Address all correspondence to: Margaret.Jordan@jcu.edu.au

References

[1] Walton C, King R, Rechtman L, Kaye W, Leray E, Marrie RA, et al. Rising prevalence of multiple sclerosis worldwide: Insights from the Atlas of MS. Multiple Sclerosis Journal. 2020;**26**(14):1816-1821

[2] Weiner HL. Multiple sclerosis is an inflammatory T-cell–mediated autoimmune disease. Archives of Neurology. 2004;**61**(10):1613-1615

[3] McDonald WI, Compston A, Edan G, Goodkin D, Hartung HP, Lublin FD, et al. Recommended diagnostic criteria for multiple sclerosis: Guidelines from the International Panel on the diagnosis of multiple sclerosis. Annals of Neurology: Official Journal of the American Neurological Association and the Child Neurology Society. 2001;**50**(1):121-127

[4] Pender MP, Greer JM. Immunology of multiple sclerosis. Current Allergy and Asthma Reports. 2007;**7**(4):285-292

[5] Cottrell D, Kremenchutzky M, Rice G, Koopman W, Hader W, Baskerville J, et al. The natural history of multiple sclerosis: A geographically based study: 5. The clinical features and natural history of primary progressive multiple sclerosis. Brain. 1999;**122**(4):625-639

[6] Tremlett H, Paty D, Devonshire V. The natural history of primary progressive MS in British Columbia, Canada. Neurology. 2005;**65**(12):1919-1923

[7] Goudarzi MH, Eadie MJ, Hollingworth SA. Disease modifying therapies for relapsing-remitting multiple sclerosis: Use and costs in Australia (1996-2019). Multiple Sclerosis and Related Disorders. 2021;**50**:102835

[8] Confavreux C, Vukusic S, Moreau T, Adeleine P. Relapses and progression of disability in multiple sclerosis. New England Journal of Medicine. 2000;**343**(20):1430-1438

[9] Confavreux C, Vukusic S. Natural history of multiple sclerosis: A unifying concept. Brain. 2006;**129**(3):606-616

[10] Kurtzke JF. Epidemiology in multiple sclerosis: A pilgrim's progress. Brain. 2013;**136**(9):2904-2917

[11] Wang Y, Marling SJ, Zhu JG, Severson KS, DeLuca HF. Development of experimental autoimmune encephalomyelitis (EAE) in mice requires vitamin D and the vitamin D receptor. Proceedings of the National Academy of Sciences. 2012;**109**(22):8501-8504

[12] Haghmorad D, Yazdanpanah E, Jadid Tavaf M, Zargarani S, Soltanmohammadi A, Mahmoudi MB, et al. Prevention and treatment of experimental autoimmune encephalomyelitis induced mice with 1, 25-dihydroxyvitamin D3. Neurological Research. 2019;**41**(10):943-957

[13] Amezcua L, Conti DV, Liu L, Ledezma K, Langer-Gould AM. Place of birth, age of immigration, and disability in Hispanics with multiple sclerosis. Multiple Sclerosis and Related Disorders. 2015;**4**(1):25-30

[14] Hammond S, English D, McLeod J. The age-range of risk of developing multiple sclerosis: Evidence from a migrant population in Australia. Brain. 2000;**123**(5):968-974

[15] Levin LI, Munger KL, O'Reilly EJ, Falk KI, Ascherio A. Primary infection with the Epstein-Barr virus and risk of

multiple sclerosis. Annals of Neurology. 2010;**67**(6):824-830

[16] Pender MP. The essential role of Epstein-Barr virus in the pathogenesis of multiple sclerosis. The Neuroscientist. 2011;**17**(4):351-367

[17] Correale J, Farez MF. Smoking worsens multiple sclerosis prognosis: Two different pathways are involved. Journal of Neuroimmunology. 2015;**281**:23-34

[18] Sadovnick AD, Baird PA, Ward RH, Optiz JM, Reynolds JF. Multiple sclerosis. Updated risks for relatives. American Journal of Medical Genetics. 1988;**29**(3):533-541

[19] Sadovnick A, Armstrong H, Rice G, Bulman D, Hashimoto L, Party D, et al. A population-based study of multiple sclerosis in twins: Update. Annals of Neurology. 1993;**33**(3):281-285

[20] Consortium IMSG. Risk alleles for multiple sclerosis identified by a genomewide study. New England Journal of Medicine. 2007;**357**(9):851-862

[21] Consortium IMSG. IL12A, MPHOSPH9/CDK2AP1 and RGS1 are novel multiple sclerosis susceptibility loci. Genes and Immunity. 2010;**11**(5):397

[22] Sawcer S, Hellenthal G, Pirinen M, Spencer CC, Patsopoulos NA, Moutsianas L, et al. Genetic risk and a primary role for cell-mediated immune mechanisms in multiple sclerosis. Nature. 2011;**476**(7359):214

[23] Consortium*† IMSG, ANZgene, IIBDGC, WTCCC2. Multiple sclerosis genomic map implicates peripheral immune cells and microglia in susceptibility. Science. 2019;**365**(6460):eaav7188

[24] Gresle MM, Jordan MA, Stankovich J, Spelman T, Johnson LJ, Laverick L, et al. Multiple sclerosis risk variants regulate gene expression in innate and adaptive immune cells. Life science alliance. 2020;**3**(7):1-11

[25] Palacios R, Goni J, Martinez-Forero I, Iranzo J, Sepulcre J, Melero I, et al. A network analysis of the human T-cell activation gene network identifies JAGGED1 as a therapeutic target for autoimmune diseases. PloS One. 2007;**2**(11):e1222

[26] Jordan MA, Baxter AG. In: Rose NR, Mackay IR, editors. The Autoimmune Diseases. Boston, MA, USA: Elsevier; 2020. pp. 383-418

[27] Rose NR. Autoimmune disease: Reflections and projections. In: Rose NR. Mackay IR, editors. The Autoimmune Diseases. Boston, MA, USA: Elsevier; 2020. pp. 3-8

[28] Harari A, Enders FB, Cellerai C, Bart P-A, Pantaleo G. Distinct profiles of cytotoxic granules in memory T cells correlate with function, differentiation stage, and antigen exposure. Journal of Virology. 2009;**83**(7):2862-2871

[29] Zahavi D, AlDeghaither D, O'Connell A, Weiner LM. Enhancing antibody-dependent cell-mediated cytotoxicity: A strategy for improving antibody-based immunotherapy. Antibody Therapeutics. 2018;**1**(1):7-12

[30] Kaufmann SH. Immunology's foundation: The 100-year anniversary of the Nobel Prize to Paul Ehrlich and Elie Metchnikoff. Nature Immunology. 2008;**9**(7):705-712

[31] Hedges JF, Lubick KJ, Jutila MA. γδ T cells respond directly to pathogen-associated molecular patterns.

The Journal of Immunology. 2005;**174**(10):6045-6053

[32] Weiner HL. A shift from adaptive to innate immunity: A potential mechanism of disease progression in multiple sclerosis. Journal of Neurology. 2008;**255**(1):3-11

[33] Gandhi R, Laroni A, Weiner HL. Role of the innate immune system in the pathogenesis of multiple sclerosis. Journal of Neuroimmunology. 2010;**221**(1-2):7-14

[34] Shin ML, Rus HG, Niculescu FI. Membrane attack by complement: Assembly and biology of terminal complement complexes. Biomembranes: A Multi-Volume Treatise. 1996;**4**:123-149

[35] Weng N-p. Aging of the immune system: How much can the adaptive immune system adapt? Immunity. 2006;**24**(5):495-499

[36] Bonilla FA, Oettgen HC. Adaptive immunity. Journal of Allergy and Clinical Immunology. 2010;**125**(2):S33-S40

[37] Castellino F, Germain RN. Cooperation between CD4+ and CD8+ T cells: When, where, and how. Annual Review in Immunology. 2006;**24**:519-540

[38] Zhu J, Jankovic D, Oler AJ, Wei G, Sharma S, Hu G, et al. The transcription factor T-bet is induced by multiple pathways and prevents an endogenous Th2 cell program during Th1 cell responses. Immunity. 2012;**37**(4):660-673

[39] Röcken M, Saurat J-H, Hauser C. A common precursor for CD4+ T cells producing IL-2 or IL-4. The Journal of Immunology. 1992;**148**(4):1031-1036

[40] Singh V, Mehrotra S, Agarwal S. The paradigm of Th1 and Th2 cytokines. Immunologic Research. 1999;**20**(3):147-161

[41] Bettelli E, Carrier Y, Gao W, Korn T, Strom TB, Oukka M, et al. Reciprocal developmental pathways for the generation of pathogenic effector TH17 and regulatory T cells. Nature. 2006;**441**(7090):235-238

[42] Veldhoen M, Hocking RJ, Atkins CJ, Locksley RM, Stockinger B. TGFβ in the context of an inflammatory cytokine milieu supports de novo differentiation of IL-17-producing T cells. Immunity. 2006;**24**(2):179-189

[43] Kisielow P, Blüthmann H, Staerz UD, Steinmetz M, Von Boehmer H. Tolerance in T-cell-receptor transgenic mice involves deletion of nonmature CD4+ 8+ thymocytes. Nature. 1988;**333**(6175):742-746

[44] Fontenot JD, Gavin MA, Rudensky AY. Foxp3 programs the development and function of CD4+ CD25+ regulatory T cells. Nature Immunology. 2003;**4**(4):330-336

[45] Hori S, Nomura T, Sakaguchi S. Control of regulatory T cell development by the transcription factor Foxp3. Science. 2003;**299**(5609):1057-1061

[46] Khattri R, Cox T, Yasayko S-A, Ramsdell F. An essential role for Scurfin in CD4+ CD25+ T regulatory cells. Nature Immunology. 2003;**4**(4):337-342

[47] Blair PJ, Bultman SJ, Haas JC, Rouse BT, Wilkinson JE, Godfrey VL. CD4+ CD8-T cells are the effector cells in disease pathogenesis in the scurfy (sf) mouse. The Journal of Immunology. 1994;**153**(8):3764-3774

[48] Lucchinetti C, Brück W, Parisi J, Scheithauer B, Rodriguez M,

Lassmann H. Heterogeneity of multiple sclerosis lesions: Implications for the pathogenesis of demyelination. Annals of Neurology: Official Journal of the American Neurological Association and the Child Neurology Society. 2000;**47**(6):707-717

[49] Wardemann H, Yurasov S, Schaefer A, Young JW, Meffre E, Nussenzweig MC. Predominant autoantibody production by early human B cell precursors. Science. 2003;**301**(5638):1374-1377

[50] Rose NR, Bona C. Defining criteria for autoimmune diseases (Witebsky's postulates revisited). Immunology Today. 1993;**14**(9):426-430

[51] Zenewicz LA, Abraham C, Flavell RA, Cho JH. Unraveling the genetics of autoimmunity. Cell. 2010;**140**(6):791-797

[52] Lerner A, Jeremias P, Matthias T. The world incidence and prevalence of autoimmune diseases is increasing. International Journal of Celiac Diseases. 2015;**3**(4):151-155

[53] Jacobson DL, Gange SJ, Rose NR, Graham NM. Epidemiology and estimated population burden of selected autoimmune diseases in the United States. Clinical Immunology and Immunopathology. 1997;**84**(3):223-243

[54] Vargas-Parada L. Research round-up: Autoimmune disease. Nature. 2021;**595**(7867):46-47

[55] Shevach EM. Mechanisms of foxp3+ T regulatory cell-mediated suppression. Immunity. 2009;**30**(5):636-645

[56] Tang Q, Bluestone JA. The Foxp3+ regulatory T cell: A jack of all trades, master of regulation. Nature Immunology. 2008;**9**(3):239-244

[57] Coffman RL. Origins of the TH1-TH2 model: A personal perspective. Nature Immunology. 2006;**7**(6):539-541

[58] Khader SA, Gaffen SL, Kolls JK. Th17 cells at the crossroads of innate and adaptive immunity against infectious diseases at the mucosa. Mucosal Immunology. 2009;**2**(5):403-411

[59] Laan M, Cui Z-H, Hoshino H, Lötvall J, Sjöstrand M, Gruenert DC, et al. Neutrophil recruitment by human IL-17 via CXC chemokine release in the airways. The Journal of Immunology. 1999;**162**(4):2347-2352

[60] Ye P, Rodriguez FH, Kanaly S, Stocking KL, Schurr J, Schwarzenberger P, et al. Requirement of interleukin 17 receptor signaling for lung CXC chemokine and granulocyte colony-stimulating factor expression, neutrophil recruitment, and host defense. The Journal of Experimental Medicine. 2001;**194**(4):519-528

[61] Romagnani S. Th1/th2 cells. Inflammatory Bowel Diseases. 1999;**5**(4):285-294

[62] Kivisäkk P, Imitola J, Rasmussen S, Elyaman W, Zhu B, Ransohoff RM, et al. Localizing central nervous system immune surveillance: Meningeal antigen-presenting cells activate T cells during experimental autoimmune encephalomyelitis. Annals of Neurology. 2009;**65**(4):457-469

[63] Brown DA, Sawchenko PE. Time course and distribution of inflammatory and neurodegenerative events suggest structural bases for the pathogenesis of experimental autoimmune encephalomyelitis. Journal of Comparative Neurology. 2007;**502**(2):236-260

[64] Frischer JM, Bramow S, Dal-Bianco A, Lucchinetti CF,

Rauschka H, Schmidbauer M, et al. The relation between inflammation and neurodegeneration in multiple sclerosis brains. Brain. 2009;**132**(5):1175-1189

[65] Fogdell-Hahn A, Ligers A, Grønning M, Hillert J, Olerup O. Multiple sclerosis: A modifying influence of HLA class I genes in an HLA class II associated autoimmune disease. Tissue Antigens. 2000;**55**(2):140-148

[66] Hohlfeld R. Biotechnological agents for the immunotherapy of multiple sclerosis. Principles, problems and perspectives. Brain: A Journal of Neurology. 1997;**120**(5):865-916

[67] Herrera BM, Ebers GC. Progress in deciphering the genetics of multiple sclerosis. Current Opinion in Neurology. 2003;**16**(3):253-258

[68] Kivisäkk P, Mahad DJ, Callahan MK, Trebst C, Tucky B, Wei T, et al. Human cerebrospinal fluid central memory CD4+ T cells: Evidence for trafficking through choroid plexus and meninges via P-selectin. Proceedings of the National Academy of Sciences. 2003;**100**(14):8389-8394

[69] Harrington LE, Hatton RD, Mangan PR, Turner H, Murphy TL, Murphy KM, et al. Interleukin 17–producing CD4+ effector T cells develop via a lineage distinct from the T helper type 1 and 2 lineages. Nature Immunology. 2005;**6**(11):1123-1132

[70] Panitch H, Haley A, Hirsch R, Johnson K. Exacerbations of multiple sclerosis in patients treated with gamma interferon. The Lancet. 1987;**329**(8538):893-895

[71] Ferber IA, Brocke S, Taylor-Edwards C, Ridgway W, Dinisco C, Steinman L, et al. Mice with a disrupted IFN-gamma gene are susceptible to the induction of experimental autoimmune encephalomyelitis (EAE). The Journal of Immunology. 1996;**156**(1):5-7

[72] Langrish CL, Chen Y, Blumenschein WM, Mattson J, Basham B, Sedgwick JD, et al. IL-23 drives a pathogenic T cell population that induces autoimmune inflammation. The Journal of Experimental Medicine. 2005;**201**(2):233-240

[73] O'Connor RA, Prendergast CT, Sabatos CA, Lau CW, Leech MD, Wraith DC, et al. Cutting edge: Th1 cells facilitate the entry of Th17 cells to the central nervous system during experimental autoimmune encephalomyelitis. The Journal of Immunology. 2008;**181**(6):3750-3754

[74] Kroenke MA, Carlson TJ, Andjelkovic AV, Segal BM. IL-12–and IL-23–modulated T cells induce distinct types of EAE based on histology, CNS chemokine profile, and response to cytokine inhibition. The Journal of Experimental Medicine. 2008;**205**(7):1535-1541

[75] Thakker P, Leach MW, Kuang W, Benoit SE, Leonard JP, Marusic S. IL-23 is critical in the induction but not in the effector phase of experimental autoimmune encephalomyelitis. The Journal of Immunology. 2007;**178**(4):2589-2598

[76] Harbo H, Lie B, Sawcer S, Celius E, Dai KZ, Oturai A, et al. Genes in the HLA class I region may contribute to the HLA class II-associated genetic susceptibility to multiple sclerosis. Tissue Antigens. 2004;**63**(3):237-247

[77] Booss J, Esiri MM, Tourtellotte WW, Mason DY. Immunohistological analysis of T lymphocyte subsets in the central nervous system in chronic progressive multiple sclerosis.

Journal of the Neurological Sciences. 1983;**62**(1-3):219-232

[78] Serafini B, Rosicarelli B, Magliozzi R, Stigliano E, Capello E, Mancardi GL, et al. Dendritic cells in multiple sclerosis lesions: Maturation stage, myelin uptake, and interaction with proliferating T cells. Journal of Neuropathology & Experimental Neurology. 2006;**65**(2):124-141

[79] Chowdhury D, Lieberman J. Death by a thousand cuts: Granzyme pathways of programmed cell death. Annual Review in Immunology. 2008;**26**: 389-420

[80] Russell JH, Ley TJ. Lymphocyte-mediated cytotoxicity. Annual Review of Immunology. 2002;**20**(1):323-370

[81] Raivich G, Jones LL, Kloss CU, Werner A, Neumann H, Kreutzberg GW. Immune surveillance in the injured nervous system: T-lymphocytes invade the axotomized mouse facial motor nucleus and aggregate around sites of neuronal degeneration. Journal of Neuroscience. 1998;**18**(15):5804-5816

[82] Oliveira AL, Thams S, Lidman O, Piehl F, Hökfelt T, Kärre K, et al. A role for MHC class I molecules in synaptic plasticity and regeneration of neurons after axotomy. Proceedings of the National Academy of Sciences. 2004;**101**(51):17843-17848

[83] Jurewicz A, Biddison WE, Antel JP. MHC class I-restricted lysis of human oligodendrocytes by myelin basic protein peptide-specific CD8 T lymphocytes. The Journal of Immunology. 1998;**160**(6):3056-3059

[84] Pouly S, Becher B, Blain M, Antel JP. Interferon-γ modulates human oligodendrocyte susceptibility to Fas-mediated apoptosis. Journal of Neuropathology & Experimental Neurology. 2000;**59**(4):280-286

[85] Rensing-Ehl A, Malipiero U, Irmler M, Tschopp J, Constam D, Fontana A. Neurons induced to express major histocompatibility complex class I antigen are killed via the perforin and not the Fas (APO-1/CD95) pathway. European Journal of Immunology. 1996;**26**(9):2271-2274

[86] Neumann H, Schmidt H, Cavalie A, Jenne D, Wekerle H. Major histocompatibility complex (MHC) class I gene expression in single neurons of the central nervous system: Differential regulation by interferon (IFN)-γ and tumor necrosis factor (TNF)-α. The Journal of Experimental Medicine. 1997;**185**(2):305-316

[87] Medana IM, Gallimore A, Oxenius A, Martinic MM, Wekerle H, Neumann H. MHC class I-restricted killing of neurons by virus-specific CD8+ T lymphocytes is effected through the Fas/FasL, but not the perforin pathway. European Journal of Immunology. 2000;**30**(12):3623-3633

[88] Dörner T, Jacobi AM, Lipsky PE. B cells in autoimmunity. Arthritis Research & Therapy. 2009;**11**(5):1-11

[89] Aung LL, Balashov KE. Decreased Dicer expression is linked to increased expression of co-stimulatory molecule CD80 on B cells in multiple sclerosis. Multiple Sclerosis Journal. 2015;**21**(9):1131-1138

[90] Barr TA, Shen P, Brown S, Lampropoulou V, Roch T, Lawrie S, et al. B cell depletion therapy ameliorates autoimmune disease through ablation of IL-6–producing B cells. Journal of Experimental Medicine. 2012;**209**(5):1001-1010

[91] Li R, Rezk A, Miyazaki Y, Hilgenberg E, Touil H, Shen P, et al. Proinflammatory GM-CSF–producing B cells in multiple sclerosis and B cell depletion therapy. Science Translational Medicine. 2015;7(310):ra166

[92] Kamburova EG, Koenen HJ, Borgman KJ, Ten Berge I, Joosten I, Hilbrands LB. A single dose of rituximab does not deplete B cells in secondary lymphoid organs but alters phenotype and function. American Journal of Transplantation. 2013;**13**(6):1503-1511

[93] Bielekova B, Catalfamo M, Reichert-Scrivner S, Packer A, Cerna M, Waldmann TA, et al. Regulatory CD56bright natural killer cells mediate immunomodulatory effects of IL-2Rα-targeted therapy (daclizumab) in multiple sclerosis. Proceedings of the National Academy of Sciences. 2006;**103**(15):5941-5946

[94] Giovannoni G, Gold R, Selmaj K, Havrdova E, Montalban X, Radue E-W, et al. Daclizumab high-yield process in relapsing-remitting multiple sclerosis (SELECTION): A multicentre, randomised, double-blind extension trial. The Lancet Neurology. 2014;**13**(5):472-481

[95] Kappos L, Wiendl H, Selmaj K, Arnold DL, Havrdova E, Boyko A, et al. Daclizumab HYP versus interferon beta-1a in relapsing multiple sclerosis. New England Journal of Medicine. 2015;**373**(15):1418-1428

[96] Sun JC, Lanier LL. NK cell development, homeostasis and function: Parallels with CD8+ T cells. Nature Reviews Immunology. 2011;**11**(10):645-657

[97] Freud AG, Becknell B, Roychowdhury S, Mao HC, Ferketich AK, Nuovo GJ, et al. A human CD34 (+) subset resides in lymph nodes and differentiates into CD56brightNatural killer cells. Immunity. 2005;**22**(3):295-304

[98] Freud AG, Yokohama A, Becknell B, Lee MT, Mao HC, Ferketich AK, et al. Evidence for discrete stages of human natural killer cell differentiation in vivo. The Journal of Experimental Medicine. 2006;**203**(4):1033-1043

[99] Eissens DN, Spanholtz J, Van Der Meer A, Van Cranenbroek B, Dolstra H, Kwekkeboom J, et al. Defining early human NK cell developmental stages in primary and secondary lymphoid tissues. PloS One. 2012;7(2):e30930

[100] Scoville SD, Mundy-Bosse BL, Zhang MH, Chen L, Zhang X, Keller KA, et al. A progenitor cell expressing transcription factor RORγt generates all human innate lymphoid cell subsets. Immunity. 2016;**44**(5):1140-1150

[101] Griffin JD, Hercend T, Beveridge R, Schlossman SF. Characterization of an antigen expressed by human natural killer cells. The Journal of Immunology. 1983;**130**(6):2947-2951

[102] Lanier LL, Testi R, Bindl J, Phillips JH. Identity of Leu-19 (CD56) leukocyte differentiation antigen and neural cell adhesion molecule. The Journal of Experimental Medicine. 1989;**169**(6):2233-2238

[103] Lanier LL, Le AM, Civin C, Loken M, Phillips J. The relationship of CD16 (Leu-11) and Leu-19 (NKH-1) antigen expression on human peripheral blood NK cells and cytotoxic T lymphocytes. The Journal of Immunology. 1986;**136**(12):4480-4486

[104] Cooper MA, Fehniger TA, Turner SC, Chen KS, Ghaheri BA, Ghayur T, et al. Human natural killer

cells: A unique innate immunoregulatory role for the CD56bright subset. Blood, The Journal of the American Society of Hematology. 2001;**97**(10):3146-3151

[105] Ferlazzo G, Pack M, Thomas D, Paludan C, Schmid D, Strowig T, et al. Distinct roles of IL-12 and IL-15 in human natural killer cell activation by dendritic cells from secondary lymphoid organs. Proceedings of the National Academy of Sciences. 2004;**101**(47):16606-16611

[106] Dalbeth N, Gundle R, Davies RJ, Lee YG, McMichael AJ, Callan MF. CD56bright NK cells are enriched at inflammatory sites and can engage with monocytes in a reciprocal program of activation. The Journal of Immunology. 2004;**173**(10):6418-6426

[107] Morandi B, Mortara L, Chiossone L, Accolla RS, Mingari MC, Moretta L, et al. Dendritic cell editing by activated natural killer cells results in a more protective cancer-specific immune response. PloS One. 2012;**7**(6):e39170

[108] Juelke K, Killig M, Luetke-Eversloh M, Parente E, Gruen J, Morandi B, et al. CD62L expression identifies a unique subset of polyfunctional CD56dim NK cells. Blood, The Journal of the American Society of Hematology. 2010;**116**(8):1299-1307

[109] Fauriat C, Long EO, Ljunggren H-G, Bryceson YT. Regulation of human NK-cell cytokine and chemokine production by target cell recognition. Blood, The Journal of the American Society of Hematology. 2010;**115**(11):2167-2176

[110] Bauer S, Groh V, Wu J, Steinle A, Phillips JH, Lanier LL, et al. Activation of NK cells and T cells by NKG2D, a receptor for stress-inducible MICA. Science. 1999;**285**(5428):727-729

[111] Bottino C, Moretta L, Moretta A. NK cell activating receptors and tumor recognition in humans. Immunobiology of Natural Killer Cell Receptors. 2006;**298**:175-182

[112] Welte S, Kuttruff S, Waldhauer I, Steinle A. Mutual activation of natural killer cells and monocytes mediated by NKp80-AICL interaction. Nature Immunology. 2006;**7**(12):1334-1342

[113] Diefenbach A, Hsia JK, Hsiung MYB, Raulet DH. A novel ligand for the NKG2D receptor activates NK cells and macrophages and induces tumor immunity. European Journal of Immunology. 2003;**33**(2):381-391

[114] Hayakawa Y, Kelly JM, Westwood JA, Darcy PK, Diefenbach A, Raulet D, et al. Cutting edge: Tumor rejection mediated by NKG2D receptor-ligand interaction is dependent upon perforin. The Journal of Immunology. 2002;**169**(10):5377-5381

[115] Wu J, Song Y, Bakker AB, Bauer S, Spies T, Lanier LL, et al. An activating immunoreceptor complex formed by NKG2D and DAP10. Science. 1999;**285**(5428):730-732

[116] Gunesch JT, Dixon AL, Ebrahim TA, Berrien-Elliott MM, Tatineni S, Kumar T, et al. CD56 regulates human NK cell cytotoxicity through Pyk2. Elife. 2020;**9**:e57346

[117] Saraste M, Irjala H, Airas L. Expansion of CD56Bright natural killer cells in the peripheral blood of multiple sclerosis patients treated with interferon-beta. Neurological Sciences. 2007;**28**(3):121-126

[118] Lu L, Ikizawa K, Hu D, Werneck MB, Wucherpfennig KW, Cantor H. Regulation of activated CD4+ T cells by NK cells via the Qa-1–NKG2A

inhibitory pathway. Immunity. 2007;**26**(5):593-604

[119] Leavenworth JW, Schellack C, Kim H-J, Lu L, Spee P, Cantor H. Analysis of the cellular mechanism underlying inhibition of EAE after treatment with anti-NKG2A F (ab′) 2. Proceedings of the National Academy of Sciences. 2010;**107**(6):2562-2567

[120] Hao J, Liu R, Piao W, Zhou Q, Vollmer TL, Campagnolo DI, et al. Central nervous system (CNS)–resident natural killer cells suppress Th17 responses and CNS autoimmune pathology. Journal of Experimental Medicine. 2010;**207**(9):1907-1921

[121] Boyman O, Kovar M, Rubinstein MP, Surh CD, Sprent J. Selective stimulation of T cell subsets with antibody-cytokine immune complexes. Science. 2006;**311**(5769):1924-1927

[122] Rouse M, Nagarkatti M, Nagarkatti PS. The role of IL-2 in the activation and expansion of regulatory T-cells and the development of experimental autoimmune encephalomyelitis. Immunobiology. 2013;**218**(4):674-682

[123] Liu Q, Sanai N, Jin W-N, La Cava A, Van Kaer L, Shi F-D. Neural stem cells sustain natural killer cells that dictate recovery from brain inflammation. Nature Neuroscience. 2016;**19**(2):243-252

[124] Taniguchi M, Seino K-i, Nakayama T. The NKT cell system: Bridging innate and acquired immunity. Nature Immunology. 2003;**4**(12):1164-1165

[125] Kawano T, Cui J, Koezuka Y, Toura I, Kaneko Y, Motoki K, et al. CD1d-restricted and TCR-mediated activation of Vα14 NKT cells by glycosylceramides. Science. 1997;**278**(5343):1626-1629

[126] Benlagha K, Weiss A, Beavis A, Teyton L, Bendelac A. In vivo identification of glycolipid antigen–specific T cells using fluorescent CD1d tetramers. The Journal of Experimental Medicine. 2000;**191**(11):1895-1904

[127] Dellabona P, Padovan E, Casorati G, Brockhaus M, Lanzavecchia A. An invariant V alpha 24-J alpha Q/V beta 11 T cell receptor is expressed in all individuals by clonally expanded CD4-8-T cells. The Journal of Experimental Medicine. 1994;**180**(3):1171-1176

[128] Mars LT, Laloux V, Goude K, Desbois S, Saoudi A, Van Kaer L, et al. Cutting edge: Vα14-Jα281 NKT cells naturally regulate experimental autoimmune encephalomyelitis in nonobese diabetic mice. The Journal of Immunology. 2002;**168**(12):6007-6011

[129] Oh SJ, Chung DH. Invariant NKT cells producing IL-4 or IL-10, but not IFN-γ, inhibit the Th1 response in experimental autoimmune encephalomyelitis, whereas none of these cells inhibits the Th17 response. The Journal of Immunology. 2011;**186**(12):6815-6821

[130] Teige A, Teige I, Lavasani S, Bockermann R, Mondoc E, Holmdahl R, et al. CD1-dependent regulation of chronic central nervous system inflammation in experimental autoimmune encephalomyelitis. The Journal of Immunology. 2004;**172**(1):186-194

[131] Furlan R, Bergami A, Cantarella D, Brambilla E, Taniguchi M, Dellabona P, et al. Activation of invariant NKT cells by αGalCer administration protects mice from MOG35-55-induced EAE: Critical roles for administration route and IFN-γ. European Journal of Immunology. 2003;**33**(7):1830-1838

[132] Miyamoto K, Miyake S, Yamamura T. A synthetic glycolipid prevents autoimmune encephalomyelitis by inducing TH2 bias of natural killer T cells. Nature. 2001;**413**(6855):531-534

[133] Parekh VV, Wu L, Olivares-Villagómez D, Wilson KT, Van Kaer L. Activated invariant NKT cells control central nervous system autoimmunity in a mechanism that involves myeloid-derived suppressor cells. The Journal of Immunology. 2013;**190**(5):1948-1960

[134] Kojo S, Seino K-i, Harada M, Watarai H, Wakao H, Uchida T, et al. Induction of regulatory properties in dendritic cells by Vα14 NKT cells. The Journal of Immunology. 2005;**175**(6):3648-3655

[135] Qian G, Qin X, Zang YQ, Ge B, Guo TB, Wan B, et al. High doses of α-galactosylceramide potentiate experimental autoimmune encephalomyelitis by directly enhancing Th17 response. Cell Research. 2010;**20**(4):480-491

[136] Araki M, Kondo T, Gumperz JE, Brenner MB, Miyake S, Yamamura T. Th2 bias of CD4+ NKT cells derived from multiple sclerosis in remission. International Immunology. 2003;**15**(2):279-288

[137] De Biasi S, Simone AM, Nasi M, Bianchini E, Ferraro D, Vitetta F, et al. iNKT cells in secondary progressive multiple sclerosis patients display pro-inflammatory profiles. Frontiers in Immunology. 2016;**7**:555

[138] Freedman MS, Devonshire V, Duquette P, Giacomini PS, Giuliani F, Levin MC, et al. Treatment optimization in multiple sclerosis: Canadian MS Working Group recommendations. Canadian Journal of Neurological Sciences. 2020;**47**(4):437-455

[139] Navikas V, Link H. Cytokines and the pathogenesis of multiple sclerosis. Journal of Neuroscience Research. 1996;**45**(4):322-333

[140] Kouwenhoven M, Teleshova N, Özenci V, Press R, Link H. Monocytes in multiple sclerosis: Phenotype and cytokine profile. Journal of Neuroimmunology. 2001;**112**(1-2):197-205

[141] Miron VE, Boyd A, Zhao J-W, Yuen TJ, Ruckh JM, Shadrach JL, et al. M2 microglia and macrophages drive oligodendrocyte differentiation during CNS remyelination. Nature Neuroscience. 2013;**16**(9):1211-1218

[142] Ziegler-Heitbrock L, Ancuta P, Crowe S, Dalod M, Grau V, Hart DN, et al. Nomenclature of monocytes and dendritic cells in blood. Blood, The Journal of the American Society of Hematology. 2010;**116**(16):e74-e80

[143] Chitnis T, Khoury SJ. Role of costimulatory pathways in the pathogenesis of multiple sclerosis and experimental autoimmune encephalomyelitis. Journal of Allergy and Clinical Immunology. 2003;**112**(5):837-849

[144] Fisher M, Levine PH, Weiner BH, Vaudreuil CH, Natale A, Johnson MH, et al. Monocyte and polymorphonuclear leukocyte toxic oxygen metabolite production in multiple sclerosis. Inflammation. 1988;**12**(2):123-131

[145] Gjelstrup MC, Stilund M, Petersen T, Møller HJ, Petersen EL, Christensen T. Subsets of activated monocytes and markers of inflammation in incipient and progressed multiple sclerosis. Immunology and Cell Biology. 2018;**96**(2):160-174

Chapter 5

Cytotoxicity Studies of Fruit-Extracted Metal Nanostructures

Emusani Ramya and D. Narayana Rao

Abstract

Biosynthesized silver (Ag) and gold nanoparticles (Au NPs) were synthesized by using *Punica granatum* and *Citrus reticulata* extracts and these extracts act as stabilizer. The average sizes of the Silver and Gold NPs were in the range of 8–10 nm and 30–40 nm, respectively. The confirmation of NPs was done by UV-visible absorption spectra, X-ray diffraction (XRD), transmission electron microscope (TEM), and Fourier transform infrared spectra (FTIR) techniques. The luminescence studies of europium and samarium complexes with NPs were studied. Emission intensities of complexes with the presence of silver enhanced and quenched with gold due to the reabsorption. The toxicity and antimicrobial studies of nanostructures were studied.

Keywords: eco-friendly green synthesis, metal nanostructures, luminescence, MTT assay, antimicrobial activity

1. Introduction

Nanomaterials exhibit chemical, physical, and biological features that are not found in other materials. In the food industry, electrical and chemistry industry, and cosmetics industry metal nanostructures have unique applications. In medicinal and industrial uses, Ag and Au nanoparticles comprise the backbone. The food sector [1, 2], catheters [3], dressings of wound [4], surgery equipment [5], synthetics [6], and optical limiter applications [7] all use silver nanostructures. Sensors [8], catalytic [9], and electrical applications [10] are only a few of the applications for gold nanoparticles. The usage of Au and Ag nanostructures as drug delivery systems demonstrated as bactericidal against Gram-positive and negative [11].

By-products from traditional synthesis techniques can be hazardous [12]. As a result, a variety of environmental and green synthesis strategies are being researched [7, 13]. Nanomedicine has had a huge influence as a result of these biosynthesized nanomaterials, and they have become the elementary units for upcoming medications to treat many ailments [14]. Plant extracts, enzymes, fungi, bacteria, and other eco-friendly synthesis processes have been using to synthesize Ag NSa and Au NSs [7, 15, 16]. Papaya, tansy, and citrus fruits are antioxidants found in fruit extracts that are used as stabilizers in the production of Ag and Au NPs [17].

IntechOpen

Pomegranate is the scientific name for *Punica granatum*, it related to the Lythraceae's family. Polyphenol, anthocyanidins, glucose, ellagic acids, cyanidin 3-glucose, and 3,5-diglucose, hydrolysable tannins, and gall tannins, are all found in pomegranate extract [18]. Anthocyanins, which are found in abundance in pomegranates, are responsible for forming, reducing, and stabilizing of nanostructures [19]. The orange, *Citrus reticulata*, is high in phytochemicals, flavonoids, phenols, vitamins, folic acids, essential oils, and pectin [20]. Flavonoids aid in the removal of metallic ions from the body.

Ag NSs have exceptional antibacterial action, which explains why so much work is being put into developing these nanostructures. The use of Ag NPs in medicine is becoming more common. Various investigations have found that Ag NPs significantly reduce mitochondrial activity, resulting in cell apoptosis or necrosis in a variety of cells [21–23]. Interactions between Ag NPs and cells, on the other hand, have been limited. The surfaces of Au NPs are modified by DNA and amino acids when they are coupled with thiol and amine groups [24, 25]. The biocompatibility of Au NPs in therapeutic applications is critical. The size, shape, and surface modification of Au NPs determine whether they are hazardous or benign to cells [26].

Due to larger Stokes shift, narrower bandwidth of emission, and longer lifetimes emission, lanthanide ions have attracted for various applications like bio-labeling and optical amplification [27], solar cells [28], light-emitting diode (LED) [29], cancer photodynamic therapy [30], and fluoroimmunoassay [31].

Many scientists are interested in the metal enhanced fluorescence (MEF) of metal NSs and rare-earth ions because they need significantly low input intensities, and hence their applications in diagnostics have increased [32, 33]. The fluorophore's emission intensity is increased by 10–103 when it is near metal particles with sub-wavelength sizes [34]. The MEF is caused by two methods: (1) fluorophores' coupling with SPR, and (2) fluorophores' coupling to the SPR. The influence of a local electric field [35]. Fluorophores with metal particles at 10 nm spacing are acted on by increasing electric fields surrounded the particles, increasing their absorption cross-section and, ultimately, enhanced radiation. The energy transmits from surface plasmonic resonance to the rare-earth ion, and vice versa, results in either enhanced reduction or reabsorption in emission in the other mechanism. The fluorescence increase is primarily influenced by the shape and diameter of nanostructures, fluorophore's dipole orientations, and the overlap of the fluorophore luminescence intensity and absorption with the plasmonic band of nanostructures [36]. Because of the rivalry of the highly enhanced field surrounding nanoparticles and nonradiative transition due to the dampening of dipole oscillators by metal surfaces, MEF is mostly dependent on the spacing among metal structures and fluorophores [37].

The SPR value is altered by different shapes, various sizes, and dielectric functions of NSs as well as the host [38]. The optimal spacing between fluorophore ions and NPS for emission enhancement exists. Based on the different energy between the rare-earth ions and the metal SPR, quenching of luminescence may occur even at relatively short distances. The number of NSs, SPR, and number of phonon states of fluorophores all affect the luminescence efficiency. The plasmon-mediated augmentation of luminescence-based on the antenna effect has been found, it boosts the excitation effect and emission rate [39]. Metal nanoparticles (donor) absorb energy, which is subsequently transferred nonradiatively to rare-earth ions (acceptor) in the Forster process [40]. One group concluded that energy transmit is the primary cause of emission amplification near metal objects [41]. Other research has found that nanoparticles can stop rare-earth ions from transferring non-radiative energy to metal nanostructures [42, 43]. These are dependent on the luminophore-nanoparticle distance, nanoparticle size, and metal particle concentration.

The goal of this research is to investigate the binding of enhanced fluorescence from Eu(TTFA)3 and Sm(TTFA)3 complexes to metal nanostructures such as silver and gold NSs in greater detail. We discovered a robust link among SPR of NSs and rare-earth molecules, as well as metal nanoparticle reabsorption by the SPR. The MTT assay is being used to investigate the cytotoxicity of metal NSs on cancer cell lines. The findings could aid in the development of biomedication delivery systems that are less hazardous than toxic ones.

2. Materials and synthesis methods

2.1 Materials

Sigma Aldrich provided silver nitrate ($AgNO_3$) and chloroaurate ($HAuCl_4$), which were used without additional purification. The *C. reticulata* and *P. granatum* were purchased from a local market. The National Centre for Cell Science (NCCS), Pune, India, provided human lung carcinoma (A549) and colorectal carcinoma (HCT116) cell lines.

2.2 Fruit extracts

Fresh *C. reticulata* and *P. granatum* peels were gathered and carefully rinsed with distilled water. A beaker was filled with 100 ml of distilled water and 6 mg of sliced *P. granatum* and *C. reticulata* fruit peels. The combination was squashed, and the extracted peel solution and strained using filter paper (Whatman no.1). Extracts were filtered again using 1.0 m filter paper to eliminate tiny particles. Then extracts were stored at 4 °C.

2.3 Synthesis of NSs

30 ml of 1 mM $AgNO_3$/$HAuCl_4$ aqueous solution was produced, and 6 ml of extracts were mixed to the 30 ml of solution at 24°C. The solution was first colorless, then progressively turned ashy, indicating the formation of Ag NPs. In the instance of Au NPs, the yellow solution turned ruby red, it showed that the forming of Au NPs. It is conducted three times to ensure that it is repeatable.

2.4 MTT assay

The MTT (3-(4,5-dimethylthiazol-2-yl)-2,5-diphenyltetrazolium bromide) test was used to investigate the cytotoxic effects of Ag NPs and Au NPs on A549 (lung) and HCT116 (colon) cancer cells. In a 96-well plate, the cell lines were plated at a density of 1 × 104 cells/well in DMEM/1% FBS. For 24 h, cells were incubated at 37°C in an environment of air and CO2 (95% + 5%). When the cells had reached around 80% confluence, they were treated with Ag NPs and Au NPs in various concentrations of 25, 50, 75, 100, and 125 M. MTT assay was used to determine cell vitality after 24 h. MTT solution was added to each cell in a volume of 20 L and incubated for 3–4 h. The formazan was then diffused in dimethyl sulfoxide (DMSO), and optical densities were measured on an ELISA plate reader at 550 nm. As a control, cells were grown without nanoparticles. The ratio of mean optical density to the control was calculated after the measurements were repeated twice. The following calculation was used to calculate cell viability for each well: CV = optical density (OD) of treated well/OD value of nontreated control well) × 100%. The cytotoxicity of Ag NPs and Au NPs was compared to a positive control drug, Oxaliplatin (OXP).

2.5 Antibacterial test

Antibacterial activities of Ag and Au NPs against Gram-negative (Acinetobacter *baumannii*) and Gram-positive (*E. coli*) bacteria were tested using the well diffusion method. Before the experiment, bacterial cultures were produced to 0.5McFarland standards. Pure bacterial cultures were subcultured on Luria-Bertani (LB) agar medium and swabbed onto individual plates in a uniform manner. The wells were filled with 10 mL, 20 mL, and 30 mL Ag NPs solutions, which were then left to diffuse and incubated for 24 h at 37°C. To examine the antibacterial activity, the diameter of the zone of inhibition was employed. For the sake of reproducibility, the experiments were repeated twice.

3. Experimental details

UV-vis absorbance spectra, X-ray diffraction (XRD), transmission electron microscopy (TEM), photoluminescence, cytotoxicity, and antibacterial methods were used. The NP solutions were placed in a 10 mL cuvette and their linear absorption spectra were found with a JASCO-V670 UV/VIS/NIR spectrometer with a wavelength range of 200–800 nm and a 1 nm resolution. Samples were drop casted onto carbon grids and cured at RT for transmission electron microscopy (TEM) examinations. To assess the size of the nanoparticles, the TEM investigation was performed on an FEI model TECHNAI G2 S-Twin. The NPS were analyzed using a Bruker D8 diffractometer at 40 kV, 30 mA, using Cu-K radiation at 1° min-1 scan rate and 0.02 step size across 2 = 20–80°. Eu and Sm complexes with Ag NPs and Au NPs had their photoluminescence spectra recorded on a HORIBA JOBIN YVON fluorescence spectrometer with excitation at 350 nm.

4. Results and analysis

4.1 UV-visible absorbance spectra

The surface plasmon resonance, which is the interaction between free electrons surrounded by metal nanostructures and radiation, was used to validate the formation and stability of Ag and Au NPs. Fruit extracts reduce aqueous silver and gold ions. From **Figure 1** shows the well-dispersed formation of Ag and Au NPs, resulted in the SPR peaks were observed in the visible region in the ranges of 350–600 nm and 500–700 nm, respectively. Due to the excitation of SPR vibrations, *P. granatum* synthesized Ag and Au NPs (**Figure 1a** and **b**) and *Citrus reticulata* Ag and Au NPs (**Figure 1c** and **d**) produced respectively. The ionization of phenolic groups in fruit extract results the formation of nanostructures. For a few months, these metal nanoparticles remain stable.

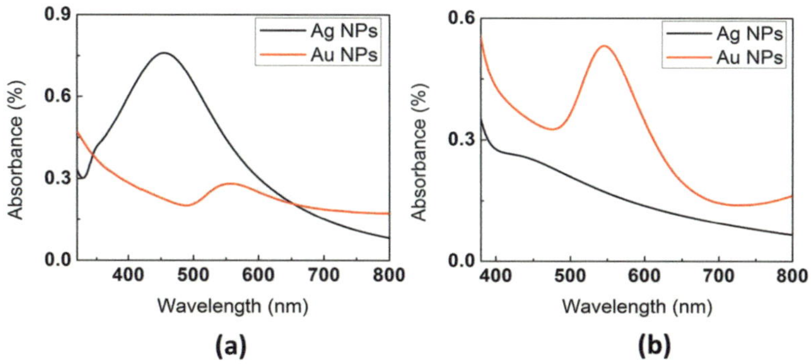

Figure 1.
Absorption spectrum of silver and gold NSs prepared by (a) Punica granatum and (b) Citrus reticulata extracts.

4.2 TEM images

The TEM images in **Figure 2a** and **b** shows that Ag nanostructures are in spherical shape and range in size from 30 to 70 nm (average size 62 nm). Ag NPs have crystalline FCC structure in their SAED (selected area electron diffraction) patterns. The circular planes corresponding to (1 1 1), (2 0 0), (2 2 0), and (3 1 1) match the XRD pattern very well. With a spherical dimension of 26 nm and edge length of a triangle 52 nm, Au NPs have spherical and triangular shape. Crystallinity of Au NPs with SAED patterns are shown in the planes (1 1 1), (2 0 0), (2 2 0), and (3 1 1). Similarly, the TEM images of Ag and Au NPs by citrus reticulata are shown in **Figure 2c** and **d**. The Ag NPs had a diameter of 56 nm and were spherical. The circular rings in the SAED pattern revealed that the produced Ag NPs were of crystalline structure. **Figure 2d** shows the presence of triangles in addition to spherical nanoparticles. Au NPs had a mean spherical particle size of 21 nm and a triangle edge size of 48 nm. The crystalline structure of Au NPs is revealed by the SAED pattern, which was indexed to (1 1 1), (2 0 0), (2 0 2) and (3 1 1) reflections.

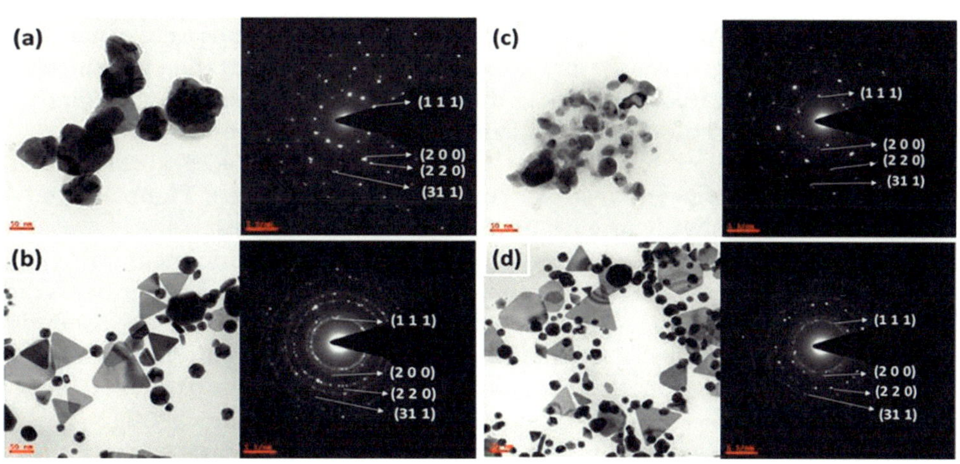

Figure 2.
TEM images and diffraction patterns of Ag and Au NSs (a) & (b) Punica granatum and (c) & (d) Citrus reticulata extract synthesized.

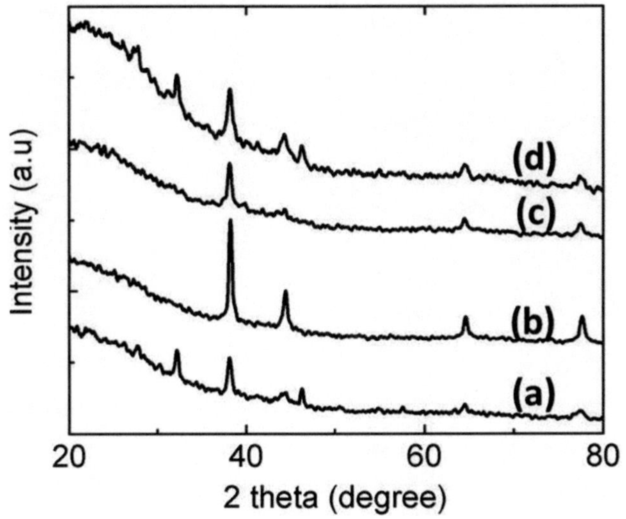

Figure 3.
XRD of Ag and Au NPs (a) & (b) Punica granatum and (c) & (d) Citrus reticulata extract synthesized.

4.3 XRD

Figure 3a–d reveals the XRD of *Punica granatum* and *Citrus reticulata* synthesized Ag and Au NSs. It shows different diffraction peaks in the range 2θ = 20°–80° at 38.5°, 44.3°, 64.2° and 77.4° corresponds to (1 1 1), (2 0 0), (2 0 2), (3 1 1) planes with FCC, respectively.

4.4 Luminescence emission

In **Figure 4a** and **c** showed that emission of europium with Ag NPs excited at 350 nm. The Ag concentrations varied from 5 to 300 μl at 15 μl of Eu. The emission at 577 nm, 590 nm (magnetic), and 614 nm (electric) are contributing to $^5D_0 \rightarrow {}^7F_0$, $^5D_0 \rightarrow {}^7F_1$, $^5D_0 \rightarrow {}^7F_2$ transitions respectively. These emission bands were enhanced but emission wavelengths were remains unaltered. With concentration increment of Ag, intensities increased and decreased with further increment of Ag. Similarly, **Figure 4b** and **d** shows the emission of Samarium with Au. The emissions at 645 nm($^4G_{5/2} \rightarrow {}^6H_{9/2}$), 566 and 602 nm ($^4G_{5/2} \rightarrow {}^6H_{5/2}$ and $^4G_{5/2} \rightarrow {}^6H_{7/2}$) correspond to electric and magnetic dipole transitions respectively. In the inset of **Figure 4b** and **d**, we show the effect of silver concentration on different transitions.

The Ag NPs' concentration dependence on the emission enhancement for Eu(TTFA)$_3$ is shown in the inset of **Figure 4a**, where the normalized intensity is the intensity ratio of Eu(TTFA)$_3$ solution containing Ag NPs to pure Eu(TTFA)$_3$ solution for the $^5D_0 \rightarrow {}^7F_2$ transition. The luminescence enhancement gets strongly affected by the Ag NPs concentration, the maximum enhancement factor was 23 at 70 μl Ag concentration. For $^5D_0 \rightarrow {}^7F_0$ and $^5D_0 \rightarrow {}^7F_1$, the enhancement factor was observed as 3. The luminescence enhancement is greater for electric dipole transitions than for magnetic dipole transitions, according to this. The hyper-sensitive transition is influenced by changes in the refractive index and ligand fields around rare-earth ions caused by interactions with metal NPs [44].

The emission of rare-earth ions near NSs are affected by the near-field environment, which induces enhancement or quenching relay on the distance between nanoparticles and rare-earth ions [45]. Rare-earth ions transfer non-radiative energy to metal particles through very low distances between NPs and molecules. Because of plasmonic resonances in metal NPs, magnetic field enhancement occurs only until a certain distance [46]. More Eu ions can be closer to NSs and the distance between Eu ions and Ag NPs can be reduced by increment in number of Ag NSs. With more NPs or reabsorption of SPR, the emission intensity is quenched by energy transmit between europium and NSs. The electric dipole transition (5D0 7F2) and magnetic dipole transitions (5D0 7F0 and 5D0 7F1), on the other hand, had distinct concentration dependence on luminescence intensities. The enhancement is considered to be caused to a hard coupling between SPR and probes. **Figure 1** shows that the absorption spectra of Ag NPs are in the wavelength range of 350–600 nm, which corresponds to the absorption of rare-earth ions. The enhancement resulted from a sensitive balance between surface Plasmon resonance reabsorption and local electromagnetic (EM) field enhancement.

Figure 4e and **g** presents the emission of 20 µl europium with Au NSs and excited at 350 nm. The emission bands at 614 nm is $^5D_0 \rightarrow {}^7F_2$ electric dipole and at 577 nm ($^5D_0 \rightarrow {}^7F_0$), and 590 nm ($^5D_0 \rightarrow {}^7F_1$) are magnetic dipole transitions. The dependency of emission of europium at 10, 20, 30 µl with gold at 5 to 300 µl is shown in the inset of

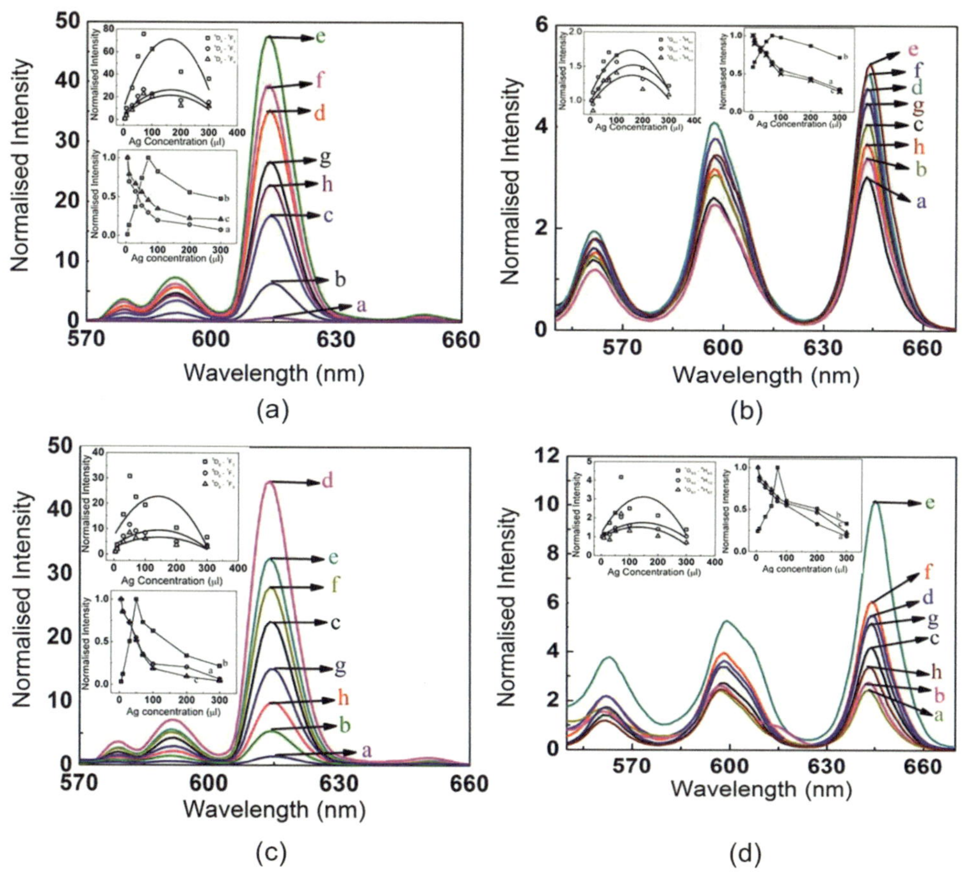

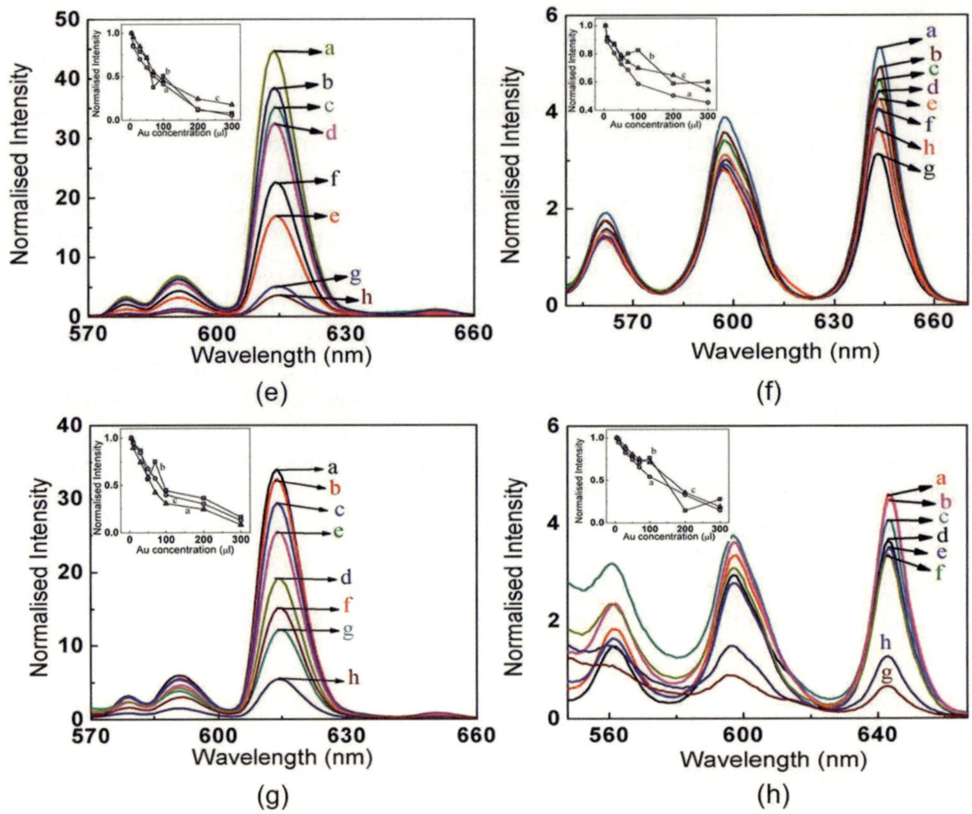

Figure 4.
((a) Punica granatum and (c) citrus reticulata) Photoluminescence of 15 μl Eu(TTFA)$_3$ with Ag NPs concentration: (a) 5, (b) 10, (c) 30, (d) 50, (e) 70, (f) 100, (g) 200 μl and (h) 300 μl. Inset figures: dependency of Ag with various dipole transitions and Ag concentration with emission at 10, 15, 30 μl of europium concentrations. ((b) P. granatum and (d) Citrus reticulata) Photoluminescence of 200 μl Sm(TTFA)$_3$ with Ag (a) 5, (b) 10, (c) 20, (d) 30, (e) 50, (f) 70, (g) 100 and (h) 200 μl concentrations. Inset figures: dependency of Ag with various dipole transitions and Ag concentration with emission at 180, 200, 220 μl of samarium concentrations. ((e) Punica granatum and (g) Citrus reticulata) Photoluminescence of 20 μl Eu(TTFA)$_3$ with Ag NPs concentration: (a) 5, (b) 10, (c) 30, (d) 50, (e) 70, (f) 100, (g) 200 μl and (h) 300 μl. Inset figures: dependency of Au with various dipole transitions and Au concentration with emission at 10, 20, 30 μl of europium concentrations. ((f) Punica granatum and (h) Citrus reticulata) Photoluminescence of 220 μl Sm(TTFA)$_3$ with Au (a) 5, (b) 10, (c) 20, (d) 30, (e) 50, (f) 70, (g) 100 and (h) 200 μl concentrations. Inset figures: dependency of Au with various dipole transitions and Au concentration with emission at 200, 220, 240 μl of samarium concentrations.

Figure 4e and **g**. Similarly, **Figure 4f** and **h** displays the emission of samarium with Au NTs. The band at 645 nm ($^4G_{5/2} \rightarrow {}^6H_{9/2}$) is an electric dipole and 566 nm ($^4G_{5/2} \rightarrow {}^6H_{5/2}$) and 602 nm ($^4G_{5/2} \rightarrow {}^6H_{7/2}$) are magnetic dipole transitions. In **Figure 4e** and **g** inset, emission of samarium 10, 20, 30 μl with Au at 5 to 300 μl is shown. The emission is quenched exponentially for Eu^{3+} at 10 and 30 μl and Sm^{3+} at 100 and 120 μl. The quenching is owing to the re-absorption of nanoparticles.

4.5 In vitro cytotoxicity

Green Ag and Au NTs were tested in MTT assays to see how they affected the growth of cancer cell lines. This is the first study to test the cytotoxicity of Ag NPs and Au NTs generated from *Punica granatum* and *Citrus reticulata* against cancer cells. **Figure 5a–d** shows the cell viability of Ag and Au nanoparticles, as well as their

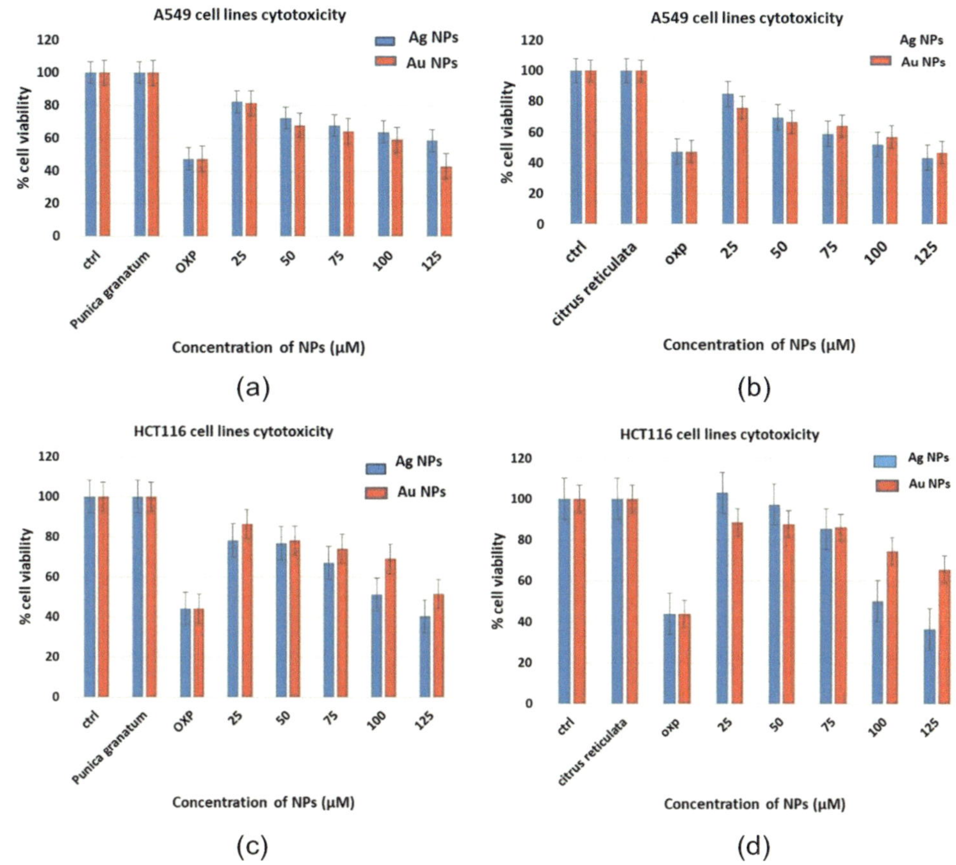

Figure 5.
Cytotoxicity studies of (a) & (b) A549 (lung cancer) cell lines of Punica granatum and Citrus reticulata and (c) & (d) HCT116 (colon cancer) cell lines of Punica granatum and Citrus reticulata of Ag NPs and Au NSs.

comparison to controls. *Punica granatum*-produced Ag and Au NPs demonstrated considerable cytotoxicity against lung and colon cancer cell lines, as shown in **Figure 5a** and **c**. *Punica granatum* produced Au NPs against lung cell lines and Ag NPs against colon cell lines, which demonstrated decreased viability of 42–40%. *Punica granatum* extract revealed no substantial toxicity, indicating that the capping agent has a minor role in toxicity. *Punica granatum*-produced Ag and Au NPs have half-maximum inhibitory concentrations (IC50) of 115 g/ml and 114 g/ml; 107 g/ml and 143 g/ml, respectively. **Figure 5b** and **d** reveals that Ag and Au NPs generated by *Citrus reticulata* are resistant to A549 and HCT116 cell lines. Cell viability decreased as the concentration of Ag and Au NPs rose. *Citrus reticulata* produced Au NPs against A549 and Ag NPs against HCT116 cell lines were found to have reduced viability, however *Citrus reticulata* extract was shown to have 100% viability. *Citrus reticulata* produced Ag and Au NPs have IC50 values of 106 g/ml and 121 g/ml; 110 g/ml and 206 g/ml, respectively.

4.6 Antimicrobial studies

Figure 6 and **Table 1** indicate the inhibitory effect of Ag and Au NSs derived from *Punica granatum* and *Citrus reticulata* extracts. Silver nanoparticles from *Citrus*

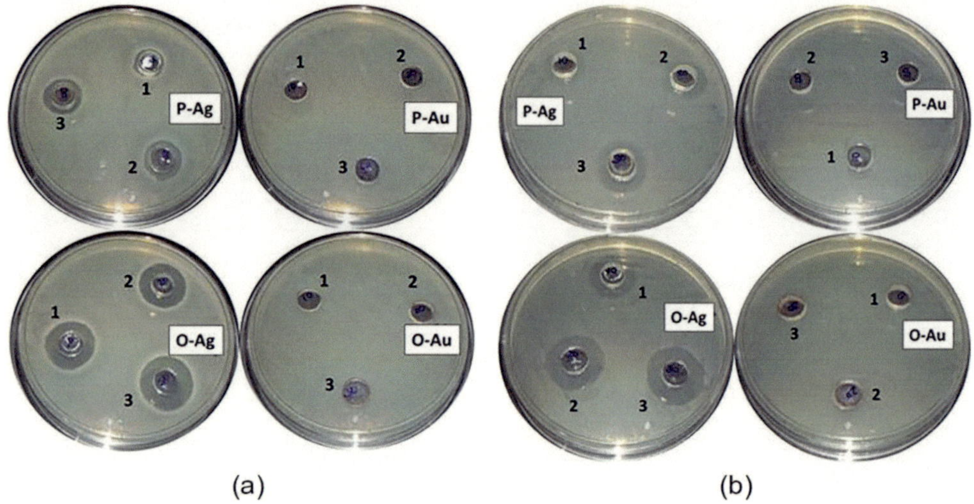

Figure 6.
Antibacterial activities of Ag and Au NPs against (a) Acinetobacter baumannii and (b) Staphylococcus aureus with (1) 10 µl, (2) 20 µl and (3) 30 µl.

Ag/Au NPs concentration (µl)		**Zone of inhibition (cm)**	
		A. baumannii	***S. aureus***
P-Ag	10	1.3	1.0
	20	1.5	1.5
	30	1.8	2
P-Au	10	0	0
	20	0	0
	30	0	0
O-Ag	10	2	1.8
	20	2	2.2
	30	2.5	2.4
O-Au	10	0	0
	20	0	0
	30	0	0

Table 1.
Inhibition zones of Ag and Au NSs.

reticulata extract (O-Ag) outperform *Punica granatum* silver nanoparticles (P-Ag) in these tests. However, Au NSs produced from *Punica granatum* (P-Au) and *Citrus reticulata* extracts (O-Au) had no antibacterial activity. The bacterial activity of the nanoparticles is determined by the inhibitory zones. Compared to *S. aureus*, silver nanoparticles have higher activity against *A. baumannii* (Gram-negative) (Gram-negative). The zone of inhibitions of P-Ag with10, 20, and 30 µl against Gram-negative and Gram-positive bacteria are 1.3, 1.5, 1.8 cm, and 1, 1.5, 2 cm, respectively. Similarly, for O-Ag are 2, 2, 2.5 cm, and 1.8, 2.2, and 2.4 cm against Gram-negative and Gram-negative bacteria, respectively, at different doses. When it comes to bacterial activity, smaller particles having large surface area have greater bacterial activity than larger particles. According to the literature, cell penetration causes intracellular loss, which leads to cell death, and this inhibition is reliant on number of nanoparticles [47, 48]. The cell death is caused by silver. Due to their thin peptidoglycan coating, Ag NPs are usually more effective for Gram-negative bacteria than Gram-positive bacteria. However, in recent studies, Ag NPs have been shown to have higher antibacterial activity against Gram-positive bacteria, indicating that these Ag NPs have significantly higher antibacterial activity than other nanoparticles. Because of their small size, nanoparticles are prone to aggregation, which limits their use. In toxicity investigations, it is significant.

5. Conclusions

The effect of metal nanostructures on emission of rare-earth compounds is reported in brief research. The emission of rare-earth ions is increased initially, but then decreased when the Ag NPs concentration was increased further. The dependence of Ag NPs on emission intensity reveals a strong linkage of luminescence centers and Plasmon resonance of NSs, as well as reabsorption of SPR of Ag NPs. The anticancer and antibacterial properties of the nanoparticles were remarkable.

Conflict of interest

The authors declare no conflict of interest.

Author details

Emusani Ramya[1]* and D. Narayana Rao[2]

1 Department of Humanities and Science, CMR Institute of Technology, Hyderabad, Telangana, India

2 School of Physics, University of Hyderabad, Hyderabad, Telangana, India

*Address all correspondence to: eramya@uohyd.ac.in

References

[1] Ali HRK, Moawad MS, Selim SA. In vitro study for comparing the cytotoxicity of silver and gold nanospheres on raw 264.7 murine macrophage cell line. Journal of Bacteriology Parasitology. 2016;**7**:1000264. DOI: 10.4172/2155-9597.1000264

[2] Cheng D, Yang J, Zhao Y. Antibacterial materials of silver nanoparticles application in medical appliances and appliances for daily use. China Medical Equipment Journal. 2004;**4**:26-32

[3] Samuel U, Guggenbichler JP. Prevention of catheter-related infections: The potential of a new nano-silver impregnated catheter. International Journal of Antimicrobial Agents. 2004;**23**:75-78. DOI: 10.1016/j.ijantimicag.2003.12.004

[4] Chen J, Han CM, Lin XW, Tang ZJ, Su SJ. Effect of silver nanoparticle dressing on second-degree burn wound. Zhonghua Wai Ke Za Zhi. 2006;**44**(1):50-52

[5] Lee HY, Park HK, Lee YM, Kim K, Park SB. A practical procedure for producing silver nanocoated fabric and its antibacterial evaluation for biomedical applications. Chemical Communications (Cambridge). 2007;**28**:2959-2961

[6] Lee HJ, Yeo SY, Jeong SH. Antibacterial effect of nanosized silver colloidal solution on textile fabrics. Journal of Materials Science. 2003;**38**:2199-2204

[7] Ramya E, Jyothi L, Rao DN. Influence of optical properties of Ag NPs from *Raphanus sativus* leaf extract on lanthanide complexes. Plasmonics. 2017;**12**:1601-1611. DOI: 10.1007/s11468-016-0424-x

[8] Chah S, Hammond MR, Zare RN. Gold nanoparticles as a colorimetric sensor for protein conformational changes. Chemistry & Biology. 2005;**12**(3):323-328

[9] Prati L, Villa A, Lupini AR, Veith GM. Gold on carbon: One billion catalysts under a single label. Physical Chemistry Chemical Physics. 2012;**14**:2969-2978. DOI: 10.1039/C2CP23405J

[10] Shipway AN, Katz E, Willner I. Nanoparticle arrays on surfaces for electronic, optical, and sensor applications. Chemical Physics Chemistry. 2000;**1**:18-52

[11] Han G, Ghosh P, Rotello VM. Functionalized gold nanoparticles for drug delivery. Nanomedicine (London, England). 2007;**1**:113-123

[12] Malina D, Agnieszka SK, Zbigniew W, Zygmunt K. Silver nanoparticles synthesis with different concentrations of polyvinylpyrrolidone. Digest Journal of Nanomaterials and Biostructures. 2012;**7**:1527-1534

[13] Ramya E, Jyothi L, Vivek Vardhan P, Sriram Gopal N, Rao DN. Optical and biomedical applications of eco-friendly biosynthesized silver nano spheres using zingiber officinale root extract. Nano Ex. 2020;**1**:010021

[14] Cruz D, Falé PL, Mourato A, Vaz PD, Serralheiro ML, Lino AR. Preparation and physicochemical characterization of Ag nanoparticles biosynthesized by Lippia citriodora (Lemon Verbena). Colloids and Surfaces. B, Biointerfaces. 2010;**81**:67-73. DOI: 10.1016/j.colsurfb.2010.06.025

[15] Kim J, Lee J, Kwon S, Jeong S. Preparation of biodegradable polymer/silver nanoparticles composite and its antibacterial efficac. Journal of Nanoscience and Nanotechnology. 2009;**9**:1098-1102. DOI: 10.1166/jnn.2009.C096

[16] Samadi N, Golkaran D, Eslamifar A, Jamalifar H, Fazeli MR, Mohseni FA. Intra/extracellular biosynthesis of silver nanoparticles by an autochthonous strain of Proteus mirabilis isolated from photographic waste. Journal of Biomedical Nanotechnology. 2009;**5**:247-253. DOI: 10.1166/jbn.2009.1029

[17] Jain D, Kumar Daima H, Kachhwaha S, Kothari SL. Synthesis of plant mediated silver nanoparticles using papaya fruit extract and evaluation of their antimicrobial activities. Digest Journal of Nanomaterials and Biostructures. 2009;**4**:557-563

[18] Gomez-Caravaca AM, Verardo V, Toselli M, Segura-Carretero A, Fernandez-Gutierrez A, Caboni MF. Determination of the major phenolic compounds in pomegranate juices by HPLC-DAD-ESI-MS. Journal of Agricultural and Food Chemistry. 2013;**61**:5328-5337. DOI: 10.1021/jf400684n

[19] Nadagouda MN, Iyanna N, Lalley J, Han C, Dionysiou DD, Varma RS. Synthesis of silver and gold nanoparticles using antioxidants from blackberry, blueberry, pomegranate, and turmeric extracts. ACS Sustainable Chemistry & Engineering. 2014;**7**:1717-1723. DOI: 10.1021/sc500237k

[20] Rafiq S, Rajkumari Kaul SA, Sofi NB, Nazir F, Nayik GA. Citrus peel as a source of functional ingredient: A review. Journal of the Saudi Society of Agricultural Sciences. 2018;**17**:351-358. DOI: 10.1016/j.jssas.2016.07.006

[21] Hussain SM, Javorina AK, Schrand AM, Duhart HM, Ali SF, Schlager JJ. The interaction of manganese nanoparticles with PC-12 cells induces dopamine depletion. Toxicological Sciences. 2006;**92**:456-463. DOI: 10.1093/toxsci/kfl020

[22] Braydich-Stolle L, Hussain S, Schlager JJ, Hofmann MC. In vitro cytotoxicity of nanoparticles in mammalian germline stem cells. Toxicological Sciences. 2005;**88**:412-419. DOI: 10.1093/toxsci/kfi256

[23] Hussain SM, Hess KL, Gearhart JM, Geiss KT, Schlager JJ. In vitro toxicity of nanoparticles in BRL 3A rat liver cells. Toxicology in Vitro. 2005;**19**:975-983. DOI: 10.1016/j.tiv.2005.06.034

[24] Jana NR, Gearheart L, Murphy CJ. Wet chemical synthesis of high aspect ratio cylindrical gold nanorods. The Journal of Physical Chemistry B. 2001;**105**:4065-4067. DOI: 10.1021/jp0107964

[25] Selvakannan PR, Mandal S, Phadtare S, Gole A, Pasricha R, Adyanthaya SD, et al. Water-dispersible tryptophan-protected gold nanoparticles prepared by the spontaneous reduction of aqueous chloroaurate ions by the amino acid. Journal of Colloid and Interface Science. 2004;**269**:97-102. DOI: 10.1016/S0021-9797(03)00616-7

[26] Devika CB, Arezou AG, Warren CWC. Determining the size and shape dependence of gold nanoparticle uptake into mammalian cells. Nano Letters. 2006;**6**:662-668

[27] Bouzigues C, Gacoin T, Alexandrou A. Biological applications of rare-earth based nanoparticles. ACS Nano. 2011;**5**:8488-8505. DOI: 10.1021/nn202378b

[28] Timmerman D, Izeddin I, Stallinga P, et al. Space-separated quantum cutting with silicon nanocrystals for photovoltaic applications. Nature Photonics. 2008;**2**:105-109. DOI: 10.1038/nphoton.2007.279

[29] McGehee MD, Bergstedt T, Zhang C, Saab AP, O'Regan MB, Bazan GC, et al. Narrow bandwidth luminescence from blends with energy transfer from semiconducting conjugated polymers to europium complexes. Advanced Materials. 1999;**11**:1349-1354

[30] Wang C, Tao H, Cheng L, Liu Z. Near-infrared light induced in vivo photodynamic therapy of cancer based on upconversion nanoparticles. Biomaterials. 2011;**32**:6145-6154. DOI: 10.1016/j.biomaterials.2011.05.007

[31] Mukkala V-M, Helenius M, Hemmilä I, Kankare J, Takalo H. Development of luminescent europium(III) and terbium(III) chelates of 2,2′:6′,2″- terpyridine derivatives for protein labelling. Helvetica Chimica Acta. 1993;**76**:1361-1378. DOI: 10.1002/hlca.19930760323

[32] Aslan K, Geddes CD. Directional surface plasmon coupled luminescence for analytical sensing applications: Which metal, what wavelength, what observation angle. Analytical Chemistry. 2009;**81**:6913-6922. DOI: 10.1021/ac900973r

[33] Eltzov E, Prilutsky D, Kushmaro A, Marks RS, Geddes CD. Metal-enhanced bioluminescence: An approach for monitoring luminescent processes. Applied Physics Letters. 2009;**94**:083901. DOI: 10.1063/1.3086283

[34] Xie F, Baker MS, Goldys EM. Enhanced fluorescence detection on homogeneous gold colloid self-assembled monolayer substrates. Chemistry of Materials. 2008;**20**:1788-1797. DOI: 10.1021/cm703121m

[35] Ray K, Chowdhury MH, Zhang J, et al. Plasmon-controlled fluorescence towards high-sensitivity optical sensing. Advances in Biochemical Engineering/Biotechnology. 2009;**116**:29-72. DOI: 10.1007/10_2008_9

[36] Geddes CD, Parfenov A, Roll D, Gryczynski I, Malicka J, Lakowicz JR. Silver fractal-like structures for metal-enhanced fluorescence: Enhanced fluorescence intensities and increased probe photostabilities. Journal of Fluorescence. 2003;**13**. DOI: 10.1023/A:1025046101335

[37] Campion A, Gallo AR, Harris CB, Robota HJ, Whitmore PM. Electronic energy transfer to metal surfaces: A test of classical image dipole theory at short distances. Chemical Physics Letters. 1980;**73**:447-450. DOI: 10.1016/0009-2614(80)80692-0

[38] Maier SA, Atwater HA. Plasmonics: Localization and guiding of electromagnetic energy in metal/dielectric structures. Journal of Applied Physics. 2005;**98**:011101. DOI: 10.1063/1.1951057

[39] Busson MP, Rolly B, Stout B, Bonod N, Wenger J, Bidault S. Photonic engineering of hybrid metal-organic chromophores. Angewandte Chemie. 2012;**51**:11083-11087

[40] Strohhöfer C, Polman A. Silver as a sensitizer for erbium. Applied Physics Letters. 2002;**81**:1414. DOI: 10.1063/1.1499509

[41] Eichelbaum M, Rademann K. Plasmonic enhancement or energy transfer? On the luminescence of gold-, silver-, and lanthanide-doped silicate glasses and its potential for light-emitting

devices. Advanced Functional Materials. 2009;**19**:2045-2052. DOI: 10.1002/adfm.200801892

[42] Anger P, Bharadwaj P, Novotny L. Enhancement and quenching of single-molecule fluorescence. Physical Review Letters. 2006;**96**:113002

[43] Kühn S, Håkanson U, Rogobete L, Sandoghdar V. Enhancement of single-molecule fluorescence using a gold nanoparticle as an optical nanoantenna. Physical Review Letters. 2006;**97**:017402

[44] Nabika H, Deki S. Surface-enhanced luminescence from Eu3+ complex nearby Ag colloids. European Physical Journal D: Atomic, Molecular, Optical and Plasma Physics. 2003;**24**:369-372. DOI: 10.1140/epjd/e2003-00165-x

[45] Aslan K, Geddesa CD. Surface plasmon coupled chemiluminescence from zinc substrates: Directional chemiluminescence. Applied Physics Letters. 2009;**94**:073104. DOI: 10.1063/1.3082175

[46] Liaw JW, Tsai HY, Huang CH. Size-dependent surface enhanced fluorescence of gold nanorod: Enhancement or quenching. Plasmonics. 2012;7:543-553. DOI: 10.1007/s11468-012-9341-9

[47] Panacek A, Kvítek L, Prucek R, Kolar M, Vecerova R, Pizúrova N, et al. Silver colloid nanoparticles: Synthesis, characterization, and their antibacterial activity. The Journal of Physical Chemistry. B. 2006;**110**:16248-16253. DOI: 10.1021/jp063826h

[48] Sondi I, Salopek-Sondi B. Silver nanoparticles as antimicrobial agent: A case study on *E. coli* as a model for gram-negative bacteria. Journal of Colloid and Interface Science. 2004;**275**:177-182. DOI: 10.1016/j.jcis.2004.02.012

Chapter 6

Cytotoxic Activity of Secondary Metabolite Compounds from Myanmar Medicinal Plants

Khun Nay Win Tun, Nanik Siti Aminah, Alfinda Novi Kristanti, Hnin Thanda Aung and Yoshiaki Takaya

Abstract

Plants are the storage place for many active ingredients that are currently used in pharmaceuticals. Proofs have been expanded for ages to indicate the profitable capacity of medicinal plants used in various infections, especially for cancer treatment. The plants, *Chromolaena odorata* (Family Asteraceae), *Croton oblongifolius* Roxb. (Family Euphorbiaceae), *Tinospora cordifolia* (Family Menispermaceae), *Melastoma malabathricum* L. (Family Melastomaceae), and *Dioscorea bulbifera* (Family Dioscoreaceae) have been used for a long time in Myanmar's traditional medicine. This study aimed to review systematically the cytotoxic activity of the whole plants and their extracts, fractions, and isolated compounds from these selected medicinal plants. This chapter also be substantiated for additional analysis on phytochemical constituents and pharmacological action of therapeutic plants species in Myanmar.

Keywords: cytotoxic activity, Myanmar, medicinal plants, isolated compounds, secondary metabolite

1. Introduction

Cancer is defined as the abnormal proliferation of cells in our bodies, which can result in death. As a result, there is currently no effective cancer therapy available, and the disease has the potential to spread globally. Plants are a rich source of novel bioactive compounds. This is because of their innate biological capabilities, which have the potential to be used in medicine and other disciplines of human health promotion. There is a significant need for new anticancer medications that are more effective and less damaging to healthy cells, such as plant-derived substances. The preliminary strategy for identifying active chemicals in plants is cytotoxic screening [1–3].

Chromolaena odorata is a tropical and subtropical flowering shrub widely distributed in various parts of the world. Fresh leaves or a decoction of *C. odorata* have been used for folk medicine in Vietnam. In other traditional medicine, it is useful for burn wounds, leech bites, skin infections, soft tissue wounds, and dento-alveolitis.

In Myanmar, local people are used for stomachache, cancer, and urinary tract infection. Chemical components of this plant have been discovered as chalcones, flavonoids, glycosides, anthraquinones, triterpenoids, fatty acids, coumarino lignoids, and phenolic compounds [4–12].

Croton oblongifolius is a medium-size tree and is widely distributed throughout Asia and Myanmar. In folk medicine, this plant is useful for dysmenorrhea, purgative, dyspepsia, and dysentery. In Myanmar, it is used to treat diarrhea (seeds) and liver disorders (barks). Phytochemical studies of *C. oblongifolius* have been reported in several different types of research and many clerodane diterpenes, cembranoid diterpenes, halimane-type diterpenoids, labdane diterpenoids, furanocembranoids, and megastigmane glycosides have been isolated from the stem barks, leaves, and roots [5–21].

Tinospora cordifolia is a huge evergreen and annual climbing plant distributed throughout the world. The plant is used to treat skin diseases, allergic conditions, jaundice, urinary disorders, anaemia, rheumatism, inflammation, and diabetes in both traditional and homeopathic medicine. In Myanmar, this plant is used for diabetes and hypertension. An important phytoconstituents reported from this plant were clerodane furano diterpene glucosides, phenylpropene disaccharides, phenylpropanoids, clerodane diterpenoid, steroids, terpenoids, and alkaloids [22–31].

Melastoma malabathricum is an evergreen shrub that belongs to the family Melastomataceae. In Malaysia, India, and Indonesia, this plant has traditionally been used to treat a variety of symptoms and disorders. In Myanmar, this plant is used to treat various kinds of cancer, toothache, diabetes, asthma, and lung disorders. Chemical constituents reported from *M. malabathricum* were flavonoids, triterpenes, alkaloids, steroids, lipids, phenolic compounds, and tannins [32–37].

Dioscorea bulbifera is a perennial vine widely distributed throughout tropical and temperate areas. It is commonly known as air potato. This herb is used in the treatment of leprosy and cancer by triple people in Bangladesh. In China, this herb is utilized for cancer and thyroid diseases. In Myanmar, this plant is used for various cancer diseases such as breast cancer, cervix cancer, etc. Phytochemical studies revealed the presence of alkaloids, anthraquinones, norclerodane diterpenoids, clerodane diterpenoids, flavonoids, phenolic compounds, steroidal saponins, steroidal sapogenins, and glycosides [38–49].

Our main focus for this chapter is secondary metabolites derived from the plant's kingdom, i.e., *C. odorata*, *C. oblongifolius*, *Tinospora cordifolia*, *M. malabathricum*, and *D. bulbifera*, and cytotoxic compounds obtained from them (**Table 1**).

Plants name	Family	Local name	Traditional/local uses
Chromolaena odorata	Asteraceae	Bhi-sub	stomachache, cancer, and urinary tract infection
Croton oblongifolius	Euphorbiaceae	Thet-yin-gyi	diarrhea and liver disorders
Tinospora cordifolia	Menispermaceae	Sin-don-ma-nwe	diabetes and hypertension
Melastoma malabathricum	Melastomaceae	Say-aoe-bote	various kinds of cancer, toothache, diabetes, asthma, and lung disorders
Dioscorea bulbifera	Dioscoreaceae	Myauk-U	breast cancer and cervix cancer

Table 1.
Myanmar medicinal plants and their traditional (or) local uses.

2. Secondary metabolites

Secondary metabolites can be found in abundance in plants. They are the focus of several research since they have a wide range of biological actions. Scientific reports have demonstrated the medicinal value of various parts of plant species (*C. odorata*, *C. oblongifolius*, *T. cordifolia*, *M. malabathricum*, and *D. bulbifera*) against a wide range of human ailments. Only the cytotoxicity of various secondary metabolites from different plant species will be discussed in this study.

3. Cytotoxic activity of asteraceae

In 2013, Kouamé and his coworkers isolated 5-hydroxy-7,4′-dimethoxyflavanone, 2′-hydroxy-4,4′,5′,6′-tetramethoxychalcone, and cadalene from the hexane soluble

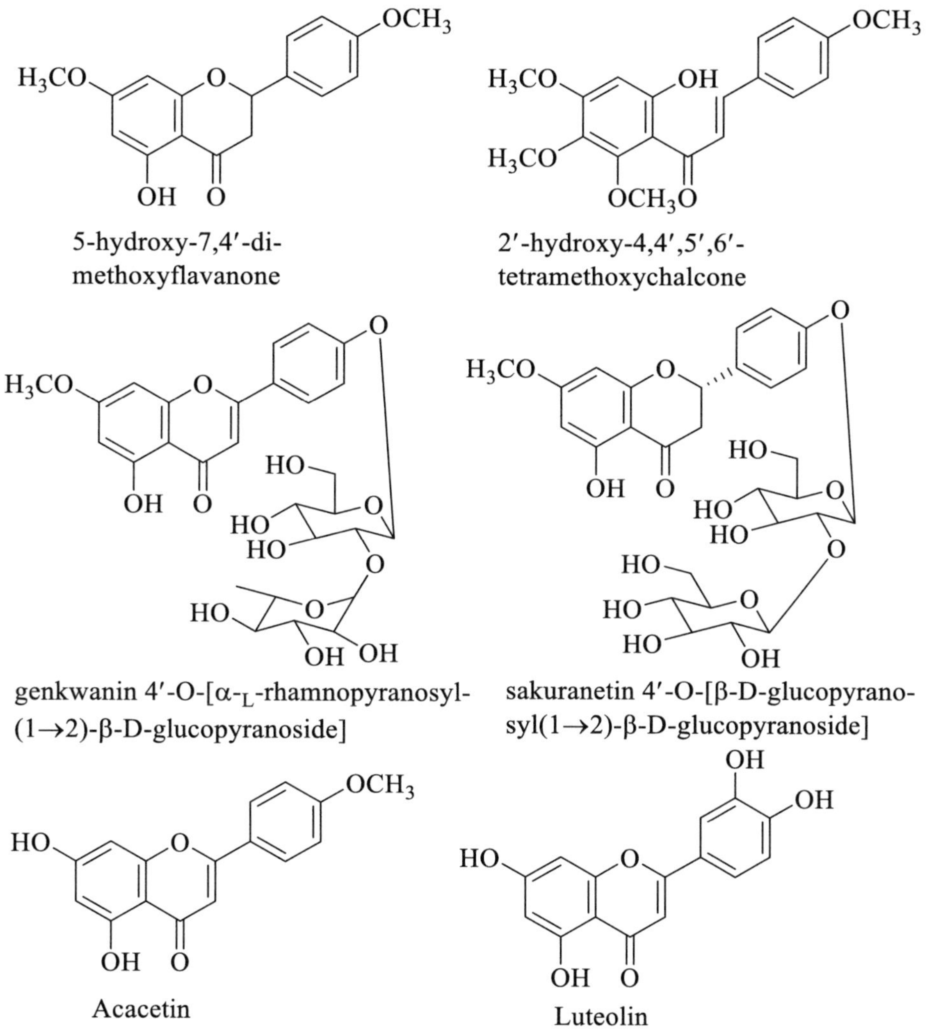

Figure 1.
Cytotoxic compounds from Chromolaena odorata.

fraction of *C. odorata*. These were screened for their cytotoxicity and anticancer properties. A chalcone, 2′-hydroxy-4,4′,5′,6′-tetramethoxychalcone has both cytotoxic and anticlonogenic actions against a wide range of cell lines (Cal51, MCF7, and MDAMB-468). Also, it enhances apoptosis in Cal51 breast cancer cells when combined with the Bcl2 inhibitor ABT737 [50]. Another research group reported that genkwanin 4′-O-[α-L-rhamnopyranosyl(1 → 2)-β-D-glucopyranoside] has cytotoxic effects on LLC (IC_{50}: 28.2 μM) and HL-60 (IC_{50}: 11.6 μM) cancer cell line. Similarly, sakuranetin 4′-O-[β-D-glucopyranosyl(1 → 2)-β-D-glucopyranoside] displayed potential cytotoxic activity in HL-60 (IC_{50}: 10.8 μM) cancer cell line [11]. Likewise, Suksamrarm and colleagues document that acacetin (IC_{50}: 24.6 μM) and luteolin (IC_{50}: 19.2 μM) possessed moderate cytotoxic activity in human small cell lung cancer cells (NCI-H187). Luteolin (IC_{50}: 38.4 μM) showed weak activity on human breast cancer cells (BC) [51]. A recent study proved that the EtOH leaves extract from *C. odorata* has anticancer and antiproliferative activity [52]. Another finding reported that EtOAc soluble fraction from EtOH leaves extract from *C. odorata* had cytotoxic and antiproliferative actions on HeLa cells [53]. In other reports, both the aqueous and EtOH extract of *C. odorata* demonstrated cytotoxic effects, with LC_{50} values of 324 and 392 μg/mL, respectively [54]. The chemical structures of some cytotoxic compounds are shown in **Figure 1**.

4. Cytotoxic activity of euphorbiaceae

Numerous investigations indicated that the *C. oblongifolius* shows cytotoxic properties. Neocrotocembranal obtained from the stem bark of *C. oblongifolius* displayed cytotoxic activity toward P-388 cells with an IC_{50} value of 6.48 μg/mL [13]. *C. oblongifolius* contain methyl 15,16-epoxy-3,13(16),14-ent-clerodatrien-18,19-olide-17-carboxylate, dimethyl 15,16-epoxy-12-oxo-3,13(16),14-ent-clerodatrien-17,18-dicarboxylate, nasimaluns A and B, levatin, (−)-hardwickiic acid, 15-hydroxy-cis-ent-cleroda-3,13-(E)-diene, and patchoulenone. These diterpenoids exhibited mild cytotoxic activity against HUCCA-1 (human cholangiocarcinoma cancer cells), KB (human epidermoid carcinoma of the mouth), HeLa (cervical adenocarcinoma cells), MDA-MB231 (human breast cells), and T47D (human mammary adenocarcinoma cells) cell lines with the IC_{50} values ranging from 10 to 50 μg/mL [14]. Additionally, the in vitro cytotoxic activity of 2,3-dihydroxy-labda-8(17),12(E),14-triene has been demonstrated non-specific moderate activities in KATO-3 (IC_{50} value 2.2 μg/mL), SW620 (IC_{50} value 2.7 μg/mL), BT474 (IC_{50} value 4.6 μg/mL), HEP-G2 (IC_{50} value 3.7 μg/mL), and CHAGO (IC_{50} value 3.3 μg/mL) [15]. Moreover, crotohalimaneic acid, crotohalimonetic acid, and 12-benzouloxy crotohalimaneic acid from *C. oblongifolius* were examined for their cytotoxic activity against human breast ductol carcinoma (BT474), lung carcinoma (CHAGO), human liver hepatoblastoma (HEP-G2), human gastric carcinoma (KATO-3), and human colon adenocarcinoma (SW620). The latter compound was inactive and the remaining two compounds displayed non-specific strong cytotoxic activity, and the IC_{50} values for crotohalimaneic acid were 7.5, 0.1, 0.2, 0.4, 0.2, and for crotohalimonetic acid were 0.1, 0.1, 5.2, 8.2, 0.1 μg/mL, respectively [17]. In another study, novel furanocembranoids, found from the stem bark of *C. oblongifolius*, have been shown broad cytotoxic effects on BT474, CHAGO, Hep-G2, KATO-3, and SW-620 [18]. Besides, croblongifolin obtained from *C. oblongifolius* was reported to have high cytotoxic activities against human hepatocarcinoma (HEP-G2) (IC_{50} 0.35 μM), breast carcinoma (BT474) (IC_{50} 0.12 μM), colon carcinoma (SW620) (IC50 0.47 μM), lung carcinoma (CHAGO) (IC50 0.24 μM), and gastric carcinoma (KATO 3)

(IC_{50} 0.35 μM) [19]. In a cytotoxic study by Sommit et al., natural labdanes (labda-7,12(E),14-triene-17-al and 17-hydroxylabda-7,12(E)14-triene) and their modified derivatives (15-hydroxylabda-7,13(*E*)diene-17,12-olide, labda-7,13(E)-diene-17,12-olide, 12,17-dihydroxylabda-7,13(*E*)-diene) were found to have non-specific moderate cytotoxicity against BT474, CHAGO, HEP-G2, KATO-3, and SW620 [20]. The ethyl acetate extract of *C. oblongifolius* was found to be cytotoxic toward the A549 lung cancer cell line [55]. In 2020, Poofery et al. studied the cytotoxic activity of ethyl acetate extract from *Bridelia ovata* and *Croton oblongifolius*. They reported that both extracts possessed anticancer properties in breast cancer cells and induce apoptosis in these cells

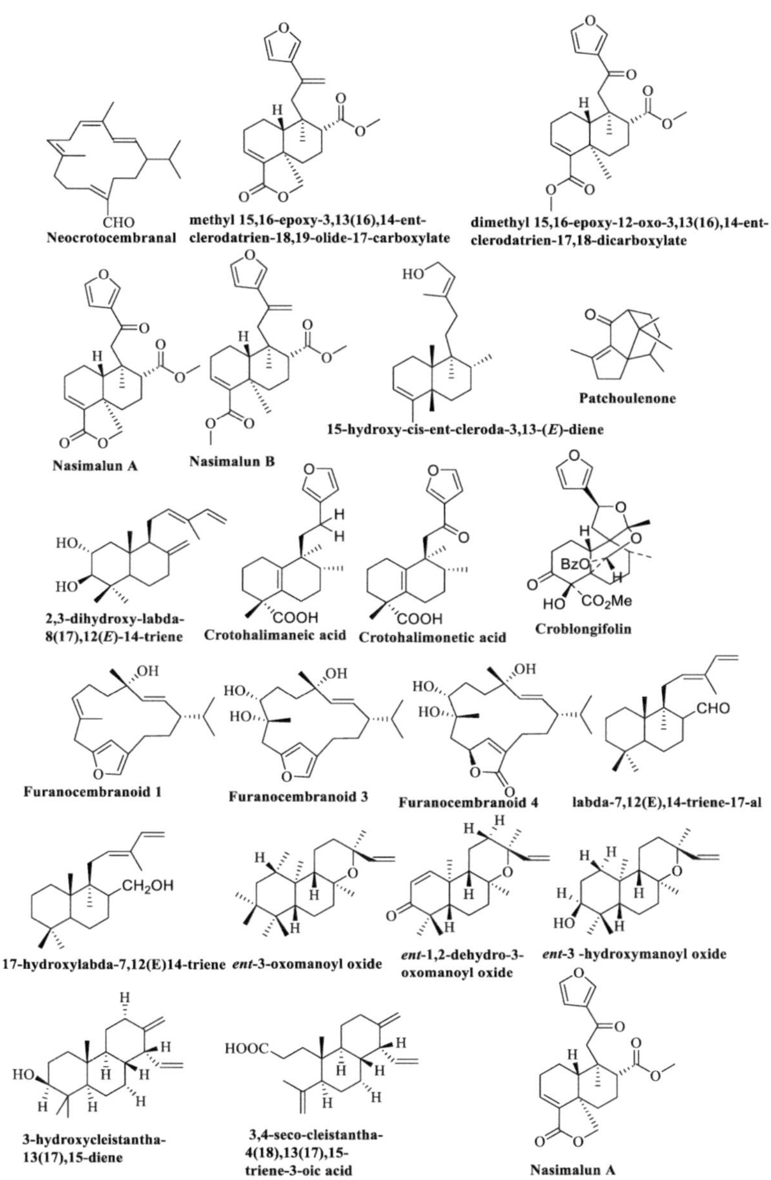

Figure 2.
Cytotoxic compounds from Croton oblongifolius.

via a mitochondrial pathway and oxidative stress [56]. In addition, Chaichantipyuth and coworkers isolated eight new labdane-type diterpenoids and a clerodane from the stem bark of *C. oblongifolius*. They examined the cytotoxic capability of specific compounds and semisynthesis products against human tumor cell lines (KATO-3, BT474, CHAGO, HEPG-2, and SW620) in vitro MTT assay. Among which isolates, ent-3-oxomanoyl oxide, ent-1,2-dehydro-3-oxomanoyl oxide, ent-3α-hydroxymanoyl oxide, nidorellol, and ent-6α,7β,8α,-trihydroxylabda-13(16),14-diene showed strong cytotoxic activity (<1 μg/mL) against KATO-3, moderated activity against BT474, and undifferentiated against CHAGO, HEPG-2, and SW620. Similarly, hardwickiic acid had a substantial effect (<1 μg/mL) on BT474, while semisynthesis products (*ent*-3β-hydroxymanoyl oxide and *ent*-3β-hydroxy-1,2-dehydromanoyl oxide) had a modest effect in all cell lines. The remaining isolates were inert [57]. The authors found, two new cleistanthane-type diterpenoids, 3-hydroxycleistantha-13(17),15-diene and 3,4-seco-cleistantha-4(18),13(17),15-triene-3-oic acid, have cytotoxic activity against various human tumor cell lines. Both were isolated from the stem bark of *C. oblongifolius*. 3-hydroxycleistantha-13(17),15-diene showed nonspecific strong cytotoxic activity (KATO-3; IC50 value 6.0 μg/mL, BT474; IC_{50} value 6.1 μg/mL, HEP-G2; IC_{50} value 0.5 μg/mL, CHAGO; IC_{50} value 5.5 μg/mL) and 3,4-seco-cleistantha-4(18),13(17),15-triene-3-oic acid showed weak activity (KATO-3; IC_{50} value 9.6 μg/mL, BT474; IC50 value 10 μg/mL, HEP-G2; IC_{50} value 8.6 μg/mL) [58]. In another study, Nasimalun A has a cytotoxic effect on the MOLT-3 cell line with an IC_{50} value of 26.44 μg/mL [59]. The chemical structures of some cytotoxic compounds are shown in **Figure 2**.

5. Cytotoxic activity of menispermaceae

In 2010, Uddin and his coworkers tested the cytotoxicity of various fractions (pet ether, CCl_4, $CHCl_3$, and aqueous soluble fraction) of *T. cordifolia* methanolic crude extract by brine shrimp lethality bioassay. They recorded that CCl_4 and $CHCl_3$ soluble fractions showed highly cytotoxic in Artemia salina with LC_{50} values of 0.402 and 0.691 μg/mL [60]. Likewise, in 2021, Modi and coworkers documented that ethanol stem and leaves extract of *T. cordifolia* had a cytotoxic effect in A. salina with LC_{50} values of 462.38 and 676.08 μg/mL [61]. Furthermore, Jagetia and Rao studied the cytotoxic effects of DCM extract of T. cordifolia on cultured HeLa cells. They suggested that the cytotoxic effect on DCM extract of *T. cordifolia* may be due to lipid peroxidation and release of lactase dehydrogenase (LDH) and decline in glutathione-S-transferase (GST) [62]. In another study, the cytotoxic activity of aqueous and hexane extract of *T. cordifolia* was evaluated against six cancer lines, including prostate (PC-3), colon (Colo-205), HCT-116), lung (A-549, NCI-H322), and breast cancer (T47D). Results showed that aqueous extract has potent cytotoxic activity (67–99%) in prostate, lung, and colon cancer lines [63]. Also, the in vitro cytotoxic activity of methanolic stem extract from *T. cordifolia* against human breast cancer cell line MDA-MB-231 and normal Vero epithelial cell line. It was revealed that methanolic stem extract displayed cytotoxic activity toward human breast cancer cells with an IC_{50} value of 59 ± 4.05 μg/mL [64]. In addition to the anti-proliferative activity, various fractions (ethanol, pet ether, DCM, butanol, and aqueous) from *T. cordifolia* were evaluated against cervical carcinoma (HeLa) cell lines by MTT and SRB methods. It was noted that both DCM (MTT; IC_{50}: 54.23 ± 0.94 μg/mL, and SRB; 48.91 ± 0.33 μg/mL) and ethanol extract (MTT; IC_{50}: 101.26 ± 1.42 μg/mL, and SRB; 87.93 ± 0.85 μg/mL) displayed significant cytotoxic effect in HeLa cell with their respective assay [65]. Moreover, Bala and his coworkers

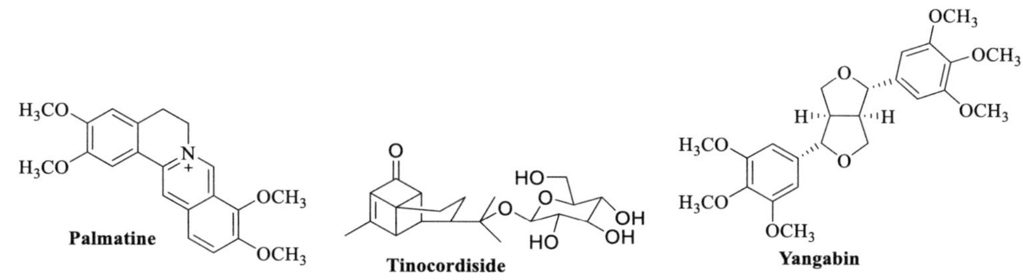

Figure 3.
Cytotoxic compounds from Tinospora cordifolia.

also reported the anti-cancer and immunomodulatory activities of different extracts, fractions, and isolated compounds from the stem of *T. cordifolia*. They were screened against four different human cancer cell lines, KB (human oral squamous carcinoma), CHOK-1 (hamster ovary), HT-29 (human colon cancer), SiHa (human cervical cancer), and murine primary cells. All extracts and fractions showed promising activity against KB and CHOK-1 cells while the isolated palmatine (KB; IC_{50}: 46.1 μM, HT-29; IC_{50}: 49.1 μM), tinocordiside (KB; IC_{50}: 45.5 μM, CHOK-1; IC_{50}: 44.9 μM), and yangambin (KB; IC_{50}: 43.8 μM) also showed promising activity with their respective cell lines [25]. The chemical structures of some cytotoxic compounds are shown in **Figure 3**.

6. Cytotoxic activity of Melastomaceae

The cytotoxic effect of the ethyl acetate fraction from *Melastoma malabathricum* in human breast and lung cancer cell lines (MCF-7 and A549) was studied by using an MTT assay. Results showed that MCF-7 and A549 cells undergo secondary necrosis 24 hours post-treatment with *M. malabathricum* extract [66]. Another observation showed that after 72 h of treatment, the *M. malabathricum* leaves extract exhibited important anticancer activity in the MCF-7 (IC_{50}: 7.14 μg/mL) cell line, although the methanol and chloroform extract from the flowers had modest activities in MCF-7 (IC_{50}: 33.63 μg/mL and 45.76 μg/mL) cell line [67]. In addition, Zakaria et al. studied the in vitro antiproliferative activity of various extracts (aqueous, chloroform, and methanol extracts) from *M. malabathricum*. The aqueous extract was found to be potent inhibitor of Caov-3 (IC_{50}: 58 μg/mL), HL-60 (IC_{50}: 58 μg/mL) cell lines, whereas $CHCl_3$ extract showed antiproliferative activity toward Caov-3 (IC_{50}: 34 μg/mL), HL-60 (IC_{50}: 96 μg/mL), and CEM-SS (IC_{50}: 22 μg/mL) cell lines. The methanol extract established antiproliferative activity in more cell lines, such as MCF-7 (IC_{50}: 87 μg/mL), HeLa (IC_{50}: 88 μg/mL), Caov-3 (IC50: 41 μg/mL), HL-60 (IC50: 13 μg/mL), CEM-SS (IC_{50}: 30 μg/mL), and MDA-MB-231 (IC_{50}: 59 μg/mL) cancer cell lines [68]. In 2019, Isnaini et al. studied the quercetin and kaempferol level in *M. malabathricum* ethanolic fruit extract. The analysis was done by HPLC MS/MS. In comparison to kaempferol (43.52 μg/g), quercetin (67.78 μg/g) had a greater concentration in the test sample. They also compared the extract to quercetin in terms of antioxidant and cytotoxic properties. The authors concluded that the extract exhibits antioxidant (IC_{50}: 16.82 ± 0.24 ppm) and cytotoxic properties (LC_{50}: 313.44 ppm) which may be attributed to the presence of quercetin [69]. The compounds such as flavonoids, naringenin, and kaempferol-3-O-(2′′,6′′-di-O-*p*-trans-coumaroyl)glucoside, isolated from the flower of *M. malabathricum*, have been proved to be the potent inhibitor of MCF7 with IC_{50} values of 0.28 μM and 1.3 μM [70].

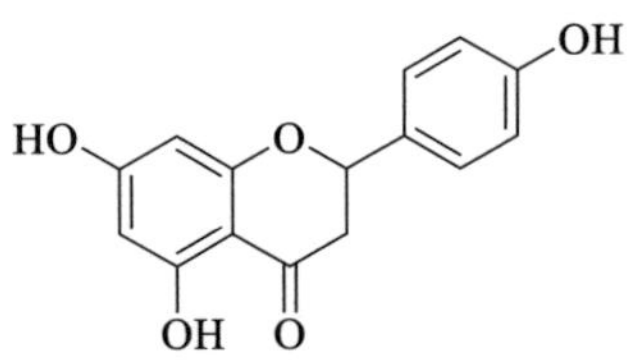

Naringenin

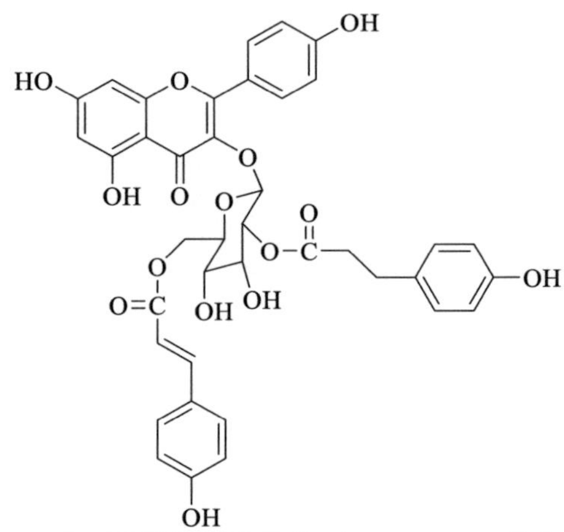

kaempferol-3-O-(2″,6″-di-*O*-*p*-trans-coumaroyl)glucoside

Figure 4.
Cytotoxic compounds from Melastoma malabathricum.

In another cytotoxic study by Hamid et al., the *M. malabathricum* leaves extract showed inhibitory effects on the MDA-MB-231 cell line after 48 and 72 h treatments [71]. The chemical structures of some cytotoxic compounds are shown in **Figure 4**.

7. Cytotoxic activity of dioscoreaceae

Several studies have been performed to examine the cytotoxic activity of constituents in *Dioscorea bulbifera*. Many studies have observed cytotoxic compounds, largely found as steroidal saponins. Isolated pennogenin-3-O-α-L-rhamnopyranosyl-(1 → 3)-[α-L-rhamnopyranosyl-(1 → 2)]-β-D-glucopyranoside and penogenin-3-O-α-L-rhamnopyranosyl-(1 → 4)-[α-L-rhamnopyranosyl-(1 → 2)]-β-D-glucopyrano-side demonstrated cytotoxic activity in two human hepatocellular carcinoma cell lines (Bel-7402; 99.1% and 92.6% inhibition, SMMC7721; IC_{50} value 4.54 μM and 4.85 μM) [47]. Diosgenin-3-O-[α-L-rhamnopyranosyl-(1 → 2)-[α-L-rhamnopyranosyl-(1 → 3)]-β-D-glucopyrano- side and diosgenin-3-O-[α-L-rhamnopyranosyl-(1 → 2)-[α-L-rhamnopyranosyl-(1 → 4)]-β-D-glucopyranoside were obtained from the rhizomes of *D. bulbifera* and displayed moderate cytotoxic activities against two human hepatocellular carcinoma cell lines (Bel-7402; IC_{50} values 10.87 and 19.10 μM, and SMMC7721; IC_{50} values 3.89 and 7.47 μM) [48]. In another study, spiroconazol A, penogenin-3-O-α-L-rhamnopyranosyl-(1 → 4)-α-L-rhamnopyranosyl-(1 → 4)-[α-L-rhamnopyranosyl-(1 → 2)]-β-D-glucopyranoside, and 26-O-β-D-glucopyranosyl-(25R)-5-en-furost-3β,17α,22α,26-tetraol-3-O-α-L- rhamno-pyranosyl-(1 → 4)-α-L-rhamnopyranosyl-(1 → 4)-[α-L-rhamnopyranosyl-(1 → 2)]-β-D-glucopyrano-side were reported to have moderate activity against human urinary bladder carcinoma cell (ECV-304) [72]. In 2018, Wang and colleagues evaluated the cytotoxic activity of three new norclerodane diterpenoids (diosbulbin E acetate, diosbulbin R, and diosbulbin S) in the SMMC-7721 cancer cell line. These compounds were isolated from the ethanol extract of *D. bulbifera*. Interestingly, the isolates showed no inhibitory effects on SMMC-7721 (IC_{50} > 40 μM) [73]. Chloroform and methanol soluble fractions from *D. bulbifera* were tested for their cytotoxic activity in T47D breast cancer cells by Rinto Muhammad Nur and Laurentius Hartanto Nugroho. The results revealed that the chloroform fraction displayed the highest activity with an IC_{50} value of 115.63 ± 86.01 μg/mL. They also screened the combined leaf chloroform fractions and the data showed that fractions F-5 (IC_{50}: 14.55 ± 8.62 μg/mL) and F-6 (IC50: 14.55 ± 8.62 μg/mL) were found to be the most toxic fractions. Nevertheless, both fractions (F-5 and F-6) were very weak when compared with the cancer drug, Doxorubicin (IC_{50}: 0.04 ± 0.02 μg/mL) [74]. Mainasara et al. reported that *D. bulbifera* has cytotoxic effects on MAD-MB-231 and MCF-7 breast cancer cell lines. The IC_{50} values for MAD-MB-231 were found to be 41.17, 15.71, and 11.53 μg/mL, at 24, 48, and 72 h of

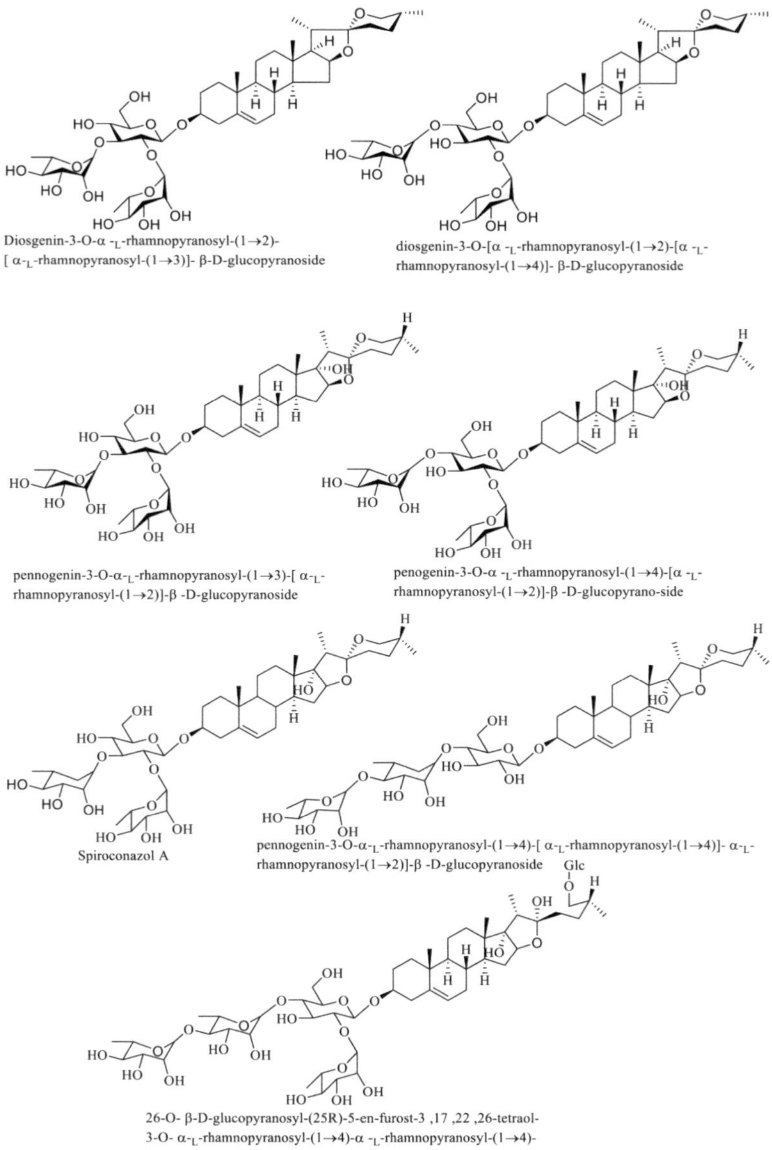

Figure 5.
Cytotoxic compounds from Dioscorea bulbifera*.*

incubation time. For MCF-7 cell lines, the IC_{50} values were 4.29, 1.86, and 1.23 μg/mL, respectively, after 24, 48, and 72 h of incubation time. They also studied cycle analysis and apoptosis by using flow cytometry analysis. It was found that D. bulbifera induced apoptosis at various phases, with a considerable drop in viable cells after 24 h and significant improvements after 48 and 72 h of treatment [75]. Wang et al. observed the antitumor activity of various extracts and compound diosbulbin B from *D. bulbifera* in mice model [76]. In another study, two new phenolic derivatives (diosbulbiol A and diosbulbiol B) from the tubers of *D. bulbifera* were tested for their cytotoxic activity against SMMC7721, MCF-7, K562 and A549 cancer cell lines. Interestingly, these two compounds showed no inhibitory effect on all selected cell lines ($IC_{50} > 40$ μM) [77]. The chemical structures of some cytotoxic compounds are shown in **Figure 5**.

8. Conclusions

Several investigations have been conducted on the bioactivities of Myanmar medicinal herbs. Novel anticancer medicines and lead compounds are still abundant in higher plants. The key benefits of anticancer natural compounds are their low toxicity, low cost, and wide range of modes of action. In this chapter, we examined and studied the potential of Myanmar medicinal plants as a source of cytotoxic compounds. Although several of these plants' metabolites exhibited potential bioactivities, some of them were determined to be inert or to have weak activities in present investigations. This chapter will surely assist academics and practitioners working with these specific plant species in determining the best application for them. Because the majority of cytotoxic effects are evaluated in vitro, their therapeutic advantages may not be adequately demonstrated. Animal models and human trials should be used in a future study to assess how effective they are at providing a platform for enhancing anticancer therapeutic approaches.

Acknowledgements

The authors are grateful to: (1) Daw Ni Ni Aung at the Department of Botany, Lashio University for her valuable support, and (2) Universitas Airlangga for the funding support.

Conflict of interest

The authors declare no conflict of interest.

Author details

Khun Nay Win Tun[1,2], Nanik Siti Aminah[1,3*], Alfinda Novi Kristanti[1,3], Hnin Thanda Aung[4] and Yoshiaki Takaya[5]

1 Faculty of Science and Technology, Department of Chemistry, Universitas Airlangga, Komplek Kampus C UNAIR, Jl. Mulyorejo, Surabaya, Indonesia

2 Department of Chemistry, Pathein University, Pathein, Myanmar

3 Biotechnology of Tropical Medicinal Plants Research Group, Universitas Airlangga, Indonesia

4 Department of Chemistry, Mandalay University, Mandalay, Myanmar

5 Faculty of Pharmacy, Meijo University, Nagoya, Japan

*Address all correspondence to: nanik-s-a@fst.unair.ac.id

References

[1] Khazir J, Riley DL, Pilcher LA, De-Maayer P, Mir BA. Anticancer agents from diverse natural sources. Natural Product Communications. 2014;**9**(11):1655-1669

[2] Wang H, Khor TO, Shu L, Su ZY, Fuentes F, Lee JH, et al. Plants vs. cancer: A review on natural phytochemicals in preventing and treating cancers and their druggability. Anti-Cancer Agents in Medicinal Chemistry. 2012;**12**(10):1281-1305. DOI: 10.2174/187152012803833026

[3] Lee JJ, Saiful Yazan L, Kassim NK, Che Abdullah CA, Esa N, Lim PC, et al. Cytotoxic activity of christia vespertilionis root and leaf extracts and fractions against breast cancer cell lines. Molecules. 2020;**25**(11):2610. DOI: 10.3390/molecules25112610

[4] Vijayaraghavan K, Rajkumar J, Bukhari SN, Al-Sayed B, Seyed MA. Chromolaena odorata: A neglected weed with a wide spectrum of pharmacological activities (Review). Molecular Medicine Reports. 2017;**15**(3):1007-1016. DOI: 10.3892/mmr.2017.6133

[5] Pel P, Chae HS, Nhoek P, Kim YM, Khiev P, Kim GJ, et al. A stilbene dimer and flavonoids from the aerial parts of Chromolaena odorata with proprotein convertase subtilisin/kexin type 9 expression inhibitory activity. Bioorganic Chemistry. 2020;**99**:103869. DOI: 10.1016/j.bioorg.2020.103869

[6] Omokhua-Uyi AG, Abdalla MA, Leonard CM, Aro A, Uyi OO, Van Staden J, et al. Flavonoids isolated from the South African weed Chromolaena odorata (Asteraceae) have pharmacological activity against uropathogens. BMC Complementary Medicine and Therapies. 2020;**20**(1):233. DOI: 10.1186/s12906-020-03024-0

[7] Zhang ML, Irwin D, Li XN, Sauriol F, Shi XW, Wang YF, et al. PPARγ agonist from Chromolaena odorata. Journal of Natural Products. 2012;**75**(12):2076-2081. DOI: 10.1021/np300386d

[8] Putri DA, Fatmawati S. A New Flavanone as a Potent Antioxidant Isolated from Chromolaena odorata L. Leaves. Evidence-Based Complementary and Alternative Medicine. 2019;**2019**:1453612. DOI: 10.1155/2019/1453612

[9] Pisutthanan N, Liawruangrath B, Liawruangrath S, Bremner JB. A new flavonoid from Chromolaena odorata. Natural Product Research. 2006;**20**(13):1192-1198. DOI: 10.1080/14786410600899050

[10] Emani L, Ravada S, Meka B, Garaga M, Golakoti T. A New Flavanone from the Leaves of Chromolaena odorata. Natural Product Communications. 2015;**10**(9):1555-1559

[11] Hung TM, Cuong TD, Dang NH, Zhu S, Long PQ, Komatsu K, et al. Flavonoid glycosides from Chromolaena odorata leaves and their in vitro cytotoxic activity. Chem Pharm Bull (Tokyo). 2011;**59**(1):129-131. DOI: 10.1248/cpb.59.129

[12] Dhar R, Kimseng R, Chokchaisiri R, Hiransai P, Utaipan T, Suksamrarn A, et al. 2′,4-Dihydroxy-3′,4′,6′-trimethoxychalcone from Chromolaena odorata possesses anti-inflammatory effects via inhibition of NF-κB and p38 MAPK in lipopolysaccharide-activated RAW 264.7 macrophages. Immunopharmacology

and Immunotoxicology. 2018;**40**(1):43-51. DOI: 10.1080/08923973.2017.1405437

[13] Roengsumran S, Singtothong P, Pudhom K, Ngamrochanavanich N, Petsom A, Chaichantipyuth C. Neocrotocembranal from Croton oblongifolius. Journal of Natural Products. 1999;**62**(8):1163-1164. DOI: 10.1021/np9900268

[14] Youngsa-ad W, Ngamrojanavanich N, Mahidol C, Ruchirawat S, Prawat H, Kittakoop P. Diterpenoids from the roots of Croton oblongifolius. Planta Medica. 2007;**73**(14):1491-1494. DOI: 10.1055/s-2007-990247

[15] Roengsumran S, Petsom A, Kuptiyanuwat N, Vilaivan T, Ngamrojnavanich N, Chaichantipyuth C, et al. Cytotoxic labdane diterpenoids from Croton oblongifolius. Phytochemistry. 2001;**56**(1):103-107. DOI: 10.1016/s0031-9422(00)00358-7

[16] Takeshige Y, Kawakami S, Matsunami K, Otsuka H, Lhieochaiphant D, Lhieochaiphant S. Oblongionosides A-F, megastigmane glycosides from the leaves of Croton oblongifolius Roxburgh. Phytochemistry. 2012;**80**:132-136. DOI: 10.1016/j.phytochem.2012.05.011

[17] Roengsumran S, Pornpakakul S, Muangsin N, Sangvanich P, Nhujak T, Singtothong P, et al. New halimane diterpenoids from Croton oblongifolius. Planta Medica. 2004;**70**(1):87-89. DOI: 10.1055/s-2004-815466

[18] Pudhom K, Vilaivan T, Ngamrojanavanich N, Dechangvipart S, Sommit D, Petsom A, et al. Furanocembranoids from the stem bark of Croton oblongifolius. Journal of Natural Products. 2007;**70**(4):659-661. DOI: 10.1021/np060520t

[19] Roengsumran S, Musikul K, Petsom A, Vilaivan T, Sangvanich P, Pornpakakul S, et al. Croblongifolin, a new anticancer clerodane from Croton oblongifolius. Planta Medica. 2002;**68**(3):274-277. DOI: 10.1055/s-2002-23138

[20] Sommit D, Petsom A, Ishikawa T, Roengsumran S. Cytotoxic activity of natural labdanes and their semi-synthetic modified derivatives from Croton oblongifolius. Planta Medica. 2003;**69**(2):167-170. DOI: 10.1055/s-2003-37708

[21] Roengsumran S, Achayindee S, Petsom A, Pudhom K, Singtothong P, Surachetapan C, et al. Two new cembranoids from Croton oblongifolius. Journal of Natural Products. 1998;**61**(5):652-654. DOI: 10.1021/np9704765

[22] Singh D, Chaudhuri PK. Chemistry and Pharmacology of Tinospora cordifolia. Natural Product Communications. 2017;**12**(2):299-308

[23] Sharma U, Bala M, Kumar N, Singh B, Munshi RK, Bhalerao S. Immunomodulatory active compounds from Tinospora cordifolia. Journal of Ethnopharmacology. 2012;**141**(3):918-926. DOI: 10.1016/j.jep.2012.03.027

[24] Singh B, Sharma P, Kumar A, Chadha P, Kaur R, Kaur A. Antioxidant and in vivo genoprotective effects of phenolic compounds identified from an endophytic Cladosporium velox and their relationship with its host plant Tinospora cordifolia. Journal of Ethnopharmacology. 2016;**194**:450-456. DOI: 10.1016/j.jep.2016.10.018

[25] Bala M, Pratap K, Verma PK, Singh B, Padwad Y. Validation of ethnomedicinal potential of Tinospora cordifolia for anticancer and immunomodulatory

activities and quantification of bioactive molecules by HPTLC. Journal of Ethnopharmacology. 2015;**175**:131-137. DOI: 10.1016/j.jep.2015.08.001

[26] Sivasubramanian A, Gadepalli Narasimha KK, Rathnasamy R, Campos AM. A new antifeedant clerodane diterpenoid from Tinospora cordifolia. Natural Product Research. 2013;**27**(16):1431-1436. DOI: 10.1080/14786419.2012.722088

[27] Pan L, Terrazas C, Lezama-Davila CM, Rege N, Gallucci JC, Satoskar AR, et al. Cordifolide A, a sulfur-containing clerodane diterpene glycoside from Tinospora cordifolia. Organic Letters. 2012;**14**(8):2118-2121. DOI: 10.1021/ol300657h

[28] Maurya R, Manhas LR, Gupta P, Mishra PK, Singh G, Yadav PP. Amritosides A, B, C and D: clerodane furano diterpene glucosides from Tinospora cordifolia. Phytochemistry. 2004;**65**(14):2051-2055. DOI: 10.1016/j.phytochem.2004.05.017

[29] Phan VK, Chau VM, Nguyen TD, La VK, Dan TH, Nguyen HN, et al. Aporphine alkaloids, clerodane diterpenes, and other constituents from Tinospora cordifolia. Fitoterapia. 2010;**81**(6):485-489. DOI: 10.1016/j.fitote.2010.01.005

[30] Ahmad F, Ali M, Alam P. New phytoconstituents from the stem bark of Tinospora cordifolia Miers. Natural Product Research. 2010;**24**(10):926-934. DOI: 10.1080/14786410802435679

[31] Patel MB, Mishra S. Isoquinoline alkaloids from Tinospora cordifolia inhibit rat lens aldose reductase. Phytotherapy Research. 2012;**26**(9):1342-1347. DOI: 10.1002/ptr.3721

[32] Sirat HM, Susanti D, Ahmad F, Takayama H, Kitajima M. Amides, triterpene and flavonoids from the leaves of Melastoma malabathricum L. Journal of Natural Medicines. 2010;**64**(4):492-495. DOI: 10.1007/s11418-010-0431-8

[33] Mazura MP, Susanti D, Rasadah MA. Anti-inflammatory action components from Melastoma malabathricum. Pharmaceutical Biology. 2007;**45**(5):372-375

[34] Wong KC, Hag Ali DM, Boey PL. Chemical constituents and antibacterial activity of Melastoma malabathricum L. Natural Product Research. 2012;**26**(7):609-618. DOI: 10.1080/14786419.2010.538395

[35] Yoshida TNF, Hosotani KNA, Okudat T. Dimeric hydrolysable tannins from melastoma Malabathricum. Phytochemistry. 1992;**31**(8):2829-2833. DOI: 10.1016/0031-9422(92)83641-b

[36] Yoshida T, Nakata F, Hosotani K, Nitta A, Okuda T. Tannins and Related Polyphenols of Melastomataceous Plants. V. Three New Complex Tannins from Melastoma malabathricum L. Chemical & Pharmaceutical Bulletin. 1992;**40**(7):1727-1732. DOI: 10.1248/cpb.40.1727

[37] Mayasari D, Murti YB, Pratiwi SUT, Sudarsono S, Hanna G, Hamann MT. TLC-Based Fingerprinting Analysis of the Geographical Variation of Melastoma malabathricum in Inland and Archipelago Regions: A Rapid and Easy-to-Use Tool for Field Metabolomics Studies. Journal of Natural Products. 2022;**85**(1):292-300. DOI: 10.1021/acs.jnatprod.1c00622

[38] Li H, Peng Y, Zheng J. Dioscorea bulbifera L.-induced hepatotoxicity and involvement of metabolic activation of furanoterpenoids. Drug Metabolism Reviews. 2020;**52**(4):568-584. DOI: 10.1080/03602532.2020.1800724

[39] Ngan NTT, Hoang NH, Truong VV, Hien NT, Lan NN, Tung NV, et al. Anti-inflammatory norclerodane diterpenoids and tetrahydrophenanthrene from the leaves and stems of Dioscorea bulbifera. Fitoterapia. 2021;**153**:104965. DOI: 10.1016/j.fitote.2021.104965

[40] Chaniad P, Tewtrakul S, Sudsai T, Langyanai S, Kaewdana K. Anti-inflammatory, wound healing and antioxidant potential of compounds from Dioscorea bulbifera L. bulbils. PLoS One. 2020;**15**(12):e0243632. DOI: 10.1371/journal.pone.0243632

[41] Liu JS, Gao WN, Zheng J, Wang GK, Yang QS. Chemical constituents from fresh tubers of Dioscorea bulbifera. Zhongguo Zhong Yao Za Zhi. 2017;**42**(3):510-516. DOI: 10D540/j.cnki.cjcmm.2017.0009

[42] Komori T. Glycosides from Dioscorea bulbifera. Toxicon. 1997;**35**(10):1531-1536. DOI: 10.1016/s0041-0101(97)00032-9

[43] Chaniad P, Wattanapiromsakul C, Pianwanit S, Tewtrakul S. Anti-HIV-1 integrase compounds from Dioscorea bulbifera and molecular docking study. Pharmaceutical Biology. 2016;**54**(6):1077-1085. DOI: 10.3109/13880209.2015.1103272

[44] Liu H, Tsim KW, Chou GX, Wang JM, Ji LL, Wang ZT. Phenolic compounds from the rhizomes of Dioscorea bulbifera. Chemistry & Biodiversity. 2011;**8**(11):2110-2116. DOI: 10.1002/cbdv.201000279

[45] Kuete V, Betrandteponno R, Mbaveng AT, Tapondjou LA, Meyer JJ, Barboni L, et al. Antibacterial activities of the extracts, fractions and compounds from Dioscorea bulbifera. BMC Complementary and Alternative Medicine. 2012;**12**:228. DOI: 10.1186/1472-6882-12-228

[46] Tang Y, Xue YB, Zhou L, Zhang JW, Yao GM, Luo ZW, et al. New norclerodane diterpenoids from the tubers of Dioscorea bulbifera. Chem Pharm Bull (Tokyo). 2014;**62**(7):719-724. DOI: 10.1248/cpb.c14-00057

[47] Liu H, Chou GX, Wu T, Guo YL, Wang SC, Wang CH, et al. Steroidal sapogenins and glycosides from the rhizomes of Dioscorea bulbifera. Journal of Natural Products. 2009;**72**(11):1964-1968. DOI: 10.1021/np900255h

[48] Liu H, Chou GX, Wang JM, Ji LL, Wang ZT. Steroidal saponins from the rhizomes of Dioscorea bulbifera and their cytotoxic activity. Planta Medica. 2011;**77**(8):845-848. DOI: 10.1055/s-0030-1250633

[49] Li SS, Iliya IA, Deng JZ, Zhao SX. Flavonoids and anthraquinone from Dioscorea bulbifera L. Zhongguo Zhong Yao Za Zhi. 2000;**25**(3):159-160

[50] Kouamé PB, Jacques C, Bedi G, Silvestre V, Loquet D, Barillé-Nion S, et al. Phytochemicals isolated from leaves of Chromolaena odorata: impact on viability and clonogenicity of cancer cell lines. Phytotherapy Research. 2013;**27**(6):835-840. DOI: 10.1002/ptr.4787

[51] Suksamrarn A, Chotipong A, Suavansri T, Boongird S, Timsuksai P, Vimuttipong S, et al. Antimycobacterial activity and cytotoxicity of flavonoids from the flowers of Chromolaena odorata. Archives of Pharmacal Research. 2004;**27**(5):507-511. DOI: 10.1007/BF02980123

[52] Yusuf H, Kamarlis RK, Yusni Y, Fahriani M. The anticancer activity of ethanol extract of Chromolaena odorata leaves in 7,12-Dimethylbenz[a]anthracene in (DMBA) induced breast cancer Wistar rats (Rattus novergicus).

Pharmacia. 2021;**68**(2):493-499. DOI: 10.3897/pharmacia.68.e63956

[53] Yusuf H, Fahriani M, Murzalina C. Anti-Proliferative and apoptotic effect of soluble ethyl acetate partition from ethanol extract of *Chromolaena odorata* Linn leaves against Hela cervical cancer cell line. Asian Pacific Journal of Cancer Prevention. 2022;**23**(1):183-189. DOI: 10.31557/APJCP.2022.23.1.183

[54] Asomugha RN, Ezejiofor AN, Okafor PN, Ijeh II. Acute and Cytotoxicity Studies of Aqueous and Ethanolic Leaf Extracts of Chromolaena odorata. Pakistan Journal of Biological Sciences. 2015;**18**(1):46-49. DOI: 10.3923/pjbs.2015.46.49

[55] Poofery J, Khaw-On P, Subhawa S, Sripanidkulchai B, Tantraworasin A, Saeteng S, et al. Potential of Thai Herbal Extracts on Lung Cancer Treatment by Inducing Apoptosis and Synergizing Chemotherapy. Molecules. 2020;**25**(1):231. DOI: 10.3390/molecules25010231

[56] Poofery J, Sripanidkulchai B, Banjerdpongchai R. Extracts of Bridelia ovata and Croton oblongifolius induce apoptosis in human MDA-MB-231 breast cancer cells via oxidative stress and mitochondrial pathways. International Journal of Oncology. 2020;**56**(4):969-985. DOI: 10.3892/ijo.2020.4973

[57] Chaichantipyuth C, Petson A, Taweechotipatr P, Muangsin N, Chaichit N, Puthong S, et al. New labdane-type diterpenoids from Croton oblongifolius and their cytotoxic activity. Heterocycles. 2005;**65**(4):809. DOI: 10.3987/com-0510327

[58] Roengsumran S, Pata P, Ruengraweewat N, Tummatorn J, Pornpakakul S, Sangvanich P, et al. New cleistanthane diterpenoids and 3,4-seco-cleistanthane diterpenoids from Croton oblongifolius. Chemistry of Natural Compounds. 2009;**45**:641. DOI: 10.1007/s10600-009-9442-7

[59] Wijesekera K. A bioactive diterpene; Nasimalum A from Croton oblongifolius Roxb. Prayog. Ras. 2017;**1**:41-44

[60] Uddin MH, Hossain MA, Kawsar MH. Antimicrobial and cytotoxic activities of Tinospora cordifolia (fam: menispermaceae). International Journal of Pharmaceutical Sciences and Research. 2011;**2**(3):656-658. DOI: 10.13040/IJPSR.0975-8232.2(3).656-58

[61] Modi B, Koirala N, Aryal SP, Shrestha J, Koirala S, Upadhyaya J, et al. Tinospora cordifolia (Willd.) Miers:phytochemical composition, cytotoxicity, proximate analysis and their biological activities. Molecular and Cellular Biology. 2021;**67**(1):50-57. DOI: 10.14715/cmb/2021.67.1.8

[62] Jagetia GC, Rao SK. Evaluation of cytotoxic effects of dichloromethane extract of guduchi (tinospora cordifolia Miers ex Hook F & THOMS) on cultured HeLa cells. Evidence-based Complementary and Alternative Medicine. 2006;**3**(2):267-272. DOI: 10.1093/ecam/nel011

[63] Sharma AK, Kumar S, Pandey AK. Ferric reducing, anti-radical and cytotoxic activities of Tinospora cordifolia stem extracts. (2014). Biochemistry & Analytical Biochemistry. 2014;**3**(2):1-5. DOI: 10.4172/2161-1009.1000153

[64] Ahmad R, Srivastava AN, Khan MA. Evaluation of in vitro anticancer activity of stem of Tinospora cordifolia against human breast cancer and Vero cell

lines. Journal of Medical Plants Studies. 2015;**3**(4):33-37

[65] Polu PR, Nayanbhirama U, Khan S, Maheswari R. Assessment of free radical scavenging and anti-proliferative activities of Tinospora cordifolia Miers (Willd). BMC Complementary and Alternative Medicine. 2017;**17**:457. DOI: 10.1186/s12906-017-1953-3

[66] Idris A, Zulkipli IN, Zulhilmi NR, Lee HF, Rajabalaya R, Chee LY, et al. *Melastoma malabathricum* ethyl acetate fraction induces secondary necrosis in human breast and lung cancer cell lines. Pharmacognosy Magazine. 2017;**13**(Suppl 3):S688-S692. DOI: 10.4103/pm.pm_465_15

[67] Roslen NA, Alewi NA, Ahamada H, Rasad MS. Cytotoxicity screening of *Melastoma malabathricum* extracts on human breast cancer cell lines in vitro. Asian Pacific Journal of Tropical Biomedicine. 2014;**4**(7):545-548. DOI: 10.12980/APJTB.4.2014C658

[68] Zakaria ZA, Rofiee MS, Mohamed AM, Teh LK, Salleh MZ. In vitro antiproliferative and antioxidant activities and total phenolic contents of the extracts of Melastoma malabathricum leaves. Journal of Acupuncture and Meridian Studies. 2011;**4**(4):248-256. DOI: 10.1016/j.jams.2011.09.016

[69] Isnaini I, Yasmina A, Nur'amin HW. Antioxidant and Cytotoxicity Activities of Karamunting (Melastoma malabathricum L.) Fruit Ethanolic Extract and Quercetin. Asian Pacific Journal of Cancer Prevention. 2019;**20**(2):639-643. DOI: 10.31557/APJCP.2019.20.2.639

[70] Susanti D, Sirat HM, Ahmad F, Ali RM, Aimi N, Kitajima M. Antioxidant and cytotoxic flavonoids from the flowers of Melastoma malabathricum L. Food Chemistry. 2007;**103**(3):710-716. DOI: 10.1016/j.foodchem.2006.09.011

[71] Hamid HA, Ramli ANM, Zamri N, Yusoff MM. UPLC-QTOF/MS-based phenolic profiling of Melastomaceae, their antioxidant activity and cytotoxic effects against human breast cancer cell MDA-MB-231. Food Chemistry. 2018;**265**:253-259. DOI: 10.1016/j.foodchem.2018.05.033

[72] Tapondjou LA, Jenett-Siems K, Böttger S, Melzig MF. Steroidal saponins from the flowers of Dioscorea bulbifera var. sativa. Phytochemistry. 2013;**95**:341-350. DOI: 10.1016/j.phytochem.2013.07.020

[73] Wang GK, Zheng J, Sun YP, Jin WF, Liu HW, Yu Y, et al. New norclerodane diterpenoids from Dioscorea bulbifera. Phytochemistry. 2018;**27**:59-62. DOI: 10.1016/j.phytol.2018.06.025

[74] Nur RM, Nugroho LH. Cytotoxic activities of fractions from Dioscorea bulbifera L. chloroform and methanol extracts on T47D breast cancer cells. Pharmacognosy Journal. 2018;**10**(1): 33-38. DOI: 10.5530/pj.2018.1.7

[75] Mainasara MM, Abu Bakar MF, Md Akim A, Linatoc AC, Abu Bakar FI, Ranneh YKH. Secondary metabolites, antioxidant, and antiproliferative activities of Dioscorea bulbifera leaf collected from Endau Rompin, Johor, Malaysia. Evidence-Based Complementary and Alternative Medicine. 2021;**2021**:8826986. DOI: 10.1155/2021/8826986

[76] Wang JM, Ji LL, Branford-White CJ, Wang ZY, Shen KK, Liu H, et al. Antitumor activity of Dioscorea bulbifera L. rhizome in vivo. Fitoterapia.

2012;**83**(2):388-394. DOI: 10.1016/j.
fitote.2011.12.001

[77] Wang G, Zheng J, Yang J, Yunpeng S, Zhang N. Liu Huiwen, Liu Jinsong. Phenolic derivatives from Dioscorea bulbifera. Records of Natural Products. 2019;**13**(4):301-306. DOI: 10.25135/rnp.114.18.10.1003

Chapter 7

Cancer Immunotherapy and Cytotoxicity: Current Advances and Challenges

Leisheng Zhang and Hui Cai

Abstract

Immunotherapies are revolutionizing strategies for cancer treatment and infectious disease administration, which thus occupy worldwide concerns and enthusiasms for conquering relapsing and refractory immunodysfuction-related diseases. Current preclinical and clinical studies have suggested the partial success and promising potential of cancer management by various immunotherapies such as cancer vaccine, lymphocyte-promoting cytokines, checkpoint inhibitors and the cellular immunotherapy. However, the precise controlled modulation of the recipient's immune system as well as the concomitant cytotoxicity remains the core challenge in the broad implementation of cancer immunotherapies. In this Chapter, we mainly focus on the latest updates of the cytotoxicity of cancer immunocytotherapy, together with the remarkable opportunities and conspicuous challenges, which represent the paradigm for boosting the immune system to enhance antitumor responses and ultimately eliminate malignancies. Collectively, we summarize and highlight the auspicious improvement in the efficacy and cytotoxicity of cancer immunotherapy and will benefit the large-scale preclinical investigations and clinical practice in adoptive immunotherapy.

Keywords: cancer immunotherapy, immune cells, hematopoietic malignancies, metastatic solid tumors, cytotoxicity

1. Introduction

To date, thousands upon thousands of people suffer and even die from various tumors with high morbidity and mortality including malignant hematologic tumors and metastatic solid tumors worldwide, which have cause tremendous physical and mental stress to the patients and their guardians [1–6]. Despite the encouraging progresses in cancer treatment, one of the core dilemmas is the current limitation in classical therapeutic modalities such as surgery, radiation and chemotherapy [7, 8]. For instance, surgery in combination with radiation treatment and chemotherapy drugs has been proved to be effective for localized cancers without metastasis and diffusion [9, 10]. Chemoradiotherapy has been considered as a synergistic anticancer remedy for locally advanced solid tumors whereas with increased damage to normal tissues and microbiota resistance [11–13]. Distinguish from the conventional cancer treatment (e.g., surgery, radiotherapy and chemotherapy), noncellular immunotherapy such as checkpoint inhibitors (e.g., PD-1, PD-L1, CTLA4), lymphocyte-promoting cytokines (e.g., IFN-γ, GM-CSF, G-CSF), and cancer vaccines (e.g., mRNAs) has been continuously developed to fulfill the goals for cancer administration as well [14–17]. Additionally, current progress has also highlighted the feasibility of nanomaterials (e.g., organic nanomaterials, inorganic nanomaterials, organic–inorganic hybrid nanomaterials) as promising agents for cancer therapy based on the knowledge of nanobiotechnology and clinical biomedicine [18–20]. However, the significant disadvantages of the aforementioned strategies are apparent and should not be ignored including drug delivery barriers, graft-versus-host disease, off-target effects and severe toxicity [5, 21, 22].

Therewith, the assumption of eradicating tumors by utilizing the human immune system has been successfully and extensively practiced for the past decades by pioneering investigators and clinicians [23]. To date, a series of cellular Immunotherapy has been identified to promote the outcomes and reduce adverse reactions of cancer patients, and in particular, those with late-stage patients with various treatment-refractory cancers [14–17]. Therefore, in this chapter, we mainly focus on the progress as well as the prospective and challenges in cellular immunotherapy for cancer treatment including the patient-specific immune cells (e.g., tumor infiltrating lymphocytes, cytokine induced killer cells), innate immunocytes (e.g., natural killer cells, dendritic cells, macrophage), adaptive immunocytes (T lymphocytes, B lymphocytes), engineered immunocytes (e.g., chimeric antigen receptor-transduced T cells, T cell receptor-transduced T cells, CAR-NK, CAR-M) [21, 24–27]. Collectively, the immune cell-mediated cancer immunotherapy has constituted a promising area of cancer biotherapy.

2. Classification of immunocytes

2.1 Patient-associated immunocytes

2.1.1 Lymphokine-activated killer cells (LAKs)

LAKs are effector cells with significant cytotoxic activity, which are induced by culturing with recombinant IL-2 and have been proven effective against NK cell-resistant allogeneic and autologous tumor cells or cell lines [28]. Generally, LAKs after regional intra-arterial perfusion mainly accumulate at tumor sites, whereas the intravenously infused LAKs first accumulate in lung tissues before migrating to the liver regions [29].

LAKs-mediated adoptive immunotherapy has been confirmed with limited efficacy in vivo ranging from 20–30% including the metastatic or advanced tumor models and clinical studies upon patients (e.g., hepatic carcinoma, renal cell carcinoma, melanoma, leukemia) [30–32]. For example, Rosenberg et al. reported the treatment of 157 patients with advanced cancer using LAKs and IL-2, and confirmed the marked tumor regression and even remission by the immunotherapeutic approach. Nevertheless, the ultimate role of LAKs-based cancer immunotherapy still awaits further improvement including the efficacy and the decrease of toxicity and complexity [32].

2.1.2 Tumor-infiltrating lymphocytes (TILs)

TILs are infiltrating lymphocytes in cancer tissues and play a critical role in mediating response to chemotherapy [33]. To date, of the tumor-infiltrating immune cells such as mast cells, macrophages, dendritic cells and leucocytes, TILs have been considered as selected heterogeneous populations of T lymphocytes with a higher and specific immunological reactivity against cancer cells than the non-infiltrating subset [34]. Despite the accumulation of TILs has been recognized as prognosis for elevated survival and clinical outcomes, yet tumors with high level of TILs also with increased PD-1 immune checkpoint expression [35]. Thus, the feasibility of TILs as biomarker for reflecting the immune responses and predicting the clinical outcomes of cancer immunotherapy still need systematic formulations and detailed explorations [34]. Generally, although ineffective for in vivo tumor elimination, yet TILs are adequate to functionate proliferation and effector capacity when separated from immunosuppressive tumor microenvironment [36]. Taken together, identification of specific subpopulations and the corresponding molecular mechanism of TILs in cancer might be of great importance for guiding prognosis and developing appropriate sequencing of immunotherapy [33].

2.1.3 Cytokine-induced killer cells (CIKs)

CIKs are a heterogeneous population of $CD3^{+}CD56^{+}$ natural killer T (NKT) cells and recognized as pharmacological tools for tumor immunotherapy, and in particular, in refractory to conventional radiotherapy and chemotherapy [37, 38]. Generally, CIKs contain two main subsets including the CD3 + CD56+ and CD3 + CD56- subpopulations endowed with higher anti-tumor activity and higher proliferation ability, respectively. Mature CIKs express active receptors of NK cells including NKG2D and DNAM-1, whereas with minimal expression of NKG2A, NKp44, NKp46 or CD94 [39].

For decades, numerous clinical trials of CIKs-based tumor immunotherapy have been registered attribute to the distinctive properties including intense MHC-independent antitumor activity, low toxicity effects and high safety on healthy cells [39, 40]. For instance, Yu et al. verified that CIKs-based immunotherapy was an effective adjuvant strategy in the early stage of hepatocellular carcinoma (HCC) rather than advanced HCC, and suggested that targeting myeloid-derived suppressor cells (MDSCs) would provide additional therapeutic benefits alongside CIKs-based therapy [23]. Moreover, CIKs can be expanded to match the clinical relevant rates cost-effectively and conveniently by a simple protocol, which is also the key issue for clinical application of adoptive immunotherapy [39].

Of the multiple modalities for cancer immunotherapy, vaccination with DC-CIKs has revealed limited therapeutic success in the administration of advanced solid tumors [41]. Therefore, considering the shortcomings of CIKs-based cellular immunotherapy alone, integrated therapy or combined modality therapy such as CIKs, cytokines, gene editing, immune checkpoints, radiotherapy and chemotherapy have better potentiality in relieving the major side effects and improving the clinical outcomes of standard treatment options [40].

2.2 Innate immunocytes

2.2.1 Natural killer (NK) cells

NK cells are heterogeneous cell population with unique characteristics of and belong to the innate lymphoid cells (ILCs), which can be divided into the cytotoxic $CD56^{dim}CD16^{high}$ and IFN-γ-producing $CD56^{bright}CD16^{low/neg}$ subsets and play an important role in both innate and adoptive immune responses dispense with preliminary antigen presentation via receptor-ligand mediated cytotoxicity, antibody-dependent cell-mediated cytotoxicity (ADCC), release of perforin and granzyme as well as cytokine-based paracrine effects (e.g., IFN-γ, GM-CSF) [21, 42].

For decades, we and other investigators have reported the generation of NK cells from various sources such as cell lines (e.g., NK-92, NK-92MI, YT), perinatal blood (e.g., cord blood, placental blood), peripheral blood, hematopoietic stem cells (HSCs) and human pluripotent stem cells (e.g., human embryonic stem cells, human induced pluripotent stem cells) [21, 43–48]. Distinguish from the chimeric antigen receptor-transduced T cells (CAR-T), NK cells with non-MHC-restricted recognition revealed reliable cytotoxicity against pathogenic microorganism and tumor cells without the significant disadvantages graft-versus-host disease (GvHD), cytokine release syndrome (CRS) and neurotoxicity [49–51]. Moreover, NK cells are supposed to eliminate non-proliferating or quiescent cancer stem cells (CSCs), which collectively highlights the possibility and preponderance of NK cell-based cytotherapy for cancer immunosurveillance and immunotherapy [52, 53].

2.2.2 Dendritic cells (DCs)

DCs are most potent antigen-presenting cells (APCs) arising from lympho-myeloid hematopoiesis and linking the innate and adaptive immunity, which are firstly identified in 1973 and capable of initiating and activating lymphocytes including T cells and B cells and thus play a unique function in cancer immunotherapy as well as the tumor microenvironment [54–57]. As recently reviewed by Stevens and the colleagues, the advent of DC-based immunotherapy and immune checkpoint inhibitors (e.g., PD-1/PD-L1, CTLA-4) has become a paradigm shift in the treatment of cancers including the small cell lung cancer (SCLC) and non-small cell lung cancer (NSCLC) [58].

Notably, DCs derived *in vitro* or naturally circulating DCs loaded with the tumor-associated antigens (TAAs) are safe and feasible for eliciting a robust and tumor-directed immune response [59]. Briefly, immature DCs are attracted by the damage-associated molecular patterns (DAMPs) and then transit to a mature phenotype with capabilities of phagocytic cargo processing and antigenic component engulfment [60]. Additionally, DCs have been reported with synergistic effect with CIKs or thoracic radiotherapy for the management of locally advanced or metastatic NSCLCs [61, 62].

2.2.3 Macrophages (MΦ)

Macrophages (MΦ), one type of innate immune cells including the M1 (inhibit growth and kill) and M2 (promote growth and repair) subtypes, possess superior antineoplastic effects due to the phenomena of engulfing pathogens and dead cells, which are pivotal modulators of tissue regeneration and hematopoietic stem cell renewal as well as cancer progression [63–66]. However, considering the purposiveness of identification of tumor-specific anticancer responses, the characteristics of MΦ in cancer as bidirectional executors have been largely overlooked for a long period [67, 68].

Current updates have indicated the biofunction of macrophages in the regression of cancers by modulating the polarization peculiarity of the predominate M2 repair-type MΦ (pro-tumor) to M1 kill-type (anti-tumor) [69]. Disturbances in MΦ function usually result in deficiency of anti-inflammatory MΦ, uncontrolled secretion of inflammatory factors (e.g., TGF-β, VEGF, EGF, PGE2, IFN-γ) and production of nitric oxide (NO) as well as poor communication with functional cells (e.g., endothelial cells, epithelial cells, lymphocytes) and cancer cells [70, 71]. For instance, the tumor-associated MΦ (TAMs) are recognized as the prominent components in tumor microenvironment (TME), which thus considered as promising and advantaged targets for developing immunotherapeutic strategy [72, 73].

Thus, MΦ/innate immunology can be orchestrated to play a critical role in indirectly or directly combating numerous hematological malignancies and metastatic solid tumors via the activity of M1 kill-type and the stimulatory effect of M1-type MΦ to cytotoxic Th1 cells and relative effector cells, respectively [69, 74]. Moreover, state-of-the-art updates have suggested the feasibility of conversion of the dominating cancer growth-promoting MΦ into growth-inhibiting type and the resultant progression in cancer immunotherapy as well [69].

2.3 Adaptive immunocytes

2.3.1 T lymphocytes

According to the tumor immunosurveillance theory, considerable attention has been caused on enhancing the effectiveness and cytotoxicity of antitumor immunity, and in particular, the T lymphocyte-based host immune response and the accompanied T cell signaling and metabolism as well [75]. T lymphocytes are $CD3^+$ mononuclear cells and are essential for eradicating tumor cells, allergens and microorganisms as well as tissue repair, and thus people with the impairment or dysfunction of T cells are at high risk of cancers and infections and eventually poor prognosis or mortality [76, 77].

Immunotherapeutic resistance remains the substantial barrier of cancer immunotherapy, and the selection of appropriate standard-of-care treatments is critical for combating tumors in preclinical and clinical studies [78, 79]. Numerous investigations have indicated the feasibility of effectively eliminate immunotherapeutic resistance by promoting priming or activation of T lymphocytes, attracting or sustaining T cell-based immune response as well as favoring the immune-promoting the aforementioned TME, which is critical for cancer immunoediting and the component phases including elimination, equilibrium and escape [80, 81]. However, the developing of an effective T cell-based cytotherapy against cancer are extremely difficult largely due to the correspondence of tumor antigens to self MHC-associated fragments of self-proteins as well as the tumor heterogeneity [75, 82]. Encouragingly, functional cytolytic T lymphocyte (CTL)-mediated immune response has been reported with self-tumor antigen expression and efficient efficacy upon some cancer models and patients (e.g., melanoma, pancreatic carcinoma, breast cancer) [83, 84]. Collectively, T lymphocytes orchestrate various aspects of adaptive immunity, while subset delineation (e.g., $TCF1^+$ progenitor subsets in exhausted $CD8^+$ T cells, memory T cells) and tissue localization as well as targeted strategies (e.g., genetically engineered CAR-T or TCR-T) are key determinants of T lymphocyte function in immune responses [75, 77, 82, 85].

2.3.2 B lymphocytes

B cells, including the fetal liver-originated B-1 and bone marrow-derived B-2 subsets, are heterogeneous lymphocytes with antibody production and cytokine release capacity, which also represent a pivotal cellular constituent of humoral immunity and function critically in the maintenance of tolerance and immune regulation [86–89]. Generally, B-1 lymphocytes, consist of B-1a and B-1b subsets, are part of innate immune system and mainly function in providing immunity to various specific pathogens via producing immunoglobulins [90]. B-2 lymphocytes, including the follicular B cells (FOB) and marginal zone B cells (MZB), have been considered as mediators of adaptive immunity and are capable of differentiating into memory cells [87].

Current studies have indicated the excessive inflammatory responses of regulatory B cells (Bregs) during autoimmune diseases, unresolved infections or chronic metabolic diseases, which also contribute to the dynamic balance of equilibrium required for tolerance by simultaneously reestablishing immune homeostasis and limiting ongoing immune responses [91–93]. Bregs encompassing all B cells for immune response suppression also play a promoting role in regulatory T cell (Treg) differentiation by accelerating anti-inflammatory cytokine (IL-10, IL-35, TGF-β) secretion and inhibiting proinflammatory cytokine production [91, 94]. Additionally, memory B cells (MBCs) have also been identified in humans and play an important role for the rapid development and response of protective immunity [95]. Collectively, with the aid of novel genetic and pharmacological technologies as well as the in-depth understanding of the phenotype, function and developmental processes, B cell-based cytotherapy has become a rapidly growing field in tumor immunotherapy [86, 94–96].

2.4 Engineered immunocytes

2.4.1 Chimeric antigen receptor-transduced T cells (CAR-Ts)

T lymphocytes with the antibody-like CAR structure expression are adequate to recognize the unique structures on the surface of tumor-associated cells or tumor cells, and in particular, the remarkable efficacy of CAR-Ts for the management of hematological malignancies [27, 82, 97]. For instance, we and other investigators have reported the remission of various hematopoietic malignancies such as refractory or relapsed B acute lymphoblastic leukemia (r/rB-ALL) and B-cell non-Hodgkin's lymphoma (NHL) by utilizing the CAR-Ts targeting CD19, CD20, CD22 [98–102]. However, the adverse effects and the regulatory mechanisms of CAR-Ts-based tumor immunotherapy cannot be ignored [97, 103]. For example, Giavridis and the colleagues have reported the involvement of MΦ and IL-1 blockade in mediating and abating CAR-Ts-induced CRS [27, 104].

Despite the encouraging progress in hematologic malignancies, yet the applications of CAR-Ts-based cell therapy to solid tumors or as candidate pharmaceutical options have been challenging and suspicious [27, 68, 105–107]. In particular, the immunosuppressive TMEs of tumors represent the most important factor for limiting the efficacy of CAR-Ts-based immunotherapy [108]. It's noteworthy that pioneering investigators have suggested the synergistic effect of oncolytic virus with CAR-Ts in improving the therapeutic effect of solid tumors [109]. For instance, Watanabe and the colleagues took advantage of the Mesothelin-redirected CAR-Ts (meso-CAR-Ts) and combined with an OAd-TNFa-IL2 oncolytic adenovirus for TNF-α and IL-2 expression, which increased CAR-Ts and host T cell infiltration to TME and thus showed enhanced efficacy upon human- pancreatic ductal adenocarcinoma (PDA)-xenograft immunodeficient mice [110].

2.4.2 T cell receptor-engineered T cells (TCR-Ts)

Besides chimeric antigen receptor (CAR) transduction, T lymphocytes can also be genetically engineered with αβ T cell receptor expression, which is capable of recognizing MHC-restricted peptide antigens and therefore poised to turn into an important pillar of cellular cancer immunotherapy [82, 111]. Generally, TCRs are capable of recognizing a relatively broad range of human leucocyte antigen (HLA)/specific peptide that can be expressed in the surface of patient's T cells [112–114]. Therefore, TCR-Ts hold the potential to redirect the recognition of tumor-associated surface antigens and the removal of cancer cells. In details, TCR-Ts-based cytotherapy are principally intended to redirect circulating $CD8^+$ T lymphocytes to the targets of expressing class I epitopes of HLA compared to those $CD4^+$/$CD8^+$ T cells with HLA Class-II epitopes. Collectively, TCR-transduced T cells hold promising prospective in targeting all intracellular proteins when peptide epitopes have been presented on the HLAs including over-expressed neoantigens and self -antigens, viral antigens [115]. Even though the characteristics of adaptive evolution in T cell immunity and the preferential expansion of T lymphocytes with high-affinity TCRs, yet whether the affinity maturation of T cells by clonal selection continues during the course of tumor development remains unresolved [114, 116].

2.4.3 CAR-transduced NK cells (CAR-NKs)

Compared to CAR-Ts, the genetically modified CAR-NKs not only inherit the splendid properties of adoptive NK cells including the aforementioned biological effects but also hold promising prospects for solid tumor administration without causing adverse effects including CRS, GvHD or immune cell- associated neurotoxicity syndrome (ICANS) [49–51, 117]. Thus, the CAR-NKs-based immunotherapy represents a virgin ground of immunotherapy innovation [118–120].

As to CAR transduction, a series of methodology has been developed to fulfill the high-efficient delivery demands via retrovirus-, lentivirus-, nonviral-mediated transfection with the range from 27–70% [121, 122]. In particular, the DNA transposon system composed of the sleeping beauty (SB) and the PiggyBac (PB) subsets is competent for delivering CAR structure into the genomes of induced pluripotent stem cells (iPSCs) or primary NK cells with high-efficient and long-lasting transgene expression [27, 123]. As to the CAR structures, a group of preclinical and clinical studies have reported the successful design and delivery of vectors carrying the cassettes of CAR-conjunct targets (e.g., CD19, CD5, CD137) into NK cells with the second- or third- or fourth-generation constructs against diverse malignancies, respectively [124–128]. Of note, Li *et al* announced the high-efficient generation of NK cells from human induced pluripotent stem cells (NK-CAR-iPSC-NKs) and superiority over T-CAR-expressing iPSC-NKs (T-CAR-iPSC-NKs), which highlighted the feasibility of the standardized and targeted CAR-NKs-based cancer immunotherapy in future [118].

2.4.4 CAR-transduced macrophages (CAR-ms)

State-of-the-art updates have indicated the feasibility of CAR-Ms for solid tumor management and virion load reduction [129, 130]. Differ from the other CAR-transduced immune cells, CAR-Ms are important sources of matrix metalloproteinase

(MMPs), which can enter tumor tissue and degrade almost all extracellular matrix (ECM) components and thus destroy malignant tumor progression [130]. Strikingly, Klichinsky *et al* has reported the successful application of CAR-Ms in two solid tumor models and confirmed the prolonged overall survival and decreased tumor burden, which indicates the possibility of CAR-Ms-focused immunotherapeutic modalities [68, 131]. Notably, the therapeutic benefit of HER-2-targeting CAR-Ms was mediated by direct antigen-specific phagocytosis and indirect pro-inflammatory effects [132]. Recently, Zhang and the colleagues derived CAR-Ms from induced pluripotent stem cells (iPSC-CAR-Ms) with splendid properties such as cytokine secretion, polarization, enhanced phagocytosis and *in vivo* antitumor activity, which for the first time made iPSCs as unlimited source for "off-the-shelf" CAR-M generation [133].

3. Preclinical investigations and clinical applications

The current clinical and preclinical successes of cellular immunotherapy represent a remarkable point in cancer management, which also underscore the consequence of decoding the underlying tumor immunology [36, 134, 135]. Accumulating evidence has indicated the safety and effectiveness of cellular immunotherapy in recognizing and eliminating transformed cancer cells during various hematologic malignancies whereas those upon metastatic solid tumors are challenging and unsatisfactory [135, 136].

Generally, according to ClinicalTrials.gov (https: //www.clinicaltrials.gov/) website, a total number of 4377 trials upon tumors by immunotherapy have been registered by the end of September, 2021. Of the aforementioned trials, there are 2463 in North America, 1089 in Europe and 834 in East Asia, respectively (**Figure 1**). As shown in **Figure 1**, among the indicated registered trials, there are 3944 interventional trials including 96 trials in early phase 1 stage, 957 in phase 1 stage, 776 in phase 1/2 stage, 1555 in phase 2 stage, 59 in phase 2/3 stage, 260 in phase 3 stage and 29 in phase 4 stage. In details, lymphoma (512 trials), leukemia (331 trials) and non-Hodgkin lymphoma (281 trials) are the top three indicates among hematologic malignancies, while lung neoplasms (822 trials), neuroectodermal tumors (751 trials) and digestive system neoplasms (696 trials) are the top three indicates among solid tumors (**Figure 1**). For instance, 26 trials of NK cell-based immunotherapy have been registered for a series of metastatic or recurrent tumor administration such as acute lymphoblastic leukemia (ALL), chronic myelogenous leukemia (CML), juvenile myelomonocytic leukemia, liver cancer, breast cancer, non-small cell lung cancer (NSCLC), pancreatic cancer, ovarian cancer, cervical cancer, tongue cancer and esophageal cancer. Also, a number of 16 CAR-T-based immunotherapy (e.g., CD19, CD22, HER2, mesothelin, PSCA, MUC1, GPC3, BCMA, SLAMF7) has also been carried out for both hematological malignancies (e.g., relapsed/refractory multiple myeloma, acute lymphocytic leukemia and refractory indolent adult non-Hodgkin lymphoma) and solid tumors (e.g., advanced lung cancer, colon cancer, esophageal carcinoma, pancreatic cancer, prostate cancer, gastric cancer and hepatic carcinoma). For instance, we recently took advantage of the CD22, CD19–22 CAR-T therapy in patients with refractory or relapsed (r/r) B acute lymphoblastic leukemia (B-ALL) and confirmed the sustained remission after sequential treatment [98–100, 137, 138]. Additionally, other cancer immunotherapy has also been taken into practice including PD-1 (NCT02843204), TCR-T therapy (NCT03778814), DC-CIKs (NCT03190811, NCT01783951), Bevacizumab (NCT02857920).

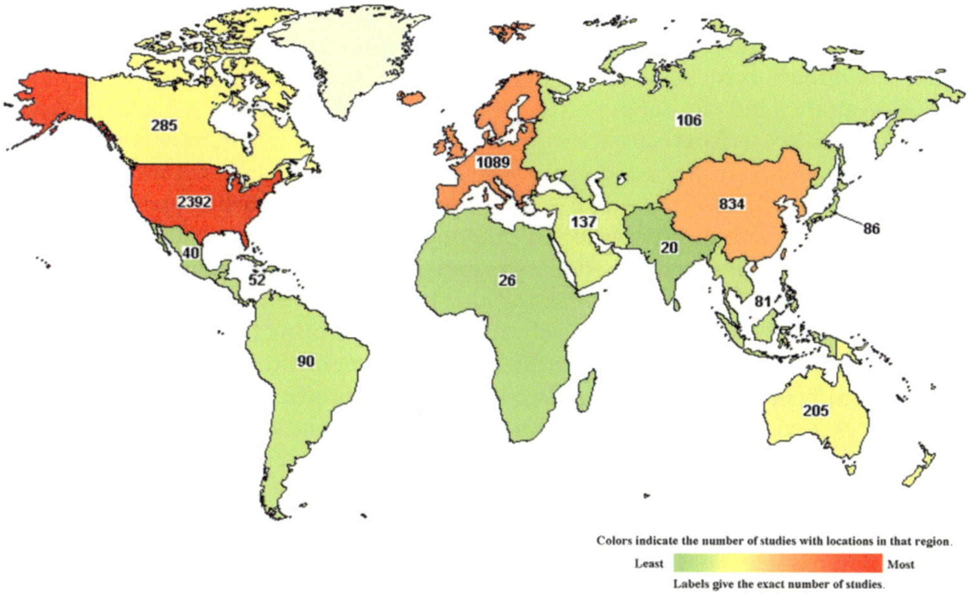

Figure 1.
The overview of registered cancer immunotherapy worldwide.

As to preclinical investigations, immunotherapy has also acted as an "off-the-shelf" strategy and a promising candidate for clinical evaluation of cancer [21, 25, 102]. For example, Sommer *et al* reported the incorporation of allogeneic BCMA-CAR-Ts and CD20 mimotope-based intra-CAR off switch for conferring lymphodepletion resistance and reducing GvHD potential with the aid of the transcription activator-like effector nuclease (TALEN)-based gene editing. Notably, the CAR-T-based immunotherapy induced sustained antitumor responses and the cells maintained intrinsic phenotype and potency after scale-up manufacturing [139]. Recently, we reported the high-efficient generation of NK cells from peripheral blood with considerable killing activity upon K562 cells and summarized the latest updates upon allogeneic NK cell- and CAR-NK cell-based immunotherapy as well [21, 45]. Interestingly, we also found the delay of disease progression and improvement of leukemic hematopoietic microenvironment (LHME) in AML mouse models by blocking the migration of regulatory T cells (Treg) [140, 141].

4. Perspectives and future directions

For decades, comprehensive strategies for eliciting anticancer immunity have been extensively explored [34]. In particular, antitumor immunotherapies involving adoptive cellular transfer or immune checkpoint inhibitors are validated as effective and promising treatment option for hematopoietic malignancies and solid tumors [78]. Of the obstacles, tumor escape and cytotoxicity are the core burning issues in oncotherapy. To overcome the shortcomings, the systematic and precise understanding of TMEs and the relevant networks including pro- and anti-inflammatory cytokines, immunosuppressive cells, tumor-associated stroma, tumor hypoxia and metabolism as well as immune inhibitory checkpoints are collectively of great importance for improving the trafficking and delivery efficacy of CAR-transduced immune cells into the tumor site and helping solve the drawback of tumor antigen heterogeneity [108].

State-of-the-art updates have indicated the rosy and powerful implement of the anti-cancer immunotherapy including adoptive cellular transplantation and immune-checkpoint inhibitors for solid tumor and hematologic malignancy management in both preclinical studies and clinical practices [142, 143]. Nevertheless, the potentially acute and chronic adverse effects caused by immunotherapy-associated cytotoxicity have led to severe outcomes in tumor patients such as neurotoxicity, aGvHD, innate or acquired resistance, autoimmunity, nonspecific inflammation, cytokine storm syndrome (CSS), and the difficulty in realizing controllable modulation of the immune response, which are also the prerequisites and key challenges in the extensive implementation of immunotherapy for tumor, and in particular, the hurdles in adoptive T cell-based immunotherapy and double-edged properties of cancer immunoediting [75, 142, 143].

As to gene-edited adoptive immune cells, improvement strategies for enhancing the transfection efficiency and target selection as well as efficiently reducing the concomitant cytotoxicity during cancer immunotherapy are the key issue. For example, even though CAR-NKs exhibit inferior baseline cytotoxicity and preferable tumor killing activity when compared with other sources, yet the occasional issues should be systematically and thoroughly overcome by subsequent stimulation with optimized cytokine cock-tails [27, 144]. Therefore, it is of great importance to explore emerging features for the efficient development of novel immunotherapies, such as the selection of ideal CARs or TCRs targeting validated antigenic epitopes with well-characterized tumor cell expression and processing, enhancing immune cell effector function, persistence, expansion, trafficking, and memory formation by strategic selection of co-stimulated substrate cells, and novel technologies for gene-engineering [21, 27].

Of note, the variations and adverse effects in cancer immunotherapy reveals the heterogeneity and instability of the current immune cell products, and thus highlight the necessity and urgency of industrialization and standardization for clinical applications [145]. Generally, a cohort of core issues both in fundamental research and clinical practice of cancer immunotherapy need to be improved before large-scale applications [21, 146]. For instance, the screening criteria of healthy tissues for generating immune cell sources, the standardized regents and procedures for cell product preparation, the dose and frequency of cell transplantation, the delivery strategies and targets for engineered cells [147–149]. Therefore, it is of great importance for the generation of clinical-grade immune cell products based on good manufacturing practice (GMP) and convenient to universally improve life quality of patients with standard supervision. Additionally, multidisciplinary research has also highlighted the feasibility and prospective of nanomaterials (e.g., surface-conjugated nanoparticles, injectable scaffolds) as promising agents for cancer therapy attribute to the rapid progresses of nanobiotechnology and clinical biomedicine [18–20].

Collectively, comprehensive treatment strategies by combining the conventional remedies (eg., laparoscopic rectal surgery, robotic surgery, radiotherapy, chemotherapy, drugs), checkpoint blockade (e.g., PD-1/PD-L1, CTLA-4), vaccines (e.g., mRNAs), biomaterials (e.g., nanoparticles) with the aforementioned cellular immunotherapy will largely benefit the malignancy management and effectively reduce the cytotoxicity [21, 108, 150–152].

5. Conclusions

To date, hematological malignancies and metastatic solid tumors have caused a prevalence of over 10 million mortalities annually [3–6]. Antitumor immunotherapy has served as a promising and alternative strategy for improving the outcomes of tumor patients as well as reducing the concomitant cytotoxicity. Objectively, there's still a long way to go and a cohort of central issues need to be solved before large-scale application in cancer immunotherapy. Overall, cancer immunotherapy has become a notable and synergistic anti-tumor remedy for a variety of hematopoietic malignancies and locally advanced solid tumors.

Acknowledgements

The authors would like to thank the members in National Postdoctoral Research Station of Gansu Provincial Hospital, Hefei Institute of Physical Science, Chinese Academy of Sciences, and Institute of Health-Biotech, Health-Biotech (Tianjin) Stem Cell Research Institute Co., Ltd. for their technical support. We also thank the staff in Beijing Yunwei Biotechnology Development Co., LTD for their language editing service. This study was supported by grants from the project Youth Fund funded by Shandong Provincial Natural Science Foundation (ZR2020QC097), the Non-profit Central Research Institute Fund of Chinese Academy of Medical Sciences (2019PT320005), Science and technology projects of Guizhou Province (QKH-J-ZK [2021]-107), the project Youth Fund funded by Jiangxi Provincial Natural Science Foundation (S2021QNJJL0277), Jiangxi Provincial Key New Product Incubation Program from Technical Innovation Guidance Program of Shangrao city (2020G002, 2020 K003), Natural Science Foundation of Tianjin (19JCQNJC12500), Jiangxi Provincial Novel Research and Development Institutions of Shangrao City (2020AB002, 2021F013).

Conflict of interest

The authors declare no conflict of interest.

Notes/thanks/other declarations

Not applicable.

Appendices and nomenclature

CAR-T	chimeric antigen receptor-transduced T cells
CAR-NK	chimeric antigen receptor-transduced NK cells
CAR-M	chimeric antigen receptor-transduced macrophage
TCR-T	T cell receptor-transduced T cells
LAKs	lymphokine-activated killer cells
TILs	tumor-infiltrating lymphocytes
CIKs	cytokine-Induced Killer cells
NKT	natural killer T
HCC	hepatocellular carcinoma
MDSCs	myeloid-derived suppressor cells
ADCC	antibody-dependent cell-mediated cytotoxicity

ILCs	innate lymphoid cells
GvHD	graft-versus-host disease
HSCs	hematopoietic stem cells
CRS	cytokine release syndrome
CSC	cancer stem cells
APCs	antigen-presenting cells
SCLC	small cell lung cancer
NSCLC	non-small cell lung cancer
DAMPs	damage-associated molecular patterns
MΦ	macrophages
NO	nitric oxide
CTL	cytolytic T lymphocyte
NHL	non-Hodgkin's lymphoma
ICANS	immune cell- associated neurotoxicity syndrome
iPSCs	induced pluripotent stem cells
MMPs	matrix metalloproteinase
ECM	extracellular matrix
CML	chronic myelogenous leukemia
ALL	acute lymphoblastic leukemia
TALEN	transcription activator-like effector nuclease
LHME	leukemic hematopoietic microenvironment

Author details

Leisheng Zhang[1,2,3,4,5]* and Hui Cai[1]

1 Key Laboratory of Molecular Diagnostics and Precision Medicine for Surgical Oncology in Gansu Province and NHC Key Laboratory of Diagnosis and Therapy of Gastrointestinal Tumor, Gansu Provincial Hospital, Lanzhou, China

2 Center for Cellular Therapies, The First Affiliated Hospital of Shandong First Medical University, Ji-nan, China

3 Key Laboratory of Radiation Technology and Biophysics, Hefei Institute of Physical Science, Chinese Academy of Sciences, Hefei, China

4 Institute of Stem Cells, Health-Biotech (Tianjin) Stem Cell Research Institute Co., Ltd., Tianjin, China

5 Jiangxi Research Center of Stem Cell Engineering, Jiangxi Health-Biotech Stem Cell Technology Co., Ltd., Shangrao, China

*Address all correspondence to: leisheng_zhang@163.com

References

[1] Salamone JM, Lucas W, Brundage SB, Holloway JN, Stahl SM, Carbine NE, et al. Promoting scientist-advocate collaborations in cancer research: Why and how. Cancer Research. 2018;**78**(20):5723-5728

[2] McEachron TA, Helman LJ. Recent advances in Pediatric cancer research. Cancer Research. 2021;**81**(23):5783-5799

[3] Zaimy MA, Saffarzadeh N, Mohammadi A, Pourghadamyari H, Izadi P, Sarli A, et al. New methods in the diagnosis of cancer and gene therapy of cancer based on nanoparticles. Cancer Gene Therapy. 2017;**24**(6):233-243

[4] Chilakamarthi U, Giribabu L. Photodynamic therapy: Past, present and future. Chemical Record. 2017;**17**(8):775-802

[5] Wang Y, Liu T, Li X, Sheng H, Ma X, Hao L. Ferroptosis-inducing nanomedicine for cancer therapy. Frontiers in Pharmacology. 2021;**12**:735965

[6] Siegel RL, Miller KD, Fuchs HE, Jemal A. Cancer statistics, 2021. CA: a Cancer Journal for Clinicians. 2021;**71**(1):7-33

[7] Bodey B, Siegel SE, Kaiser HE. Human cancer detection and immunotherapy with conjugated and non-conjugated monoclonal antibodies. Anticancer Research. 1996;**16**(2):661-674

[8] Chen M, Hu S, Li Y, Jiang TT, Jin H, Feng L. Targeting nuclear acid-mediated immunity in cancer immune checkpoint inhibitor therapies. Signal Transduction and Targeted Therapy. 2020;**5**(1):270

[9] Holyoake DLP, Smyth EC. Chemoradiotherapy or surgery for very early Esophageal squamous cancer: Can a nonrandomized trial give us the answer? Gastroenterology. 2021;**161**(6):1793-1795

[10] Solaini L, Perna F, Cavaliere D, Vaccaro C, Avanzolini A, Cucchetti A, et al. Average treatment effect of robotic versus laparoscopic rectal surgery for rectal cancer. International Journal of Medical Robotics. 2021;**17**(2):e2210

[11] Rallis KS, Lai Yau TH, Sideris M. Chemoradiotherapy in cancer treatment: Rationale and clinical applications. Anticancer Research. 2021;**41**(1):1-7

[12] He Y, Liu QW, Liao HX, Xu WW. Microbiota in cancer chemoradiotherapy resistance. Clinical and Translational Medicine. 2021;**11**(1):e250

[13] Conibear J, AstraZeneca UKL. Rationale for concurrent chemoradiotherapy for patients with stage III non-small-cell lung cancer. British Journal of Cancer. 2020;**123**(Suppl 1):10-17

[14] Bordon Y. An RNA vaccine for advanced melanoma. Nature Reviews. Immunology. 2020;**20**(9):517

[15] Sahin U, Oehm P, Derhovanessian E, Jabulowsky RA, Vormehr M, Gold M, et al. An RNA vaccine drives immunity in checkpoint-inhibitor-treated melanoma. Nature. 2020;**585**(7823):107-112

[16] Rohatgi A, Kirkwood JM. Cancer vaccine induces potent T cell responses - but is it enough? Nature Reviews. Clinical Oncology. 2020;**17**(12):721-722

[17] Sahin U, Derhovanessian E, Miller M, Kloke BP, Simon P, Lower M, et al. Personalized RNA mutanome vaccines mobilize poly-specific therapeutic

immunity against cancer. Nature. 2017;**547**(7662):222-226

[18] Ning L, Zhu B, Gao T. Gold nanoparticles: Promising agent to improve the diagnosis and therapy of cancer. Current Drug Metabolism. 2017;**18**(11):1055-1067

[19] Nazir S, Hussain T, Ayub A, Rashid U, MacRobert AJ. Nanomaterials in combating cancer: Therapeutic applications and developments. Nanomedicine. 2014;**10**(1):19-34

[20] Liang P, Mao L, Dong Y, Zhao Z, Sun Q, Mazhar M, et al. Design and application of near-infrared nanomaterial-liposome hybrid Nanocarriers for cancer Photothermal therapy. Pharmaceutics. 2021;**13**(12):2070

[21] Zhang L, Liu M, Yang S, Wang J, Feng X, Han Z. Natural killer cells: Of-the-shelf cytotherapy for cancer immunosurveillance. American Journal of Cancer Research. 2021;**11**(4):1770-1791

[22] Daher M, Melo Garcia L, Li Y, Rezvani K. CAR-NK cells: The next wave of cellular therapy for cancer. Clinical & Translational Immunology. 2021;**10**(4):e1274

[23] Yu SJ, Ma C, Heinrich B, Brown ZJ, Sandhu M, Zhang Q, et al. Targeting the crosstalk between cytokine-induced killer cells and myeloid-derived suppressor cells in hepatocellular carcinoma. Journal of Hepatology. 2019;**70**(3):449-457

[24] Lu SJ, Feng Q. CAR-NK cells from engineered pluripotent stem cells: Off-the-shelf therapeutics for all patients. Stem Cells Translational Medicine. 2021;**10**(Suppl 2):S10-S17

[25] Depil S, Duchateau P, Grupp SA, Mufti G, Poirot L. 'Off-the-shelf' allogeneic CAR T cells: Development and challenges. Nature Reviews. Drug Discovery. 2020;**19**(3):185-199

[26] Rana A, Bhatnagar S. Advancements in folate receptor targeting for anti-cancer therapy: A small molecule-drug conjugate approach. Bioorganic Chemistry. 2021;**112**:104946

[27] Zhang L, Meng Y, Feng X, Han Z. CAR-NK cells for cancer immunotherapy: From bench to bedside. Biomarker Research. 2022;**10**(1):12

[28] Subauste CS, Dawson L, Remington JS. Human lymphokine-activated killer cells are cytotoxic against cells infected with toxoplasma gondii. The Journal of Experimental Medicine. 1992;**176**(6):1511-1519

[29] Nakano K, Eura M, Chikamatsu K, Masuyama K, Ishikawa T. Characterization of transendothelial migratory lymphokine-activated killer cells. Japanese Journal of Cancer Research. 1996;**87**(4):395-400

[30] Lafreniere R, Rosenberg SA. Successful immunotherapy of murine experimental hepatic metastases with lymphokine-activated killer cells and recombinant interleukin 2. Cancer Research. 1985;**45**(8):3735-3741

[31] Rosenberg SA, Lotze MT, Muul LM, Leitman S, Chang AE, Ettinghausen SE, et al. Observations on the systemic administration of autologous lymphokine-activated killer cells and recombinant interleukin-2 to patients with metastatic cancer. The New England Journal of Medicine. 1985;**313**(23):1485-1492

[32] Rosenberg SA, Lotze MT, Muul LM, Chang AE, Avis FP, Leitman S, et al. A progress report on the treatment of 157 patients with advanced cancer using

lymphokine-activated killer cells and interleukin-2 or high-dose interleukin-2 alone. The New England Journal of Medicine. 1987;**316**(15):889-897

[33] Stanton SE, Disis ML. Clinical significance of tumor-infiltrating lymphocytes in breast cancer. Journal for Immunotherapy of Cancer. 2016;**4**:59

[34] Badalamenti G, Fanale D, Incorvaia L, Barraco N, Listi A, Maragliano R, et al. Role of tumor-infiltrating lymphocytes in patients with solid tumors: Can a drop dig a stone? Cellular Immunology. 2019;**343**:103753

[35] Lin B, Du L, Li H, Zhu X, Cui L, Li X. Tumor-infiltrating lymphocytes: Warriors fight against tumors powerfully. Biomedicine & Pharmacotherapy. 2020;**132**:110873

[36] Yang Y. Cancer immunotherapy: Harnessing the immune system to battle cancer. The Journal of Clinical Investigation. 2015;**125**(9):3335-3337

[37] Gao X, Mi Y, Guo N, Xu H, Xu L, Gou X, et al. Cytokine-induced killer cells As pharmacological tools for cancer immunotherapy. Frontiers in Immunology. 2017;**8**:774

[38] Zhang Y, Ellinger J, Ritter M, Schmidt-Wolf IGH. Clinical studies applying cytokine-induced killer cells for the treatment of renal cell carcinoma. Cancers (Basel). 2020;**12**(9):2471

[39] Giraudo L, Gammaitoni L, Cangemi M, Rotolo R, Aglietta M, Sangiolo D. Cytokine-induced killer cells as immunotherapy for solid tumors: Current evidence and perspectives. Immunotherapy. 2015;**7**(9):999-1010

[40] Shirjang S, Alizadeh N, Mansoori B, Mahmoodpoor A, Kafil HS, Hojjat-Farsangi M, et al. Promising immunotherapy: Highlighting cytokine-induced killer cells. Journal of Cellular Biochemistry. 2019;**120**(6):8863-8883

[41] Mohsenzadegan M, Peng RW, Roudi R. Dendritic cell/cytokine-induced killer cell-based immunotherapy in lung cancer: What we know and future landscape. Journal of Cellular Physiology. 2020;**235**(1):74-86

[42] Bryceson YT, March ME, Ljunggren HG, Long EO. Activation, coactivation, and costimulation of resting human natural killer cells. Immunological Reviews. 2006;**214**: 73-91

[43] Xu J, Niu T. Natural killer cell-based immunotherapy for acute myeloid leukemia. Journal of Hematology & Oncology. 2020;**13**(1):167

[44] Rezvani K, Rouce R, Liu E, Shpall E. Engineering natural killer cells for cancer immunotherapy. Molecular Therapy. 2017;**25**(8):1769-1781

[45] Liu M, Meng Y, Zhang L, Han Z, Feng X. High-efficient generation of natural killer cells from peripheral blood with preferable cell vitality and enhanced cytotoxicity by combination of IL-2, IL-15 and IL-18. Biochemical and Biophysical Research Communications. 2021;**534**:149-156

[46] Cany J, van der Waart AB, Tordoir M, Franssen GM, Hangalapura BN, de Vries J, et al. Natural killer cells generated from cord blood hematopoietic progenitor cells efficiently target bone marrow-residing human leukemia cells in NOD/SCID/IL2Rg(null) mice. PLoS One. 2013;**8**(6):e64384

[47] Kang L, Voskinarian-Berse V, Law E, Reddin T, Bhatia M, Hariri A, et al. Characterization and ex vivo expansion of human placenta-derived natural

killer cells for cancer immunotherapy. Frontiers in Immunology. 2013;**4**:101

[48] Hauswirth AW, Florian S, Printz D, Sotlar K, Krauth MT, Fritsch G, et al. Expression of the target receptor CD33 in CD34+/CD38-/CD123+ AML stem cells. European Journal of Clinical Investigation. 2007;**37**(1):73-82

[49] Marofi F, Rahman HS, Thangavelu L, Dorofeev A, Bayas-Morejon F, Shirafkan N, et al. Renaissance of armored immune effector cells, CAR-NK cells, brings the higher hope for successful cancer therapy. Stem Cell Research & Therapy. 2021;**12**(1):200

[50] Hermanson DL, Kaufman DS. Utilizing chimeric antigen receptors to direct natural killer cell activity. Frontiers in Immunology. 2015;**6**:195

[51] Geller MA, Cooley S, Judson PL, Ghebre R, Carson LF, Argenta PA, et al. A phase II study of allogeneic natural killer cell therapy to treat patients with recurrent ovarian and breast cancer. Cytotherapy. 2011;**13**(1):98-107

[52] Luna JI, Grossenbacher SK, Murphy WJ, Canter RJ. Targeting cancer stem cells with natural killer cell immunotherapy. Expert Opinion on Biological Therapy. 2017;**17**(3):313-324

[53] Hu Y, Tian ZG, Zhang C. Chimeric antigen receptor (CAR)-transduced natural killer cells in tumor immunotherapy. Acta Pharmacologica Sinica. 2018;**39**(2):167-176

[54] Kohli K, Pillarisetty VG. Dendritic cells in the tumor microenvironment. Advances in Experimental Medicine and Biology. 2020;**1273**:29-38

[55] Tran Janco JM, Lamichhane P, Karyampudi L, Knutson KL. Tumor-infiltrating dendritic cells in cancer pathogenesis. Journal of Immunology. 2015;**194**(7):2985-2991

[56] Gardner A, de Mingo PA, Ruffell B. Dendritic cells and their role in immunotherapy. Frontiers in Immunology. 2020;**11**:924

[57] Waisman A, Lukas D, Clausen BE, Yogev N. Dendritic cells as gatekeepers of tolerance. Seminars in Immunopathology. 2017;**39**(2):153-163

[58] Stevens D, Ingels J, Van Lint S, Vandekerckhove B, Vermaelen K. Dendritic cell-based immunotherapy in lung cancer. Frontiers in Immunology. 2020;**11**:620374

[59] Palucka K, Banchereau J. Cancer immunotherapy via dendritic cells. Nature Reviews. Cancer. 2012;**12**(4):265-277

[60] Zhou J, Wang G, Chen Y, Wang H, Hua Y, Cai Z. Immunogenic cell death in cancer therapy: Present and emerging inducers. Journal of Cellular and Molecular Medicine. 2019;**23**(8):4854-4865

[61] Zhang L, Yang X, Sun Z, Li J, Zhu H, Li J, et al. Dendritic cell vaccine and cytokine-induced killer cell therapy for the treatment of advanced non-small cell lung cancer. Oncology Letters. 2016;**11**(4):2605-2610

[62] Arbour KC, Riely GJ. Systemic therapy for locally advanced and metastatic non-small cell lung cancer: A review. JAMA. 2019;**322**(8):764-774

[63] Cavaillon JM. The historical milestones in the understanding of leukocyte biology initiated by Elie Metchnikoff. Journal of Leukocyte Biology. 2011;**90**(3):413-424

[64] Wynn TA, Vannella KM. Macrophages in tissue repair,

regeneration, and fibrosis. Immunity. 2016;**44**(3):450-462

[65] Luo Y, Shao L, Chang J, Feng W, Liu YL, Cottler-Fox MH, et al. M1 and M2 macrophages differentially regulate hematopoietic stem cell self-renewal and ex vivo expansion. Blood Advances. 2018;**2**(8):859-870

[66] Zhang H, Li Z, Li W. M2 macrophages serve as critical executor of innate immunity in chronic allograft rejection. Frontiers in Immunology. 2021;**12**:648539

[67] Goswami KK, Ghosh T, Ghosh S, Sarkar M, Bose A, Baral R. Tumor promoting role of anti-tumor macrophagesintumormicroenvironment. Cellular Immunology. 2017;**316**:1-10

[68] Anderson NR, Minutolo NG, Gill S, Klichinsky M. Macrophage-based approaches for cancer immunotherapy. Cancer Research. 2021;**81**(5):1201-1208

[69] Mills CD, Lenz LL, Harris RA. A breakthrough: Macrophage-directed cancer immunotherapy. Cancer Research. 2016;**76**(3):513-516

[70] Hibbs JB Jr, Vavrin Z, Taintor RR. L-arginine is required for expression of the activated macrophage effector mechanism causing selective metabolic inhibition in target cells. Journal of Immunology. 1987;**138**(2):550-565

[71] Mills CD. Anatomy of a discovery: m1 and m2 macrophages. Frontiers in Immunology. 2015;**6**:212

[72] Pei H, Qin J, Wang F, Tan B, Zhao Z, Peng Y, et al. Discovery of potent ureido tetrahydrocarbazole derivatives for cancer treatments through targeting tumor-associated macrophages. European Journal of Medicinal Chemistry. 2019;**183**:111741

[73] Wang H, Tian T, Zhang J. Tumor associated macrophages (TAMs) in colorectal cancer (CRC): From mechanism to therapy and prognosis. International Journal of Molecular Sciences. 2021;**22**(16):8470

[74] Baek SH, Lee HW, Gangadaran P, Oh JM, Zhu L, Rajendran RL, et al. Role of M2-like macrophages in the progression of ovarian cancer. Experimental Cell Research. 2020;**395**(2):112211

[75] de Aquino MT, Malhotra A, Mishra MK, Shanker A. Challenges and future perspectives of T cell immunotherapy in cancer. Immunology Letters. 2015;**166**(2):117-133

[76] Levite M. T cells plead for rejuvenation and amplification; with the Brain's neurotransmitters and neuropeptides we can make it happen. Frontiers in Immunology. 2021;**12**:617658

[77] Kumar BV, Connors TJ, Farber DL. Human T cell development, localization, and function throughout life. Immunity. 2018;**48**(2):202-213

[78] O'Donnell JS, Teng MWL, Smyth MJ. Cancer immunoediting and resistance to T cell-based immunotherapy. Nature Reviews. Clinical Oncology. 2019;**16**(3):151-167

[79] Seliger B, Massa C. Immune therapy resistance and immune escape of Tumors. Cancers (Basel). 2021;**13**(3):551

[80] Smyth MJ, Ngiow SF, Ribas A, Teng MW. Combination cancer immunotherapies tailored to the tumour microenvironment. Nature Reviews. Clinical Oncology. 2016;**13**(3):143-158

[81] Mittal D, Gubin MM, Schreiber RD, Smyth MJ. New insights into cancer immunoediting and its three component

phases--elimination, equilibrium and escape. Current Opinion in Immunology. 2014;**27**:16-25

[82] Restifo NP, Dudley ME, Rosenberg SA. Adoptive immunotherapy for cancer: Harnessing the T cell response. Nature Reviews. Immunology. 2012;**12**(4):269-281

[83] Dudley ME, Rosenberg SA. Adoptive-cell-transfer therapy for the treatment of patients with cancer. Nature Reviews. Cancer. 2003;**3**(9):666-675

[84] Ryschich E, Notzel T, Hinz U, Autschbach F, Ferguson J, Simon I, et al. Control of T-cell-mediated immune response by HLA class I in human pancreatic carcinoma. Clinical Cancer Research. 2005;**11**(2 Pt 1):498-504

[85] Beltra JC, Manne S, Abdel-Hakeem MS, Kurachi M, Giles JR, Chen Z, et al. Developmental relationships of four exhausted CD8(+) T cell subsets reveals underlying transcriptional and epigenetic landscape control mechanisms. Immunity. 2020;**52**(5):825-841 e828

[86] Baba Y, Saito Y, Kotetsu Y. Heterogeneous subsets of B-lineage regulatory cells (Breg cells). International Immunology. 2020;**32**(3):155-162

[87] Montecino-Rodriguez E, Dorshkind K. B-1 B cell development in the fetus and adult. Immunity. 2012;**36**(1):13-21

[88] Nothelfer K, Sansonetti PJ, Phalipon A. Pathogen manipulation of B cells: The best defence is a good offence. Nature Reviews. Microbiology. 2015;**13**(3):173-184

[89] Fillatreau S, Manfroi B, Dorner T. Toll-like receptor signalling in B cells during systemic lupus erythematosus. Nature Reviews Rheumatology. 2021;**17**(2):98-108

[90] Wang Y, Liu J, Burrows PD, Wang JY. B cell development and maturation. Advances in Experimental Medicine and Biology. 2020;**1254**:1-22

[91] Mauri C, Bosma A. Immune regulatory function of B cells. Annual Review of Immunology. 2012;**30**:221-241

[92] Zhu Q, Rui K, Wang S, Tian J. Advances of regulatory B cells in autoimmune diseases. Frontiers in Immunology. 2021;**12**:592914

[93] Catalan D, Mansilla MA, Ferrier A, Soto L, Oleinika K, Aguillon JC, et al. Immunosuppressive mechanisms of regulatory B cells. Frontiers in Immunology. 2021;**12**:611795

[94] Akkaya M, Kwak K, Pierce SK. B cell memory: Building two walls of protection against pathogens. Nature Reviews. Immunology. 2020;**20**(4):229-238

[95] Laidlaw BJ, Cyster JG. Transcriptional regulation of memory B cell differentiation. Nature Reviews. Immunology. 2021;**21**(4):209-220

[96] Botia-Sanchez M, Alarcon-Riquelme ME, Galicia G. B cells and microbiota in autoimmunity. International Journal of Molecular Sciences. 2021;**22**(9):4846

[97] Rodriguez-Garcia A, Palazon A, Noguera-Ortega E, Powell DJ Jr, Guedan S. CAR-T cells hit the tumor microenvironment: Strategies to overcome tumor escape. Frontiers in Immunology. 2020;**11**:1109

[98] Pan J, Niu Q, Deng B, Liu S, Wu T, Gao Z, et al. CD22 CAR T-cell therapy

in refractory or relapsed B acute lymphoblastic leukemia. Leukemia. 2019;**33**(12):2854-2866

[99] Pan J, Zuo S, Deng B, Xu X, Li C, Zheng Q, et al. Sequential CD19-22 CAR T therapy induces sustained remission in children with r/r B-ALL. Blood. 2020;**135**(5):387-391

[100] Pan J, Tan Y, Deng B, Tong C, Hua L, Ling Z, et al. Frequent occurrence of CD19-negative relapse after CD19 CAR T and consolidation therapy in 14 TP53-mutated r/r B-ALL children. Leukemia. 2020;**34**(12):3382-3387

[101] Roselli E, Faramand R, Davila ML. Insight into next-generation CAR therapeutics: Designing CAR T cells to improve clinical outcomes. The Journal of Clinical Investigation. 2021;**131**(2):e142030

[102] Frigault MJ, Maus MV. State of the art in CAR T cell therapy for CD19+ B cell malignancies. The Journal of Clinical Investigation. 2020;**130**(4):1586-1594

[103] Hay KA. Cytokine release syndrome and neurotoxicity after CD19 chimeric antigen receptor-modified (CAR-) T cell therapy. British Journal of Haematology. 2018;**183**(3):364-374

[104] Giavridis T, van der Stegen SJC, Eyquem J, Hamieh M, Piersigilli A, Sadelain M. CAR T cell-induced cytokine release syndrome is mediated by macrophages and abated by IL-1 blockade. Nature Medicine. 2018;**24**(6):731-738

[105] Maude SL, Laetsch TW, Buechner J, Rives S, Boyer M, Bittencourt H, et al. Tisagenlecleucel in children and Young adults with B-cell lymphoblastic Leukemia. The New England Journal of Medicine. 2018;**378**(5):439-448

[106] Kingwell K. CAR T therapies drive into new terrain. Nature Reviews. Drug Discovery. 2017;**16**(5):301-304

[107] Hernandez-Lopez A, Tellez-Gonzalez MA, Mondragon-Teran P, Meneses-Acosta A. Chimeric antigen receptor-T cells: A pharmaceutical scope. Frontiers in Pharmacology. 2021;**12**:720692

[108] Tahmasebi S, Elahi R, Esmaeilzadeh A. Solid Tumors challenges and new insights of CAR T cell engineering. Stem Cell Reviews and Reports. 2019;**15**(5):619-636

[109] Wing A, Fajardo CA, Posey AD Jr, Shaw C, Da T, Young RM, et al. Improving CART-cell therapy of solid Tumors with oncolytic virus-driven production of a bispecific T-cell engager. Cancer Immunology Research. 2018;**6**(5):605-616

[110] Watanabe K, Luo Y, Da T, Guedan S, Ruella M, Scholler J, et al. Pancreatic cancer therapy with combined mesothelin-redirected chimeric antigen receptor T cells and cytokine-armed oncolytic adenoviruses. JCI Insight. 2018;**3**(7):e99573

[111] Ecsedi M, McAfee MS, Chapuis AG. The anticancer potential of T cell receptor-engineered T cells. Trends Cancer. 2021;**7**(1):48-56

[112] Attaf M, Roider J, Malik A, Rius Rafael C, Dolton G, Predergast AJ, et al. Cytomegalovirus-mediated T cell receptor repertoire perturbation is present in early life. Frontiers in Immunology. 2020;**11**:1587

[113] Toya T, Taguchi A, Kitaura K, Misumi F, Nakajima Y, Otsuka Y, et al. T-cell receptor repertoire of cytomegalovirus-specific cytotoxic T-cells after allogeneic stem cell transplantation. Scientific Reports. 2020;**10**(1):22218

[114] Schober K, Voit F, Grassmann S, Muller TR, Eggert J, Jarosch S, et al.

Reverse TCR repertoire evolution toward dominant low-affinity clones during chronic CMV infection. Nature Immunology. 2020;**21**(4):434-441

[115] Li D, Li X, Zhou WL, Huang Y, Liang X, Jiang L, et al. Genetically engineered T cells for cancer immunotherapy. Signal Transduction and Targeted Therapy. 2019;**4**:35

[116] Redeker A, Arens R. 'Reverse evolution' in T cell biology. Nature Immunology. 2020;**21**(4):360-362

[117] Berrien-Elliott MM, Cashen AF, Cubitt CC, Neal CC, Wong P, Wagner JA, et al. Multidimensional analyses of donor memory-like NK cells reveal new associations with response after adoptive immunotherapy for Leukemia. Cancer Discovery. 2020;**10**(12):1854-1871

[118] Li Y, Hermanson DL, Moriarity BS, Kaufman DS. Human iPSC-derived natural killer cells engineered with chimeric antigen receptors enhance anti-tumor activity. Cell Stem Cell. 2018;**23**(2):181-192 e185

[119] Elahi R, Heidary AH, Hadiloo K, Esmaeilzadeh A. Chimeric antigen receptor-engineered natural killer (CAR NK) cells in cancer treatment; recent advances and future prospects. Stem Cell Reviews and Reports. 2021;**17**(6):2081-2106

[120] Basar R, Daher M, Rezvani K. Next-generation cell therapies: The emerging role of CAR-NK cells. Blood Advances. 2020;**4**(22):5868-5876

[121] Rostovskaya M, Fu J, Obst M, Baer I, Weidlich S, Wang H, et al. Transposon-mediated BAC transgenesis in human ES cells. Nucleic Acids Research. 2012;**40**(19):e150

[122] Vargas JE, Chicaybam L, Stein RT, Tanuri A, Delgado-Canedo A, Bonamino MH. Retroviral vectors and transposons for stable gene therapy: Advances, current challenges and perspectives. Journal of Translational Medicine. 2016;**14**(1):288

[123] Kim A, Pyykko I. Size matters: Versatile use of PiggyBac transposons as a genetic manipulation tool. Molecular and Cellular Biochemistry. 2011;**354**(1-2):301-309

[124] Xu Y, Liu Q, Zhong M, Wang Z, Chen Z, Zhang Y, et al. 2B4 costimulatory domain enhancing cytotoxic ability of anti-CD5 chimeric antigen receptor engineered natural killer cells against T cell malignancies. Journal of Hematology & Oncology. 2019;**12**(1):49

[125] Chen KH, Wada M, Pinz KG, Liu H, Lin KW, Jares A, et al. Preclinical targeting of aggressive T-cell malignancies using anti-CD5 chimeric antigen receptor. Leukemia. 2017;**31**(10):2151-2160

[126] Quintarelli C, Sivori S, Caruso S, Carlomagno S, Falco M, Boffa I, et al. Efficacy of third-party chimeric antigen receptor modified peripheral blood natural killer cells for adoptive cell therapy of B-cell precursor acute lymphoblastic leukemia. Leukemia. 2020;**34**(4):1102-1115

[127] Xia J, Minamino S, Kuwabara K. CAR-expressing NK cells for cancer therapy: A new hope. Bioscience Trends. 2020;**14**(5):354-359

[128] Daher M, Basar R, Gokdemir E, Baran N, Uprety N, Nunez Cortes AK, et al. Targeting a cytokine checkpoint enhances the fitness of armored cord blood CAR-NK cells. Blood. 2021;**137**(5):624-636

[129] Fu W, Lei C, Ma Z, Qian K, Li T, Zhao J, et al. CAR macrophages for SARS-CoV-2 immunotherapy. Frontiers in Immunology. 2021;**12**:669103

[130] Zhang W, Liu L, Su H, Liu Q, Shen J, Dai H, et al. Chimeric antigen receptor macrophage therapy for breast tumours mediated by targeting the tumour extracellular matrix. British Journal of Cancer. 2019;**121**(10):837-845

[131] Klichinsky M, Ruella M, Shestova O, Lu XM, Best A, Zeeman M, et al. Human chimeric antigen receptor macrophages for cancer immunotherapy. Nature Biotechnology. 2020;**38**(8):947-953

[132] Mukhopadhyay M. Macrophages enter CAR immunotherapy. Nature Methods. 2020;**17**(6):561

[133] Zhang L, Tian L, Dai X, Yu H, Wang J, Lei A, et al. Pluripotent stem cell-derived CAR-macrophage cells with antigen-dependent anti-cancer cell functions. Journal of Hematology & Oncology. 2020;**13**(1):153

[134] Wang W, Jiang J, Wu C. CAR-NK for tumor immunotherapy: Clinical transformation and future prospects. Cancer Letters. 2020;**472**:175-180

[135] Morgan MA, Buning H, Sauer M, Schambach A. Use of cell and genome modification technologies to generate improved "off-the-shelf" CAR T and CAR NK cells. Frontiers in Immunology. 1965;**2020**:11

[136] Siegler EL, Zhu Y, Wang P, Yang L. Off-the-shelf CAR-NK cells for cancer immunotherapy. Cell Stem Cell. 2018;**23**(2):160-161

[137] Pan J, Deng B, Ling Z, Song W, Xu J, Duan J, et al. Ruxolitinib mitigates steroid-refractory CRS during CAR T therapy. Journal of Cellular and Molecular Medicine. 2021;**25**(2):1089-1099

[138] Tan Y, Pan J, Deng B, Ling Z, Song W, Xu J, et al. Toxicity and effectiveness of CD19 CAR T therapy in children with high-burden central nervous system refractory B-ALL. Cancer Immunology, Immunotherapy. 2021;**70**(7):1979-1993

[139] Sommer C, Boldajipour B, Kuo TC, Bentley T, Sutton J, Chen A, et al. Preclinical evaluation of allogeneic CAR T cells targeting BCMA for the treatment of multiple myeloma. Molecular Therapy. 2019;**27**(6):1126-1138

[140] Wang R, Feng W, Wang H, Wang L, Yang X, Yang F, et al. Blocking migration of regulatory T cells to leukemic hematopoietic microenvironment delays disease progression in mouse leukemia model. Cancer Letters. 2020;**469**:151-161

[141] Ren J, Han L, Tang J, Liu Y, Deng X, Liu Q, et al. Foxp1 is critical for the maintenance of regulatory T-cell homeostasis and suppressive function. PLoS Biology. 2019;**17**(5):e3000270

[142] Riley RS, June CH, Langer R, Mitchell MJ. Delivery technologies for cancer immunotherapy. Nature Reviews. Drug Discovery. 2019;**18**(3):175-196

[143] Kennedy LB, Salama AKS. A review of cancer immunotherapy toxicity. CA: a Cancer Journal for Clinicians. 2020;**70**(2):86-104

[144] Balassa K, Rocha V. Anticancer cellular immunotherapies derived from umbilical cord blood. Expert Opinion on Biological Therapy. 2018;**18**(2):121-134

[145] Koepsell SA, Miller JS, McKenna DH Jr. Natural killer cells: A review of manufacturing and clinical utility. Transfusion. 2013;**53**(2):404-410

[146] Pahl JHW, Koch J, Gotz JJ, Arnold A, Reusch U, Gantke T, et al. CD16A activation of NK cells promotes NK cell proliferation and memory-like cytotoxicity against cancer cells. Cancer Immunology Research. 2018;**6**(5):517-527

[147] Lu C, Guo C, Chen H, Zhang H, Zhi L, Lv T, et al. A novel chimeric PD1-NKG2D-41BB receptor enhances antitumor activity of NK92 cells against human lung cancer H1299 cells by triggering pyroptosis. Molecular Immunology. 2020;**122**:200-206

[148] Wrona E, Borowiec M, Potemski P. CAR-NK cells in the treatment of solid Tumors. International Journal of Molecular Sciences. 2021;**22**(11):5899

[149] Jayaraman J, Mellody MP, Hou AJ, Desai RP, Fung AW, Pham AHT, et al. CAR-T design: Elements and their synergistic function. eBioMedicine. 2020;**58**:102931

[150] Grosser R, Cherkassky L, Chintala N, Adusumilli PS. Combination immunotherapy with CAR T cells and checkpoint blockade for the treatment of solid Tumors. Cancer Cell. 2019;**36**(5):471-482

[151] Zhang F, Stephan SB, Ene CI, Smith TT, Holland EC, Stephan MT. Nanoparticles that reshape the tumor milieu create a therapeutic window for effective T-cell therapy in solid malignancies. Cancer Research. 2018;**78**(13):3718-3730

[152] Yong SB, Chung JY, Song Y, Kim J, Ra S, Kim YH. Non-viral nano-immunotherapeutics targeting tumor microenvironmental immune cells. Biomaterials. 2019;**219**:119401

Chapter 8

Trypan Blue Exclusion Assay, Neutral Red, Acridine Orange and Propidium Iodide

Arinzechukwu Ude, Kaiyven Afi-Leslie, Kelechi Okeke and Emmanuel Ogbodo

Abstract

Cytotoxicity and cell viability assessments are very important parameters that are widely used in fundamental research and drug development to determine the safety profile of toxic compounds. These assays measure the degree to which a substance can cause toxic damage to cells or cell death. There are different assays that have been employed to determine the cytotoxicity of substances. These assays either determine enzymatic function, cell viability, mitochondrial activity, lipid metabolism, cell proliferation and/or cell death. These assays entail use of different kinds of dyes such as trypan blue exclusion dye, neutral red, acridine orange and propidium iodide to stain the cells. Trypan blue dye permeates compromised cell membrane to stain necrotic cells. However, this can lead to false positive and false negative results as it does not provide information on sub-lethal injury. As a result, neutral red and acridine orange can be used as counterstains for trypan blue to stain the lysosome of live cells. Acridine orange can also be used to stain nucleic acids in living cells and is usually co-stained with propidium iodide or ethidium bromide. This is because propidium iodide and ethidium bromide permeate only compromised plasma membrane thus co-staining cells with these dyes can provide vital information that can be used to differentiate between live and dead cells.

Keywords: cytotoxicity, trypan, acridine, iodide, neutral, blue, red, orange, propidium

1. Introduction

Today, humans consume food and vegetables rich in phenolic compounds and are continuously exposed to increased volume of xenobiotics. These chemicals such as phenolic compounds as well as pharmaceutical agents induce cellular toxicity and/or genomic instability most especially in the liver [1]. These chemicals are absorbed and biotransformed by the liver to their metabolites that may be less or more toxic to cells. Consequently, the evaluation of hepatocytotoxicity and erythropoietic cytotoxicity are key components of safety assessment during drug development [2]. Cytotoxicity is the ability of chemicals to destroy the functioning of living cells [3, 4]. Cytotoxic chemicals cause damages to the components of DNA and thus induce mutations that may increase the risk of cancer development. Cytotoxicity and cell viability assessments are based on targeting of functions of different cells, including the liver and bone marrow [5, 6].

Cytotoxicity assays are experimental methods used in pharmacology and *in vitro* toxicology studies to measure loss of cellular functions [5–7]. These assays generally screen xenobiotics and predict human toxicity. These assays predict the risk of these xenobiotics towards human health by relating the cytotoxic effects between the *in vitro* and *in vivo* systems. There are different *in vitro* cytotoxic assays that have been employed in this stead with different results [5–7]. In these *in vitro* systems commonly used, cells are treated or exposed to a chemical or test compound and incubated for some period. Afterwards, a marker is measured to reflect the number of viable cells present compared to the positive (toxin) and negative (vehicle) control treatments. In addition, dead cells are also evaluated to distinguish between cytotoxicity and growth arrest. Knowledge of the number of live and dead cells present during or after the end of the experiment enhances the statistical robustness of these assays [5–8]. These assays commonly estimate dead cells via two common ways: (a) penetration of an otherwise non-permeable dye into the cells due to loss of cell membrane integrity; (b) the ability of indicator molecules to partition into a compartment not achievable if the cell membrane is intact. Therefore, these cytotoxicity assays directly or indirectly assess cytotoxicity by providing information on cell membrane integrity, cell metabolism, cell machinery, cell mortality and cell proliferation [5–8]. Some of these sensitive and reproducible assays are used to obtain colorimetric, luminometric and fluorometric measurements hence entails the use of different dyes such as trypan blue exclusion dye, neutral red, acridine orange and propidium iodide [1, 9–12]. These dyes offer different advantages and disadvantages hence a combination of two or more dyes is usually employed in *in vitro* studies to avoid overestimation or underestimation of the toxicity of a chemical substance thereby increasing the reliability of the results obtained.

2. Experimental dyes used in measuring cytotoxicity

2.1 Trypan blue dye exclusion assay

Trypan blue dye exclusion assay is one of the most frequently used routine methods to determine cell viability [9, 13]. It involves the selective staining of dead cells with trypan blue and microscopic examination on haemocytometer [9]. It was developed in 1975 to measure viable cells and provide information on cell mortality [2]. Trypan blue, a non-permeable cell membrane dye, is an azo dye derived from toluidine. This is a vital stain used in bioscience to exclusively stain necrotic (dead) cells (**Figure 1**). Therefore, this assay is based on the principle that viable cells have intact cell membranes, which trypan blue cannot penetrate thus, trypan blue is excluded from viable cells [5].

In contrast, trypan blue dye penetrates cell membrane of necrotic (dead) cells due to loss of cell membrane integrity and enters the cytoplasm. Under light microscope, only necrotic (dead) cells absorb this blue colour [15]. The trypan blue staining technique is usually performed on a single sample or relatively small number of samples from simple experiments [6].

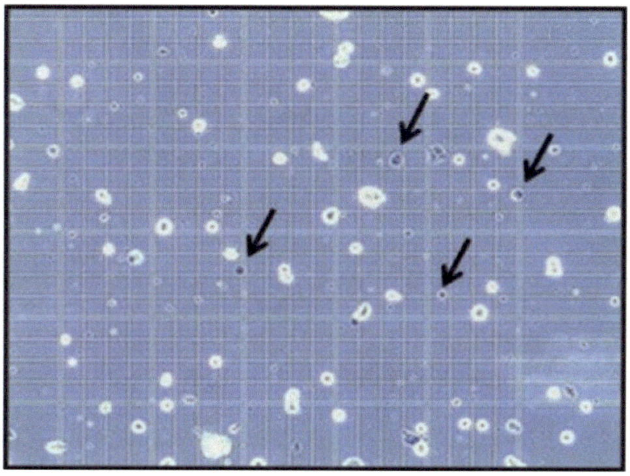

Figure 1.
Cells stained with trypan blue. Neubauer chamber counter (Adapted from Hong et al. [14]).

However, there have been some questions raised about the integrity of the trypan blue technique. These include:

1. There is the possibility of false-positive or false-negative results. Trypan blue exclusion assay can only demonstrate either viable or dead cells without detecting cells with sub-lethal injury [6]. Compromised viable cells may be counted as dead cells whilst the cell is still alive or cells with intact membranes may be counted as live cells even though they are no longer functioning as viable cells. This is because sometimes a cell may have an intact membrane and at the same time, the cell will be unable to grow or function as a viable cell [6, 16]. On the contrary, a cell may lose its cell membrane integrity and still be able to repair itself to become viable again [16]. Longer incubations with trypan blue dye may also result in faint staining of viable cells possibly due to slow uptake of the dye [16].

2. False-positive and/or false-negative results may also be due to subjective judgement of the user to determine what is a dead cell or stained debris [6]. Cellular uptake of trypan blue dye is determined by the human eye, and it is possible to overlook any small amount of the dye uptake following cellular damage due to stress or injury [9, 13]. Cells are counted manually using a haemocytometer hence a false result may arise due to human errors especially when processing many cell counts. Also, it takes a lot of time and manual labour to count and measure multiple samples [6]. Thus, if different operators count the same sample, there may be a lot of inconsistencies.

3. Trypan blue dye can also function as a mutagenic and carcinogenetic agent [6, 17]. Therefore, high concentration and prolonged exposure of the dye to cells may lead to toxic effects on the cells thus cause gradual cell damage and thence leading to false cell counts [18]. Due to its toxicity, trypan blue can also cause changes in intracellular protein expression, which may require some time to be observed/reflected in the cell count [18]. The amount of trypan blue dye exposed

to a cell with time may influence the expression of *P53* and *P21* genes, which are crucial in cell cycle. *P53* helps in making intracellular protein located in the nucleus and plays a major role in cell cycle; controlling cell division and cell death whereas *P21* is a cell cycle inhibitor including cell cycle arrest, apoptosis, and DNA repair [19–21]. Therefore, the impact of prolonged exposure of trypan blue dye to a cell on *P53* and *P21* genes expression imposes a challenge that contradicts the use of this assay [18].

In conclusion, there is need for optimisation of concentration and incubation time depending on the type of cell involved to eliminate the possibility of this dye causing a gradual damage on the cells. Also, trypan blue dye may likely underestimate cellular damage when performed immediately after treating cells with a cytotoxic agent. As a result, many researchers combine this technique with another dye to ensure the cytotoxic results are robust. Benchtop instruments have also been designed to automate imaging and improve the biased analysis steps of this assay.

2.2 Neutral red

Neutral red assay was first described by Finter in 1969 as a chemosensitivity assay [22, 23]. This assay provides information on cell machinery and quantifies cell viability and measures cell replication. In addition to hepatocytes and erythrocytes, this assay can also be done in non-adherent cells such as fibroblasts. Cells with biotransformation capacity are recommended to be used when neutral red is used to assess the cytotoxicity of chemicals requiring metabolic activation to toxic metabolites [1, 24].

Its principle is based on the ability of viable cells to absorb neutral red, a weak cationic dye, which penetrates the cell membrane and concentrates in the lysosomes of the cell [24]. The mechanism by which the dye penetrates the cell membrane and accumulates in the lysosomes is not well understood. Micropinocytosis with subsequent fusion of vesicles with secondary lysosome was first suggested however entry by non-ionic passive diffusion has always been postulated. Once inside the lysosome, it binds to the anionic and/or phosphate groups of the lysosomal matrix by electrostatic hydrophobic bonds [24–26]. The dye is then extracted from viable cells using an acidified ethanol solution, and the absorbance of the solubilised dye is quantified using a spectrophotometer at 540 nm wavelength.

The uptake of neutral red by viable cells depends on their capacity to maintain pH gradients through the production of ATP [1, 8]. The dye presents a net charge close to zero at physiological pH that enables it to penetrate the cell membrane. Once inside the lysosome, the dye becomes charged and is retained inside the lysosome due to a proton gradient in the lysosome that maintains a pH lower than that of the cytoplasm [1, 8]. As a result, the amount of retained dye is proportional to the number of viable cells (**Figure 2**).

However, alterations in cell surface or lysosomal membranes can modify the uptake of neutral red by viable cells. For example, a variety of chemicals or pharmaceutical agents induce damage to cell surface or lysosomal membrane that may alter or decrease dye uptake and subsequent retention [27]. Due to specific lysosomal capacities in different cells for taking up the dye, neutral red can be used to differentiate between viable, damaged, or dead cells. Viable cells have intact lysosome and tend to absorb more neutral red dye more than dead cells or cells undergoing apoptosis.

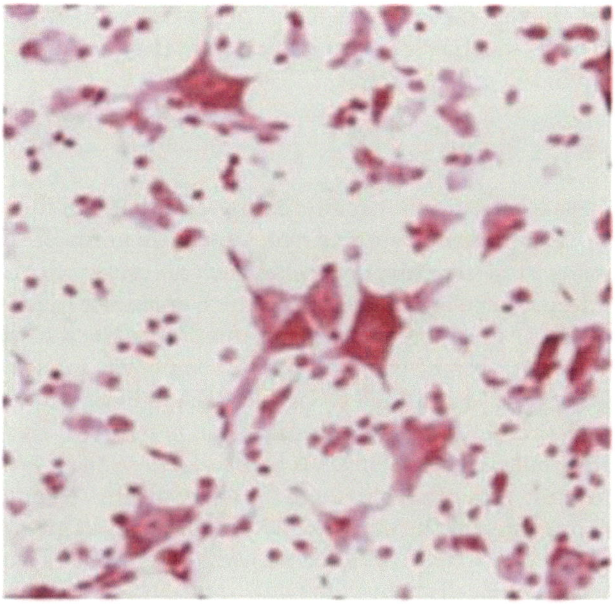

Figure 2.
Cells stained with neutral red (Adapted from Repetto et al. [24]).

Therefore, uptake of the dye as well as lysosomal integrity are highly sensitive indicators of cell viability. In addition, the results of neutral red assay are dependent on (a) the degree of acute toxicity (b) number of viable cells in the culture which determine the timing of the assay [25, 26, 28].

Neutral red dye is non-specific and non-toxic and is often used as a counterstain for dyes such as trypan blue dye. Therefore, neutral red assay poses some advantages over other cytotoxicity assays as it is very sensitive, simple, cheap, readily quantifiable, presents less interference and does not require equipment or unstable reagents such as tetrazolium salts that measure lactate dehydrogenase enzyme activity by the chemical reduction of the salts to formazans [22, 29, 30]. The neutral red assay also has advantages, which include speed and reproducibility of data. Also, neutral red estimates can be done on the same cell culture alongside protein determination.

Nevertheless, there are limitations of the neutral red assay such as underestimation of the toxicity of chemicals, which require metabolic activation to a toxic product and of substances, which bind to serum proteins [22, 29, 30]. On the contrary, some chemicals may induce irreversible precipitation of the neutral red dye into fine, needle-like crystals, which may result in an overestimation of the toxic effects [27]. Some chemicals have a localised effect on lysosomes, and this may result in low or high uptake thus leading to overestimation or underestimation of cellular toxicity. Therefore, this assay is suitable for detecting chemicals such as chloroquine that selectively target lysosome and alter its pH thus inducing a greater effect of neutral red uptake than most chemicals [24]. This causes an overestimation of the toxic effects induced by these chemicals. Changes in cell proliferation may also interfere with the estimation of lysosomal function [26, 28, 29]. To prevent this, the assay can be performed in conjunction with other cytotoxicity assays. However, performing neutral red estimates followed by determination of enzymatic function such as lactate dehydrogenase and glucose-6-phosphate dehydrogenase or total protein determination in the same cell culture may lead to a reduction in the amount of protein estimated to be present in the cell culture.

2.3 Acridine orange

Acridine orange is a heterocyclic organic compound that was first extracted from coal tar. It acts as a weak basic nucleic acid dye, which can permeate the cell membrane and accumulate in these acidic organelles such as lysosome in a pH dependent manner [31]. Acridine orange is hydrophobic, which enables it to permeate the cell membrane quickly, enter the cytoplasm and accumulate in the lysosome [31–33]. Therefore, acridine orange can be used to stain lysosome, vacuoles, and nucleus where it specifically binds to double stranded DNA and RNA in living cells by intercalation or electrostatic attractions (**Figure 3**). Acridine orange stabilises the pigment-DNA complexes via charge neutralisation of DNA backbone phosphate group [33, 35].

Under acridine orange staining, lysosome fluoresce bright-red or orange red at a wavelength of 590 nm whereas the nucleus and cytoplasm emit green fluorescence at a wavelength of 525 nm. Due to its low molecular weight (256 g/mol), it rapidly diffuses into the cytoplasm of living cells to bind to DNA and RNA [36, 37]. Once bound to single stranded DNA or RNA, red fluorescence is emitted whereas green fluorescence is emitted when bound to double stranded DNA. As a result, acridine orange can be used in apoptotic studies to stain apoptotic cells orange or red depending on the degree of loss of membrane integrity. This dye also has the capacity to label dead cells thus differentiating the apoptotic cells into early apoptosis (green) and late apoptosis (orange red) hence it offers superior accuracy than the older methods [10, 11, 38].

However, fluorescence response of acridine orange is dependent upon the concentration of acridine orange used, the solvent used in dissolving the dye, fixation, time of staining, ions present, ionic strength of the medium, pH, temperature, and complexing substrate [37, 39, 40]. Lowering the concentration of acridine orange causes a decrease in fluorescence whilst an increase in acridine orange concentration will cause a metachromatic shift. In addition, acridine orange is usually co-stained with other

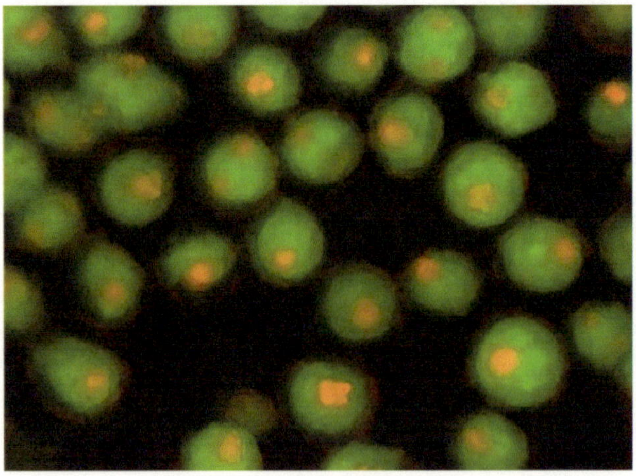

Figure 3.
Stromal cell line, HS-5 cells, stained with acridine orange, which stained the nucleus of the cells (Adapted from Okeke [34]).

dyes such as ethidium bromide and propidium iodide [10, 37–41]. The stock solution of acridine orange (100 mg/ml) and ethidium bromide/propidium iodide (100 mg/ml) is usually made in phosphate buffered saline (PBS) or distilled water and stored in a foil-wrapped bottle at 4°C. This is because acridine orange is light-sensitive and can degrade upon exposure to light.

Co-staining of cells with acridine orange and ethidium bromide or propidium iodide provides information on nuclear morphology (perinuclear chromatin condensation, nuclear collapse, and eventual fragmentation) [10, 37, 40, 41]. Ethidium bromide and propidium iodide can bind with core histones of DNA nucleosome structure but lack the metachromatic property of acridine orange and only stain dead cells when combined with acridine orange [10, 37, 40, 41]. This enables earlier identification of damaged cells by suppressing the DNA-specific green fluorescence induced by acridine orange. Ethidium bromide and propidium iodide are impermeable to intact cell membranes and intercalate the DNA and emits red or orange fluorescence when cells lose their membrane integrity [10, 12, 37–40]. In the presence of two dyes, the cells stain red or orange because the molar concentration of propidium iodide/ethidium bromide is more than 100× greater than that of acridine orange thus propidium iodide/acridine orange has greater affinity and specificity for nucleic acids than acridine orange. In addition, the active accumulation of acridine orange in the cells stops upon death, which reduces the concentration of acridine orange in the cells thereby making the red fluorescence of ethidium bromide/propidium iodide is more apparent [12, 38, 40].

Therefore, dual staining of cells with acridine orange and propidium iodide/ethidium bromide aids the detection of four main types of cells: (a) Viable cells with uniform green nuclei with organised chromatin structure (b) Necrotic cells with uniform orange red or red nuclei with organised structure due to loss of membrane integrity (c) Early apoptotic cells with irregular structured green nucleic but chromatin is condensed as apoptotic bodies or green patches/fragments (d) late apoptotic cells, with orange or red nuclei with extremely condensed or fragmented chromatin [10, 11, 38, 40–42]. Thus, since the cellular detail and nuclear outline are distinct, co-staining cells with acridine orange and ethidium bromide/propidium iodide offers a rapid, stable, sensitive, and easy-to-perform way to simultaneously visualise and identify all the possible nuclear stages with increased accuracy and ease of interpretation. In addition, light microscopy of cells co-stained with these dyes also provide visualisation of a complete morphological profile of an apoptotic cell [12, 37, 41–43]. Therefore, this assay also permits the staining and scoring of multiple specimens in batches thus it is cost-effective and labour saving. It also works well at room temperature and not subject to interference by extracellular enzymes.

However, fluorescence microscope is required to perform this dual staining thus expertise in this field is paramount. In addition, propidium iodide and ethidium bromide are carcinogenic and can cause debilitating effects to the DNA [11, 12, 38]. Interestingly, acridine orange is also carcinogenic and has been used as an anti-tumour agent in targeting different cancer cells via photosensitization as it selectively binds to malignant cells compared to normal cells [40–42]. Accumulation of acridine orange in the lysosome at low pH is also crucial for photosensitization thus resulting in the release of oxygen radicals, especially in malignant cells.

2.4 Propidium iodide

Propidium iodide is an analogue of ethidium bromide. It is an exclusion dye that binds to short non-specific double stranded RNA or DNA at excitation wavelength of

525 nm and an emission wavelength of 491–495 nm to produce red or orange fluorescence [44]. Propidium iodide is impermeable to the plasma membrane in living cells however it easily permeates compromised plasma membrane [45–47]. Once inside the cell, the dye binds to DNA, which leads to a 30-fold increase in fluorescence shifting the excitation maximum by 30–40 nm up to 525 nm to the red and emission maximum by 15 nm to the blue [44–46]. Thus, propidium iodide provides an objective number of dead or necrotic cells in a cytotoxicity assay when observed with a fluorescent microscope or flow cytometer (**Figure 4**). Therefore, assessing cytotoxicity with propidium iodide is easy, cost-effective and aids in measuring a large set of samples using automated flow cytometer.

In addition to acridine orange, propidium iodide is often co-stained with annexin V [12, 38, 41]. Annexin V binds to phosphatidylserine (PS) exposed on the outer membrane of necrotic cell surface. PS is usually situated in the inner membrane leaflet in viable cells and only translocate to the outer membrane leaflet in these cells upon induction of apoptosis [12, 48]. Co-staining propidium iodide with annexin V offers the ability to simultaneously detect and measure viable and non-viable cells. The cells are grouped into four groups: (a) viable cells (annexin V-negative/propidium iodide-negative), (b) necrotic cells (annexin V-negative/propidium iodide-positive), (c) early apoptotic (annexin V-positive/propidium iodide-negative), (d) late apoptotic (annexin V-positive/propidium iodide-negative) [3, 10, 12, 49]. Therefore, propidium iodide co-staining with annexin V or acridine orange provides reliable, reproducible results and distinguishes subpopulation of apoptotic cells with accuracy.

However, in some cases, late apoptosis can be characterised by some loss of membrane integrity therefore flow cytometric analysis of cytotoxicity using propidium iodide and annexin V cannot differentiate between late apoptosis and necrosis, which can both be annexin V-positive and propidium iodide-positive. In addition, apoptotic cells stained with propidium iodide present different hypodiploid peaks and sizes on the red fluorescent channels, which could be representative of the sub G1 (debris) in the sample [46, 49]. Therefore, despite propidium iodide being a universal cell death indicator, results can be skewed by nuclear material in solution although this

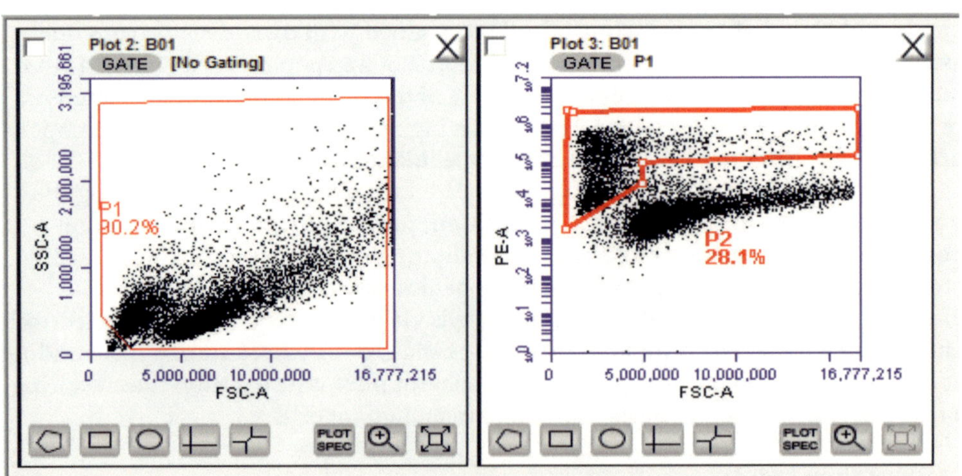

Figure 4.
Cells treated with 20 μg/mL of PI/FACS buffer for 10 min on Ice and analysed on the flow cytometer. Exclude debris using (P1 left panel) then isolate non-viable cells on the PE-A vs. FSC-A using P2 (right panel).

can be avoided by excluding the debris by size in the forward and side scatter plots or using RNAse A to digest RNA. Propidium binding to the DNA can be affected by the chromosome structure and this can be exploited to study the effects of drugs and xenobiotics on cell cycle [46, 49]. This is because propidium shares similar properties to some chemotherapy agents such as mitoxantrone, which can mask its reading and lead to underestimation of toxic effects [50, 51].

3. Conclusion

Cytotoxicity assessment is an important assessment in pharmaceutical and environmental industry to produce novel drugs and identify potentially harmful substances. In recent years, experimental methods to evaluate cytotoxicity have improved due to the progress of modern biology. However, there is no uniform cytotoxicity assay and the most popular methods currently used still entail tracking changes in cell morphology, cell proliferation and differentiation by labelling cells with colorimetric or fluorescent dyes that target compromised membranes of dead cells or specific organelles in live cells. Trypan blue dye is the most common dye used in cytotoxicity assays however due to its lack of specificity and sensitivity, it should be made mandatory to counterstain cells with dyes like neutral red, acridine orange and propidium iodide to provide a robust result. This would help eliminate potential drastic patient outcomes that may result from pushing a toxic compound into the market.

Author details

Arinzechukwu Ude[1]*, Kaiyven Afi-Leslie[2], Kelechi Okeke[3] and Emmanuel Ogbodo[4]

1 University of Brighton, Brighton, UK

2 University of Exeter, Exeter, UK

3 London South Bank University, London, UK

4 Manchester Metropolitan University, Manchester, UK

*Address all correspondence to: arinze_ude@yahoo.co.uk

References

[1] Bacanli M, Anlar HG, Başaran AA, Başaran N. Fenoliklerin sitotoksisite profillerinin değerlendirilmesi: Farklı hücrelerde farklı zaman aralıklarında nötral kırmızı ve MTT Yöntemlerinin karşılaştırılması. Turkish Journal of Pharmaceutical Sciences. 2017;**14**(2):95-107. DOI: 10.4274/tjps.07078

[2] Kamiloglu S, Sari G, Ozdal T, Capanoglu E. Guidelines for cell viability assays. Food Frontiers. 2020;**1**(3):332-329. DOI: 10.1002/fft2.44

[3] Darzynkiewicz Z, Juan G, Li X, Gorczyca W, Murakami T, Traganos F. Cytometry in cell necrobiology: Analysis of apoptosis and accidental cell death (necrosis). Cytometry. 1997;**27**(1):1-20

[4] Mullick Chowdhury S, Lalwani G, Zhang K, Yang JY, Neville K, Sitharaman B. Cell specific cytotoxicity and uptake of graphene nanoribbons. Biomaterials. 2013;**34**(1):283-293. DOI: 10.1016/j.biomaterials.2012.09.057

[5] Abhishek K. Jain, Divya Singh, Kavita Dubey, Renuka Maurya, Sandeep Mittal, & Alok K. Pandey. Chapter 3—Models and methods for in vitro toxicity. In Vitro Toxicology. 2018;**1**:45-65. ISBN 9780128046678. DOI: 10.1016/B978-0-12-804667-8.00003-1

[6] Aslantürk ÖS. In vitro cytotoxicity and cell viability assays: Principles, advantages, and disadvantages. Genotoxicity-A Predictable Risk to Our Actual World. 2018;**2**:64-80

[7] Maier K, Schmitt-Landgraf R, Siegemund B. Development of an in-vitro test system with human skin cells for evaluation of phototoxicity. Toxicology in Vitro. 1991;**5**(5)

[8] Fotakis G, Timbrell JA. In vitro cytotoxicity assays: Comparison of LDH, neutral red, MTT and protein assay in hepatoma cell lines following exposure to cadmium chloride. Toxicology Letters. 2006;**160**(2):171-177. DOI: 10.1016/j.toxlet.2005.07.001

[9] Avelar-Freitas BA, Almeida VG, Pinto MC, Mourão FA, Massensini AR, Martins-Filho OA, et al. Trypan blue exclusion assay by flow cytometry. Brazilian Journal of Medical and Biological Research = Revista brasileira de pesquisas medicas e biologicas. 2014;**47**(4):307-315. DOI: 10.1590/1414-431X20143437

[10] Mascotti K, McCullough J, Burger SR. HPC viability measurement: trypan blue versus acridine orange and propidium iodide. Transfusion. 2000;**40**(6):693-696. DOI:10.1046/j.1537-2995.2000.40060693.x

[11] Leles SM, Factori R, Rocha CLMSC. In situ analysis of apoptosis in *Aspergillus nidulans* with ethidium bromide and acridine orange. Genetics and Molecular Research. 2013;**12**(3):2895-2901. DOI: 10.4238/2013.August.12.5

[12] Baskić D, Popović S, Ristić P, Arsenijević NN. Analysis of cycloheximide-induced apoptosis in human leukocytes: Fluorescence microscopy using annexin V/propidium iodide versus acridine orange/ethidium bromide. Cell Biology International. 2006;**30**(11):924-932. DOI: 10.1016/j.cellbi.2006.06.016

[13] Tran SL, Puhar A, Ngo-Camus M, Ramarao N. Trypan blue dye enters viable cells incubated with the pore-forming toxin HlyII of Bacillus cereus. PLoS One. 2011;**6**(9):e22876. DOI: 10.1371/journal.pone.0022876

[14] Hong C, Lee J, Zheng H, Hong S-S. Porous silicon nanoparticles for cancer photothermotherapy. Nanoscale Research Letters. 2011;**6**:321. DOI: 10.1186/1556-276X-6-321

[15] Chan LL, Rice WL, Qiu J. Observation and quantification of the morphological effect of trypan blue rupturing dead or dying cells. PLoS One. 2020;**15**(1):e0227950. DOI: 10.1371/journal.pone.0227950

[16] Strober W. Trypan blue exclusion test of cell viability. Current Protocols in Immunology. 2015;**111**:A3.B.1-A3.B.3. DOI: 10.1002/0471142735.ima03bs111

[17] Ford RJ, Becker FF. The characterization of trypan blue-induced tumors in Wistar rats. The American Journal of Pathology. 1982;**106**(3):326-331

[18] Kwok AK, Yeung CK, Lai TY, Chan KP, Pang CP. Effects of trypan blue on cell viability and gene expression in human retinal pigment epithelial cells. The British Journal of Ophthalmology. 2004;**88**(12):1590-1594. DOI: 10.1136/bjo.2004.044537

[19] Karimian A, Ahmadi Y, Yousefi B. Multiple functions of p21 in cell cycle, apoptosis and transcriptional regulation after DNA damage. DNA Repair (Amst). 2016;**42**:63-71. DOI: 10.1016/j.dnarep.2016.04.008. Epub 2016 Apr 22

[20] Tonnessen-Murray CA, Lozano G, Jackson JG. The regulation of cellular functions by the p53 Protein: Cellular senescence. Cold Spring Harbor Perspectives in Medicine. 2017;**7**(2):a026112. DOI: 10.1101/cshperspect.a026112

[21] Ozaki T, Nakagawara A. Role of p53 in cell death and human cancers. Cancers. 2011;**3**(1):994-1013. DOI: 10.3390/cancers3010994

[22] Borenfreund E, Puerner JA. Toxicity determined in vitro by morphological alterations and neutral red absorption. Toxicology Letter. 1985;**24**(2-3):119-124. DOI: 10.1016/0378-4274(85)90046-3

[23] Babich H, Borenfreund E. Cytotoxicity of T-2 toxin and its metabolites determined with the neutral red cell viability assay. In Applied and Environmental Microbiology. 1991;**57**(7)

[24] Repetto G, del Peso A, Zurita JL. Neutral red uptake assay for the estimation of cell viability/cytotoxicity. Nature Protocols. 2008;**3**(7):1125-1131. DOI: 10.1038/nprot.2008.75

[25] Elliott WM, Auersperg N. Comparison of the neutral red and methylene blue assays to study cell growth in culture. Biotechnic & Histochemistry;**68**(1):29-35. DOI: 10.3109/10520299309105573

[26] Langbeen A, Jorssen EP, Granata N, Fransen E, Leroy JL, Bols PE. Effects of neutral red assisted viability assessment on the cryotolerance of isolated bovine preantral follicles. Journal of Assisted Reproduction and Genetics. 2014;**31**(12):1727-1736. DOI: 10.1007/s10815-014-0340-y

[27] Homolya L, Orbán TI, Csanády L, Sarkadi B. Mitoxantrone is expelled by the ABCG2 multidrug transporter directly from the plasma membrane. Biochimica et Biophysica Acta. 2011;**1808**(1):154-163

[28] Gray DW, Millard PR, McShane P, Morris PJ. The use of the dye neutral red as a specific, non-toxic, intra-vital stain of islets of Langerhans. British Journal of Experimental Pathology. 1983;**64**(5):553-558

[29] Zhang S, Lipsky MM, Trump BF, Hsu I. Neutral red (NR) assay for

cell viability and xenobiotic-induced cytotoxicity in primary cultures of human and rat hepatocytes. Cell Biology and Toxicology. 1990;**6**(2):219-234

[30] Valdivieso-Garcia A, Clarke RC, Rahn K, Durette A, Macleod DL, Gyles CL. Neutral red assay for measurement of quantitative vero cell cytotoxicity. Applied and Environmental Microbiology. 1993;**59**(6):1981-1983. DOI: 10.1128/aem.59.6.1981-1983.1993

[31] El-Garawani IM, Hassab El-Nabi SE. Increased sensitivity of apoptosis detection using direct DNA staining method and integration of acridine orange as an alternative safer fluorescent dye in agarose gel electrophoresis and micronucleus test. Canadian Journal of Pure & Applied Science. 2016;**10**(2):3865-3871. Àvailable from: https://www.researchgate.net/publication/320016082

[32] Iessi E, Logozzi M, Lugini L, Azzarito T, Federici C, Spugnini EP, et al. Acridine orange/exosomes increase the delivery and the effectiveness of acridine orange in human melanoma cells: A new prototype for theranostics of tumors. Journal of Enzyme Inhibition and Medicinal Chemistry. 2017;**32**(1):648-657. DOI: 10.1080/14756366.2017.1292263

[33] Lin YC, Lin JF, Tsai TF, Chen HE, Chou KY, Yang SC, et al. Acridine orange exhibits photodamage in human bladder cancer cells under blue light exposure. Scientific Reports. 2017;**7**(1):14103. DOI: 10.1038/s41598-017-13904-0

[34] Okeke K. Chemotherapy induces a genotoxic bystander effect in cell lines from the human bone marrow, evaluation of the role of the redox microenvironment [thesis]. Bristol, UK: University of the West of England; 2019. Available from: https://uwe-repository.worktribe.com/output/2812709

[35] Liu CH, Sahoo SL, Tsao MH. Acridine orange coated magnetic nanoparticles for nucleus labeling and DNA adsorption. Colloids and Surfaces B: Biointerfaces. 2014;**115**:150-156. DOI: 10.1016/j.colsurfb.2013.11.003

[36] Traganos F, Darzynkiewicz Z. Lysosomal proton pump activity: Supravital cell staining with acridine orange differentiates leukocyte subpopulations. Methods in Cell Biology. Chapter. 1994;**12**:185-194

[37] Osman H, Elsahy D, Saadatzadeh MR, Pollok KE, Yocom S, Hattab EM, et al. Acridine orange as a novel photosensitizer for photodynamic therapy in glioblastoma. World Neurosurgery. 2018;**114**:e1310-e1315. DOI: 10.1016/j.wneu.2018.03.207

[38] Hussain H, Santhana RL, Ahmad S, Abd Razak MF, Wan Mohamud WN, Bakar J, et al. (Reviewing Editor). Determination of cell viability using acridine orange/propidium iodide dual-spectrofluorometry assay. Cogent Food & Agriculture. 2019;**5**:1. DOI: 10.1080/23311932.2019.1582398

[39] Nafisi S, Saboury AA, Keramat N, Neault JF, Tajmir-Riahi HA. Stability and structural features of DNA intercalation with ethidium bromide, acridine orange and methylene blue. Journal of Molecular Structure. 2007;**827**(1-3):35-43. DOI: 10.1016/j.molstruc.2006.05.004

[40] Criswell KA, Krishna G, Zielinski D, Urda GA, Theiss JC, Juneau P, et al. Use of acridine orange in: Flow cytometric evaluation of erythropoietic cytotoxicity. 1998;**414**(1-3):49-61. DOI: 10.1016/s1383-5718(98)00041-2

[41] Valentini A, Pucci D, Crispini A, Federici G, Bernardini S. Acridine orange based platinum(II) complexes inducing cytotoxicity and cell cycle perturbation

in spite of GSTP1 up-regulation. Chemico-Biological Interactions. 2006;**161**(3):241-250. DOI: 10.1016/j.cbi.2006.03.011

[42] Bank HL. Rapid assessment of islet viability with acridine orange and propidium iodide. In Vitro Cellular & Developmental Biology. 1988;**24**:266-273. DOI: 10.1007/BF02628826

[43] Wallen CA, Higashikubo R, Dethlefsen LA, Wallen CA. Comparison of two flow cytometric assays for cellular RNA-acridine orange and propidium iodide. Cytometry. 1980;**3**(3):155-160

[44] Information NCFB. PubChem Compound Summary for CID 104981. Propidium iodide: PubChem; 2022. Available from: https://pubchem.ncbi.nlm.nih.gov/compound/Propidium-iodide

[45] Crowley LC, Scott AP, Marfell BJ, Boughaba JA, Chojnowski G, Waterhouse NJ. Measuring cell death by propidium iodide uptake and flow cytometry. 2016;**2016**(7). DOI: 10.1101/pdb.prot087163

[46] Riccardi C, Nicoletti I. Analysis of apoptosis by propidium iodide staining and flow cytometry. Nature Protocols. 2006;**1**(3):1458-1461

[47] Deitch AD, Law H, deVere WR. A stable propidium iodide staining procedure for flow cytometry. The Journal of Histochemistry and Cytochemistry. 1982;**30**(9):967-972

[48] Pollack A, Ciancio G. Chapter 3 cell cycle phase-specijric analysis of cell viability using hoechst 33342 and propidium iodide after ethanol preservation. Methods in Cell Biology. 1990;**33**:19-24

[49] Silva MT. Secondary necrosis: The natural outcome of the complete apoptotic program. FEBS Letters. 2010;**584**(22):4491-4499

[50] Hussein RA, Mohsin AJ. Trypan blue exclusion verifies in vitro cytotoxicity of new cis-platinum (II) complex in human cells. Baghdad Science. 2019;**16**(3):555-558. DOI: 10.21123/bsj.2019.16.3.0555

[51] Nowak E, Kammerer S, Küpper JH. ATP-based cell viability assay is superior to trypan blue exclusion and XTT assay in measuring cytotoxicity of anticancer drugs Taxol and Imatinib, and proteasome inhibitor MG-132 on human hepatoma cell line HepG2. Clinical Hemorheology and Microcirculation. 2018;**69**(1-2):327-336. DOI: 10.3233/CH-189120

Chapter 9

In Vitro Cytotoxic and Apoptosis Induction Potential of Two Plant Extracts on HeLa Cells

Özlem Dağdeviren Özsöylemez and Gül Özcan Arican

Abstract

Natural products are commonly used for developing anticancer drugs that are beneficial for various cancer types. The aim of this study is to apply *Colchicum umbrosum* Steven and *Colchicum baytopiorum* CD Brickell (one of the endemic species in Turkey) extracts on HeLa cell lines and determine changes in cytotoxicity and viability. For this aim, kinetic parameters such as proliferation rate have been determined by MTT assay, and apoptotic index (AI) has been researched by fluorescence microscopies using DAPI staining. Also, some apoptosis-related genes have been examined by the RT-PCR method. Five different concentrations of both extracts from the two Colchicum species have cytotoxic effects and it has been understood that HeLa cells were more sensitive to the most effective concentration of the *C. baytopiorum* extract, which is 0.1 mg/ml, and it showed antitumor effects by causing apoptosis for 48 h. The cytotoxic activity and apoptotic effects of *Colchicum umbrosum* Steven and *Colchicum baytopiorum* (Colchicaceae/Liliaceae) have been studied for the first time on HeLa cell lines. We suggested that the medicines derived from natural products seem to be a new promising treatment for cancer.

Keywords: apoptosis, *Bcl-2* gene family, cancer, *Colchicum baytopiorum*, *Colchicum umbrosum*, HeLa cells

1. Introduction

Cancer is a disease characterized by constant clonal proliferation of the somatic cells due to which the homeostatic feedback mechanism gets out of control through genetic and epigenetic alterations and kills by invading normal tissues [1–4]. Natural products extracted from plants, microbes, and other marine organisms have secondary or non-essential metabolism. Natural products have gained importance in novel anticancer agents' discovery and investigation of more effective drugs [5, 6]. Colchicine, which is obtained from plants of the genus Colchicum belonging to Liliaceae (Colchicaceae) family, is one of the natural product. Colchicine is an anti-neoplastic drug used in cancer therapy as a mitotic inhibitor [7].

Apoptosis plays a critical role in the elimination of damaged or undesirable cells and the development and maintenance of homeostasis. Disturbances in these processes cause various diseases ranging from cancer and autoimmune disorders to ischemic injuries and neurodegenerative diseases [8–14].

Apoptosis is encouraged by cell signaling, lack of growth factors, the release of granzyme, specific receptor-ligand interactions, and various stress factors (chemotherapy, radiation, etc.). Then, the cell death process is initiated. Some of the original proteins play a role in this process. As a consequence thereof, signals are transferred to the execution mechanism with the help of adaptor proteins and are regulated the mitochondrial permeability by bcl-2 family proteins. An increase in permeability of the outer mitochondrial membrane is caused by the release of cytochrome c. Cytochrome c is connected to apoptosis protease activation factor (Apaf-1) in the cytosol. Apaf-1 binds to the pro-caspase-9 molecule by oligomerization in the presence of adenosine triphosphate (ATP) and constitutes apoptosome. The activated caspase-9 plays a role in the activation of caspase-3 and caspase-7. Nevertheless, the *bcl-2* in the mitochondrial membrane prevents activation of caspases by preventing the increase of mitochondrial permeability and making stable Apaf-1 and so on proteins. Thus, apoptosis does not occur. *bcl-xL,* which is a member of the bcl-2 family, and *bcl-2* play the anti-apoptotic role. *bax*, *bad,* and *bak* are triggered by apoptosis [15–22].

In recent years, investigating the molecular mechanisms of apoptosis by using agents has become extremely important in cancer treatment [23–26]. Our study is aimed to investigate the cytotoxic effects of different concentrations of two extracts that belong to the genus *Colchicum*, one of them is endemic in Turkey (*Colchicum umbrosum* Steven (S.) and *Colchicum baytopiorum* CD Brickell (CD) is an endemic species in Turkey [27, 28], on HeLa cells and also show the molecules that play a role in the regulation of molecular mechanisms of the resultant apoptosis.

2. Materials and methods

2.1 Cell line and culture conditions

Human cervical cancer cell line (HeLa) cells were obtained from American Type Culture Collection, HeLa (CCL-2). The cells were cultured in a 25-cm^2 flask following Minimum Essential Medium (Sigma, MEM) containing 10% FBS (Gibco Lab.), penicillin (100 unit/ml), and streptomycin (50 mg/ml). The cells were incubated at 37°C in a humidified %5 CO2 incubator [29].

2.2 Plant material and extraction

Corms of *Colchicum baytopiorum* (ISTE: 81438, Antalya-Termessos, July 5, 2005) and *Colchicum umbrosum* (ISTE: 85333, Bolu-Abant, June 25, 2008) were dried. All dried plant materials were extracted with methanol. Methanol extracts were evaporated in a rotavapor. Plant extracts were prepared in five different concentrations (0.01 mg/ml, 0.05 mg/ml, 0.1 mg/ml, 0.5 mg/ml, 1 mg/ml) in MEM, supplemented with %10 FBS. Prepared concentrations were treated to HeLa cells in time period of 24, 48, and 72 h [30, 31].

2.3 Assays for in vitro cytotoxicity

Cytotoxic effects of plant extracts were measured by using the MTT (3-[4, 5-dimethylthiazol-2-yl]-2, 5-diphenyltetrazolium bromide; Sigma) assay. Cells were seeded onto a 96 well plate (30,000 cells/well), allowed to attach overnight, and then exposed to five different concentrations of plant extracts for 24, 48, and 72 h. MTT solution dissolved in PBS (40 μl/well) was added. Cells were incubated at 37°C for 4 h. Formazan crystals were dissolved in 160 μl DMSO (dimethyl sulfoxide) at 37°C for 1 h. The plate was read on ELISA reader at 570 nm reference of 690 nm (μQuant, Bio-tek) [32].

2.4 Apoptotic Index

Apoptotic cells showed characteristic fragmented nuclei, whereas survivors showed an intact nuclear morphology. These cells were fixed with cold methanol. The cells were counterstained with 4′, 6-diamidino-2-phenylindole (DAPI) nuclear staining mounting solution for the determination of the apoptotic index (AI) [33]. Then, the slides were washed in phosphate-buffered saline (PBS). The AI was calculated by scoring the percentage of apoptotic cells in control and the extract treatment (0.1 mg/ml) group of 1000 cells and at least 30 areas/each slide under the fluorescence microscope [34].

2.5 Phase-contrast microscopies

After the treatment with C. baytopiorum extract (0.1mg/ml), the morphological changes in HeLa cells were examined after 24, 48, and 72 h under phase contrast microscopies (x200).

2.6 Light microscopies

The morphological changes that occurred in HeLa cells were examined with Giemsa stain and Feulgen method under a light microscope (×1000) after 24, 48, and 72 h [35].

2.7 The isolation of total RNA

Total RNA was isolated from cells in the control and the extract treatment (0.1 mg/ml) group, for 48 h by using a total RNA isolation kit (Invitrogen PureLink Micro-to-Midi Total RNA Isolation Kit, Cat. No: 12183-180). For the determination of the amount of isolated total RNA (diluted 1: 200), spectrophotometer (Cintra 20, GBC) was used at 260 nm [36].

2.8 Reverse transcriptase (RT)-PCR

A total of 21 μl RNase-free water was put in each PCR tube. In this mix, 25 μl of the 2x Reaction mix (0.4 mM MSO_4 of each deoxynucleoside triphosphate) and 2 l of RT/Platinum ™ Taq Max were added. Then, 1l of *bcl-x, bik, mcl-1, bfl-1*, and *β-actin* (Takara BCL-2 familly, Cat no.6623) and 1l of total RNA of each experimental group was added to the tube for one-step RT-PCR (The Super Script™ One-Step RT-PCR Kit Invitrogen, Cat no. 10928-042).

Reverse transcription and amplification were performed in TC412 Techne. The RT-PCR profile used was as follows: a reverse transcription stage at 55°C for 30 min, followed by an initial denaturation stage at 94°C for 2 min. This was then followed by 40 amplification cycles at 94°C for 15 s, at 60°C for 30 s, and at 68°C for 1 min, and a final extension step at 68°C for 5 min. The resulting PCR products were visualized by electrophoresis through 1.8% agarose gel containing ethidium bromide and then visualized under UV light [37].

2.9 Statistical analysis

Apoptotic index and cytotoxicity were analyzed by one-way ANOVA, followed by Dunnett's test for separate comparisons with the control group and the T-test for separate comparisons with each of the groups. Differences were considered significant at $p < 0.05$ (GraphPad Prism version 4.00, GraphPad Software, San Diego California USA).

3. Results

In our study, the cytotoxicity of all extracts was determined by MTT assay. Accordingly Cytotoxicity calculation method, it has been determined that the cytotoxicity and % vitality (calculated by assuming the control group as 100%) in HeLa cells, treated with the extracts from two species of Colchicum plant, are being increased in concentration compared to the control group ($p < 0.05$). Different concentrations of both extracts from the two Colchicum species have cytotoxic effects ($p < 0.05$) and HeLa cells are more sensitive to the species of *C. baytopiorum*. The meaningful cytotoxic effect of the 0.001 mg/ml concentration extracted from *C. umbrosum* has been occurred in only 72 h, compared to the control group ($p < 0.01$) (**Figures 1** and **2**).

Determined viability% values of HeLa cells treated with *C. baytopiorum* are respectively 70% for B1, 74% for B2, 58% for B3, 75% for B4, 77% for B5 in 24 h; 28% for B1, 23% for B2, 20% for B3, 26% for B4, 20% for B5 in 48 h; 15% for B1, 19% for B2, 27% for B3, 17% for B4, 14% for B5 in 72 h. Vitality % values of HeLa cells treated with *C. umbrosum* are respectively 101% for U1, 70% for U2, 82% for U3, 52% for U4, 77% for U5 in 24 h; 99% for U1, 28% for U2, 19% for U3, 33% for U4, 21% for U5 in 48 h; 63% for U1, 20% for U2, 21% for U3, 20% for U4, 23% for U5 in 72 h.

Apoptotic index is determined by fluorescence microscopy using DAPI staining (**Figure 3**). DAPI staining was also conducted to show the apoptosis of HeLa cells. The AI was calculated as follows:

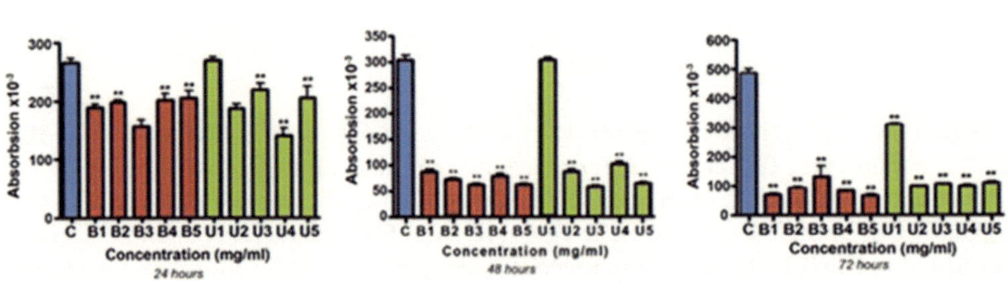

Figure 1.
*MTT results of HeLa cells treated with C. baytopiorum and C. umbrosum (B1 and U1: 0.001, B2 and U2: 0.05, B3 and U3: 0.1, B4 and U4: 0.5, B5 and U5: 1 mg/ml) for 24, 48, and 72 h. *$p < 0.05$, **$p < 0.01$ (in comparison to control).*

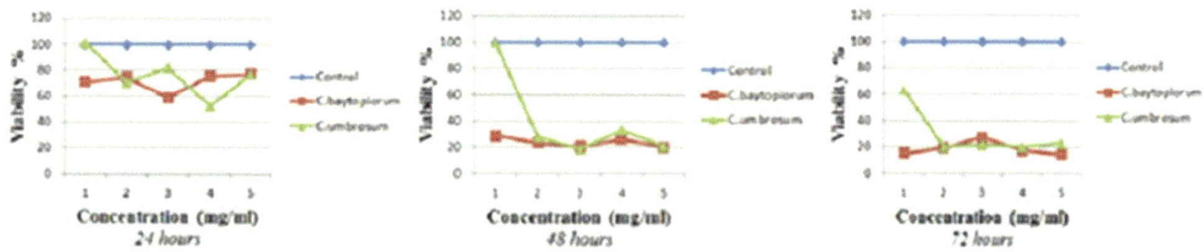

Figure 2.
Viability % values of HeLa cells treated with C. baytopiorum and C. umbrosum (1: 0.001, 2: 0.05, 3: 0.1, 4: 0.5, 5: 1 mg/ml) for 24, 48, and 72 h.

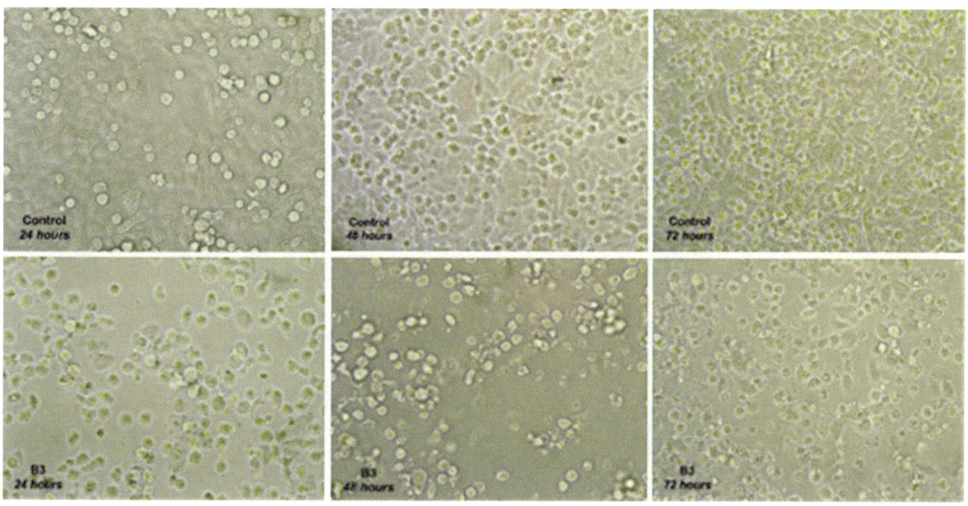

Figure 3.
Phase-contrast microscopy of control and B3 (0.1 mg/ml) concentration of the C. baytopiorum extract treatments groups on HeLa cells for 24, 48, and 72 h (×200).

AI = (number of apoptotic cells/total number counted) × 100%.

Apoptotic index values of HeLa cells treated with *C. baytopiorum* and *C. umbrosum* for 24, 48, and 72 h are shown in **Figure 4**. AI values were determined as 2.3% for control group, 18.3% for B1, 16.4% for B2, 12.37% for B3, 25.12% for B4, 18.02% for B5 in the application of the *C. baytopiorum* extract; 8.66% for U1, 6.78% for U2, 14.49% for U3, 15.88% for U4, 18.91% for U5 in the application of the *C. umbrosum* extract in 24 h.

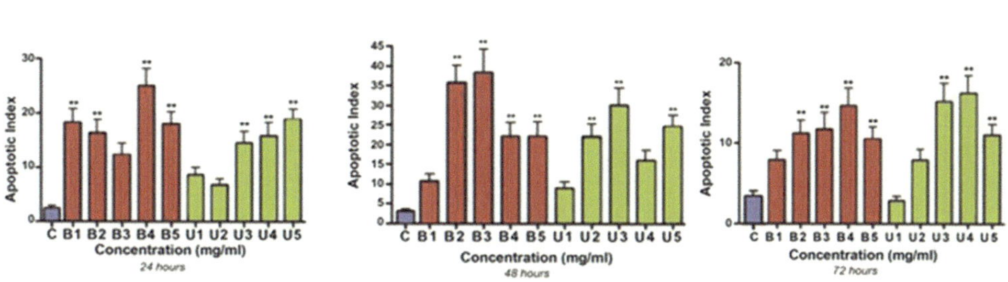

Figure 4.
*Determined apoptotic index (AI) values of HeLa cells treated with C. baytopiorum and C. umbrosum (B1 and U1: 0.001, B2 and U2: 0.05, B3 and U3: 0.1, B4 and U4: 0.5, B5 and U5: 1 mg/ml) for 24, 48, and 72 h. **p < 0.01 (in comparison to control).*

In our experiments, in 24 h, it was determined significant cytotoxic effect according to the control ($p < 0.01$) in the experimental groups B1, B2, B4 and B5 applied the extract of *C. baytopiorum* and the experimental groups U3, U4 and U5 applied *C. umbrosum* extract.

AI values were determined as 2.84% for control group, 10.67% for B1, 35.91% for B2, 38.38% for B3, 22.04% for B4, 22.03% for B5 in the application of the C. *baytopiorum* extract; 8.9% for U1, 22.05% for U2, 30.19% for U3, 16.01% for U4, 23.58% for U5 in the application of the *C. umbrosum* extract in 48 h.

In our experiments in B2, B3, B4, and B5 experimental groups applied *C. baytopiorum* extract and U2, U3, U5 experimental groups applied *C. umbrosum*

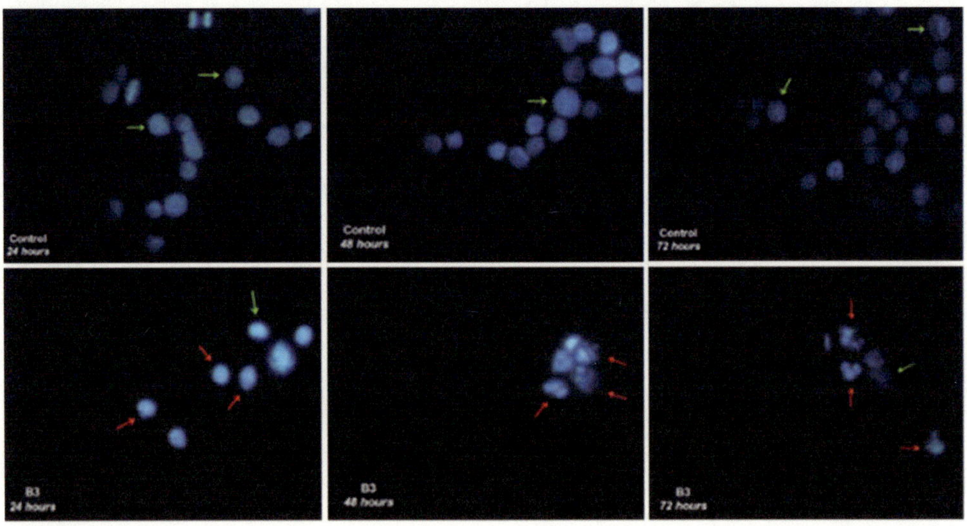

Figure 5.
Fluorescence microscopy of control and B3 (0.1 mg/ml) concentration of the C. baytopiorum extract treatments groups on HeLa cells for 24, 48, and 72 h (×1000; DAPI) (→: non-apoptotic cell, →: apoptotic cell).

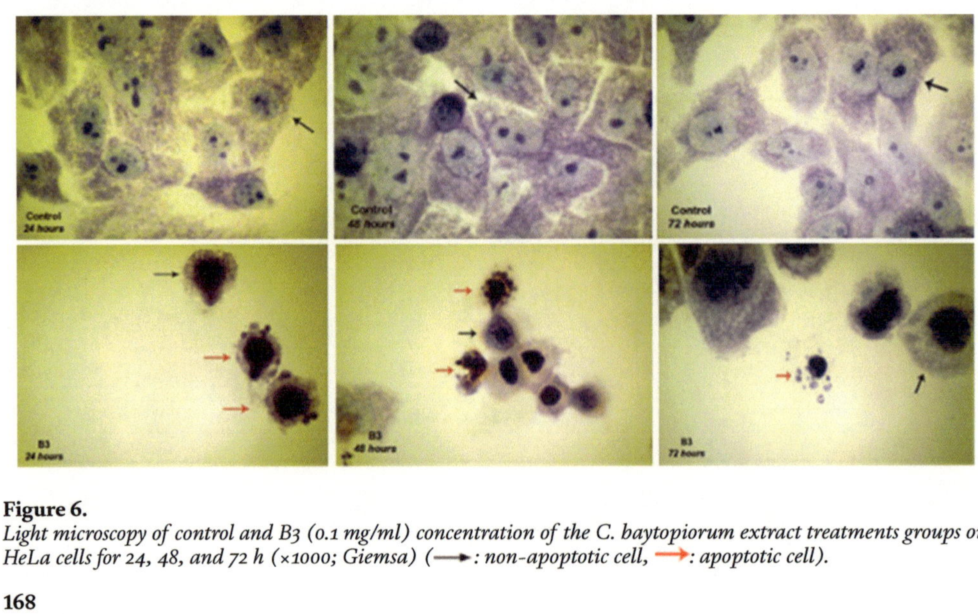

Figure 6.
Light microscopy of control and B3 (0.1 mg/ml) concentration of the C. baytopiorum extract treatments groups on HeLa cells for 24, 48, and 72 h (×1000; Giemsa) (→: non-apoptotic cell, →: apoptotic cell).

DOI: http://dx.doi.org/10.5772/intechopen.105696

extract were determined to have a significant cytotoxic effect according to control ($p < 0.01$) in 48 h.

AI values were determined as 2.92% for control group, 7.89% for B1, 11.19% for B2, 11.76% for B3, 14.72% for B4, 10.5% for B5 in the application of the *C. baytopiorum*

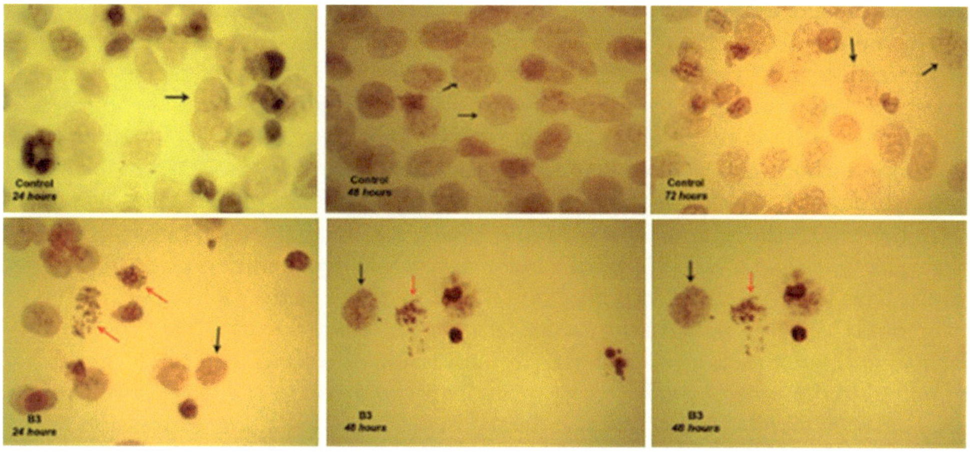

Figure 7.
Light microscopy of control and B3 (0.1 mg/ml) concentration of the C. baytopiorum extract treatments groups on HeLa cells for 24, 48, and 72 h (×1000; Feulgen) (→: non-apoptotic cell, →: apoptotic cell).

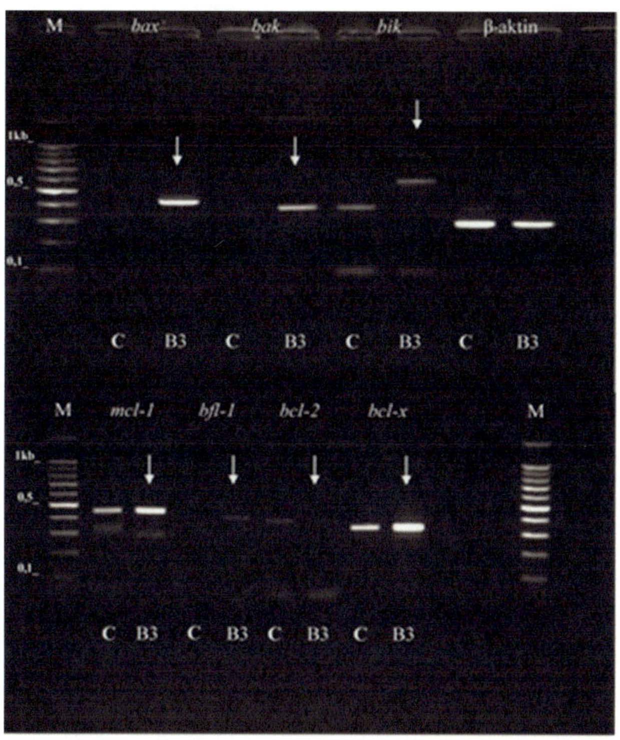

Figure 8.
RT-PCR analysis of bcl-2 members of control and B3 (0.1 mg/ml) concentration of the C. baytopiorum extract treatments groups on HeLa cells for 48 h (M: 1.5 kb marker, C: control, B3: 0.1 mg/ml).

extract; 2.82% for U1, 7.91% for U2, 15.22% for U3, 16.23% for U4, 11% for U5 in the application of the *C. umbrosum* extract in 72 h.

In our experiments, in 72 h, it was determined significant cytotoxic effect according to the control ($p < 0.01$) in the experimental groups B1, B2, B4 and B5 applied the extract of *C. baytopiorum* and the experimental groups U3, U4 and U5 applied *C. umbrosum* extract.

The methanol extract of *C. baytopiorum* has been treated with the most effective concentration of 0.1 mg/ml in 48 h experiment group (after the evaluation of cytotoxicity and AI value).

The morphological changes belonging to apoptosis such as blebbing of membrane and apoptotic body including cell organelles and chromatin parts [32, 33] have been shown through phase-contrast, fluorescence, and light microscopies (**Figures 3** and **5–7**).

To determine the apoptosis at the molecular level, the expression rates of the genes of the *bax*, *bak*, *bik*, *mcl-1*, *bfl-1*, *bcl2*, and *bcl-x*, which are members of the *bcl-2* gene family, have been searched through the RT-PCR method. In the 48 h experimental group of HeLa cells which were treated with the concentration of 0.1 mg/ml of *C. baytopiorum* extract, an increase in the expression of the *mcl-1* and *bcl-x* genes by compared to the control group was observed, and an increase of the expression of *bax, bak, bik* and *bfl-1* genes was also determined . In contrast, any expression was observed in the *bcl-2* gene (**Figure 8**).

4. Discussion

Some studies reported that crude extracts of natural products containing a variety of molecules that exhibit antitumoral activity are highly effective in cancer cell death [38–40].

However, some reports suggested that natural product mixtures often claim that they are more effective than purified compounds due to synergistic, additive, or antagonistic interactions [41]. Also, different species of the same genus may have different concentrations of the same compounds [42].

In our study, cytotoxicity of all extracts was determined by MTT assay. Accordingly, it has been determined that the cytotoxicity and % vitality (calculated by assuming the control group as 100%) in HeLa cells, treated with the extracts from two species of Colchicum plant, are being increased in concentration compared to the control group ($p < 0.05$). Different concentrations of both extracts from the two Colchicum species have cytotoxic effects ($p < 0.05$) and HeLa cells are more sensitive to the species of *C. baytopiorum*. The meaningful cytotoxic effect of the 0.001 mg/ml concentration extracted from *C. umbrosum* has been occurred in only 72 h, compared to the control group ($p < 0.01$) (**Figures 1** and **2**).

Determined viability % values of HeLa cells treated with *C. baytopiorum* are, respectively, 70% for B1, 74% for B2, 58% for B3, 75% for B4, 77% for B5 in 24 h; 28% for B1, 23% for B2, 20% for B3, 26% for B4, 20% for B5 in 48 h; and 15% for B1, 19% for B2, 27% for B3, 17% for B4, 14% for B5 in 72 h. Vitality % values of HeLa cells treated with *C. umbrosum* are, respectively, 101% for U1, 70% for U2, 82% for U3, 52% for U4, 77% for U5 in 24 h; 99% for U1, 28% for U2, 19% for U3, 33% for U4, 21% for U5 in 48 h; and 63% for U1, 20% for U2, 21% for U3, 20% for U4, 23% for U5 in 72 h.

Several compounds derived from natural products, such as Genistein, Puerarin, Formononetin, Cyanidin 3-O-glucoside, Epigallocatechin gallate, Quercetin, Kaempferol, Hesperidin, Naringin, Apigenin 7-glucoside, Wogonin, Eugenol,

Anthocyanins, Curcumin, Emodin, Epifriedelanol, Mitomycin C, and Piperine, induce apoptosis in cervical cancers [6, 43]. Activation of apoptosis is used as an anticancer mechanism in cancer therapy research [6].

The determination of the AI is the method that helps to predict and diagnose patients by considering the tumor, its stage, the course of the disease, its consequences, the patient's strength of resistance, and the possibilities of helping the patient [44].

AI has been determined by DAPI respectively the fluorescence microscopies (**Figure 3**). DAPI staining was also conducted to show the apoptosis of HeLa cells. AI was calculated as follows (**Table 1**):

AI = (number of apoptotic cells/Total number counted) × 100%

Table 1.
AI calculation formula.

AI values of HeLa cells treated with *C. baytopiorum* and *C. umbrosum* for 24, 48, and 72 h are shown in **Figure 4**. AI values were determined as 2.3% for control group, 18.3% for B1, 16.4% for B2, 12.37% for B3, 25.12% for B4, 18.02% for B5 in the application of the *C. baytopiorum* extract; 8.66% for U1, 6.78% for U2, 14.49% for U3, 15.88% for U4, 18.91% for U5 in the application of the *C. umbrosum* extract in 24 h.

Bungu et al. [23] applied the prepared extracts from the leaves and bulbs of *Tulbaghia violacea* on HeLa, HT29, MCF-7, and WHCO3 cells. After 24 h, a significant number of apoptotic cells were detected. If after 48 h, a further increase in the number of apoptotic cells was observed. The results of our study are consistent with this and other similar studies.

In our experiments, in 24 h, it was determined significant cytotoxic effect according to the control ($p < 0.01$) in the experimental groups B1, B2, B4 and B5 applied the extract of *C. baytopiorum* and the experimental groups U3, U4 and U5 applied *C. umbrosum* extract.

AI values were determined as 2.84% for control group, 10.67% for B1, 35.91% for B2, 38.38% for B3, 22.04% for B4, 22.03% for B5 in the application of the C. *baytopiorum* extract; 8.9% for U1, 22.05% for U2, 30.19% for U3, 16.01% for U4, 23.58% for U5 in the application of the *C. umbrosum* extract in 48 h.

In our experiments, in 48 h, it was determined significant cytotoxic effect according to the control ($p < 0.01$) in the experimental groups B1, B2, B4 and B5 applied the extract of *C. baytopiorum* and the experimental groups U3, U4 and U5 applied *C. umbrosum* extract.

AI values were determined as 2.92% for control group, 7.89% for B1, 11.19% for B2, 11.76% for B3, 14.72% for B4, and 10.5% for B5 in the application of the *C. baytopiorum* extract; 2.82% for U1, 7.91% for U2, 15.22% for U3, 16.23% for U4, and 11% for U5 in the application of the *C. umbrosum* extract in 72 h.

In our experiments in B2, B3, B4, and B5 experimental groups applied *C. baytopiorum* extract and the U3, U4, and U5 experimental groups applied *C. umbrosum* extract was determined to have a significant cytotoxic effect according to control ($p < 0.01$) in 72 h.

The methanol extract of *C. baytopiorum* has been treated with the most effective concentration of 0.1 mg/ml in 48 h experiment group (after the evaluation of cytotoxicity and AI value).

The morphological changes belonging to the apoptosis such as blebbing of membrane and apoptotic body including cell organelles and chromatin parts [26, 45, 46] have been shown through phase-contrast, fluorescence, and light microscopies (**Figures 3** and 5–7).

Most studies on apoptosis of plant extracts were confirmed by mechanisms such as caspase and *bcl-2* gene family [6]. A study showed that *Dendrobium chrysanthum* ethanol extracts induct apoptosis by upregulated Bax, and p53 and downregulated Bcl-2 in HeLa cell lines at the density of 450 μg/ml for 24 h [47]. Another study reported that 0–100 μmol/L Wogonin induces apoptosis via elevating the Bax/Bcl-2 expression ratio, activating Caspase 3 and 9 in SiHa and CaSki cells [48].

To determine the apoptosis at the molecular level, the expression rates of the genes of the *bax*, *bak*, *bik*, *mcl-1*, *bfl-1*, *bcl2*, and *bcl-x*, which are members of the *bcl-2* gene family, have been searched through the RT-PCR method. In the 48 h experiment group of the HeLa cells that were treated with the 0.1 mg/ml concentration of the *C. baytopiorum* extract, an increase has seen in expressions of *mcl-1* and *bcl-x* genes compared to the control group, and some increase has been determined in the *bax*, *bak*, *bik,* and *bfl-1* genes expressions. On the other hand, no expression was observed in *bcl-2* gene (**Figure 8**).

Acknowledgements

This work was supported by the Research Fund of The University of Istanbul, Project no: 3495. The authors of this manuscript are grateful to Prof. Dr. Nurhayat Sütlüpınar and Dr. Sevda Pırıldar for providing the facilities to prepare the plant extraction.

Conflict of interest

The authors declare no conflict of interest.

Author details

Özlem Dağdeviren Özsöylemez[1]* and Gül Özcan Arican[2]

1 Institute of Gratuate Studies in Science and Engineering, Radiobiology Programme, Istanbul University, Istanbul, Turkey

2 Faculty of Science, Department of Biology, Istanbul University, Istanbul, Turkey

*Address all correspondence to: o_dagdeviren@hotmail.com

References

[1] Evan GI, Vousden KH. Proliferation, cell cycle and apoptosis in cancer. Nature. 2001;**411**:342-347

[2] Mitchell RN, Kumar V. In: Çevikbaş U, editor. Robbins Basic Pathology. Istanbul: Nobel Medicine Press; 2003. pp. 166-210

[3] Sitorus RS, Gumay S, Valk P. In: Diederich M, editor. Natural Compounds and Their Role in Apoptotic Cell Signaling Pathways. New York: Annals of the New York Akademi of Sciences; 2009. pp. 77-86

[4] Ghiraldini FG, Filipescu D, Bernstein E. Solid tumours hijack the histone variant network. Natural Review Cancer. 2021;**21**(4):257-275

[5] Dasari S, Njiki S, Mbemi A, Yedjou CG, Tchounwou PB. Pharmacological effects of cisplatin combination with natural products in cancer chemotherapy. International Journal of Molecular Science. 2022;**23**:1532

[6] Park SH, Kim M, Lee S, Jung W, Kim B. Therapeutic potential of natural products in treatment of cervical cancer: A review. Nutrients. 2021;**13**:154

[7] Hadfield JA, Ducki S, Hirst N, et al. Tubulin and microtubules as targets for anticancer drugs. Progress in Cell Cycle Research. 2003;**5**:309-325

[8] Cohen GM. Caspases: The executioners of apoptosis. Biochemical Journal. 1997;**326**:1-16

[9] Chang HY, Yang X. Proteases for cell suicide: Functions and regulation of caspases. Microbiology and Molecular Biology Reviews. 2000;**64**(4):821-826

[10] Özören N, El-Deiry WS. In: DNA Alterations in Cancer: Genetic and Epigenetic Changes, Ehrlic M (Ed). Natick MA: Eaton Publishing; 2000. pp. 13-43

[11] Collazo C, Chacón O, Borras O. Programmed cell death in plants resembles apoptosis of animals. Biotecnologia Aplicada. 2006;**23**(1):1-10

[12] Shimazu T, Degenhardt K, Nur-E-Kamal A, et al. NBK/BIK antagonizes MCL-1 and BCL-XL and activates BAK-mediated apoptosis in response to protein synthesis inhibition. Genes and Development. 2007;**21**:929-941

[13] Sankari SL, Masthan KMK, Babu NA, et al. Apoptosis in cancer—An update. Asian Pacific Journal of Cancer Prevention. 2012;**13**(10):4873-4878

[14] Kalkan H. In: Tutar Y, editor. Regulation and Dysfunction of Apoptosis. London, United Kingdom: IntechOpen; 2021. pp. 1-25

[15] Shi Y. In: Yin XM, Dong Z, editors. Essential of Apoptosis: A Guide for Basic and Clinical Research. Totowa, New Jersey: Humana Press Inc.; 2003. pp. 47-65

[16] Yin XM, Ding WX, Zhao Y. In: Yin XM, Dong Z, editors. Essential of Apoptosis: A Guide for Basic and Clinical Research. Totowa, New Jersey: Humana Press Inc.; 2003. pp. 13-14

[17] Krakstad C, Chekenya M. Survival signalling and apoptosis resistance in glioblastomas: Opportunities for targeted therapeutics. Molecular Cancer. 2010;**9**:135

[18] Galluzzi L, Vitale I, Abrams JM, et al. Molecular definitions of cell

death subroutines: Recommendations of the nomenclature committee on cell death. Cell Death and Differentiation. 2012;**2011**:1-14

[19] Raemy E, Martinou JC. Involvement of cardiolipin in tBID-induced activation of BAX during apoptosis. Chemistry and Physics of Lipids. 2014;**179**:70-74

[20] Czabotar PE, Lessene G, Strasser A, et al. Control of apoptosis by the BCL-2 protein family: Implications for physiology and therapy. Molecular Cell Biology. 2014;**15**:49-63

[21] Sulistina DR, Ratnawati R, Arsana Wiyasa IW. Rhodamine B increases hypothalamic cell apoptosis and disrupts hormonal balance in rats. Asian Pac Journal of Reproduction. 2014;**3**(3):180-183

[22] Lalier L, Vallette F, Manon S. Bcl-2 family members and the mitochondrial import machineries: The roads to death. Biomolecules. 2022;**12**:162

[23] Bungu L, Frost CL, Brauns SC, et al. Tulbaghia violacea inhibits growth and induces apoptosis in cancer cells in vitro. African Journal of Biotechnology. 2006;**5**(20):1936-1943

[24] Karagöz A, Turgut Kara N, Çakır Ö, et al. Cytotoxic activity of crude extracts from Astragalus chrysochlorus (Leguminosae). Biotechnology and Biotechnoogy Equipment. 2007;**21**(2):220-222

[25] Song HY, Deng XH, Yuan GY, et al. Expression of bcl-2 and p 53 in induction of esophageal cancer cell apoptosis by ECRG2 in combination with cisplatin. Asian Pacific Journal of Cancer Prevention. 2014;**15**(3):1397-1401

[26] Chainumnim S, Saenkham A, Dolsophon K, Chainok K, Suksamrarn S, Tanechpongtamb W. Stem Extract from Momordica cochinchinensis Induces Apoptosis in Chemoresistant Human Prostate Cancer Cells (PC-3). Molecules. 2022;**27**:1313

[27] Brickell CD. In: Davis PH, editor. Flora of Turkey and the East Aegean Islands. Vol. 8. Edinburgh: University Press; 1984. pp. 329-351

[28] Persson K. In: Güner A, Özhatay N, Ekim T, Başer KHC, editors. Flora of Turkey and the East Aegean Islands. Vol. 11. Edinburgh: University Press; 2000. pp. 246-265

[29] Freshney RI. In: Freshney RI, editor. Culture of Animal Cells: A Manual of Basic Technique. New York: Wiley-Liss; 2000. pp. 177-194

[30] Pırıldar S, Sütlüpınar N, Atasever B, et al. Chemical constituents of the different parts of Colchicum baytopiorum and their cytotoxic activities on K562 and HL60 cell lines. Pharmaceutical Biology. 2010;**48**(1):32-39

[31] Özcan Arican FG, Çakır Ö, Arican E, et al. Effects of Geven root extract on proliferation of HeLa cells and Bcl-2 gene expressions. African Journal of Biotechnology. 2012;**11**:4296-4304

[32] Kurashige S, Akuzawa Y, Endo F. Effects of astragali radix extract on carcinogenesis, cytokine production, and cytotoxicity in mice treated with a carcinogen, N-butyl-N'-butanolnitrosoamine. Cancer Investment. 1999;**17**:30-35

[33] Khalilia W. The Effect of Gamma Radiation on Molecular Mechanisms of Apoptosis on Cell Culture. İstanbul: Istanbul University; 2015

[34] Özcan Arıcan G, Khalilia W, Serbes U, et al. The effect of hypobaric conditions

on the apoptosis signalling pathways in HeLa cells. Asian Pacific Journal of Cancer Prevention. 2014;**15**:20-28

[35] Dagdeviren Ozsoylemez O, Ozcan G. Effects of Colchicum baytopiorum leaf extract on cytotoxicity and cell death pathways in C-4 I and Vero cell lines. JBUON. 2021;**26**(3):1135-1147

[36] Serbes U, Dağdeviren Özsöylemez Ö, Özcan G. evaluation of paclitaxel effects in the pattern of expression of survival and apoptotic genes regulators in HeLa cells. Current Pharmaceutical Biotechnology. 2016;**17**:1058-1067

[37] Benito A, Silva M, Grillot D, et al. Apoptosis induced by erythroid differentiation of human leukemia cell lines is inhibitied by bcl-xL. Blood. 1996;**87**(9):3837-3843

[38] Patel K, Patel DK. Medicinal importance, pharmacological activities, and analytical aspects of hispidulin: A concise report. Journal of Trade Complete Medicine. 2016;7:360-366

[39] Chicca A, Tebano M, Adinolfi B, Ertugrul K, Flamini G, Nieri P. Anti-proliferative activity of aguerin B and a new rare nor-guaianolide lactone isolated from the aerial parts of Centaurea deflexa. European Journal of Medical and Chemistry. 2011;**46**:3066-3070

[40] Forgo P, Zupko I, Molnar J, Vasas A, Dombi G, Hohmann J. Bioactivity-guided isolation of antiproliferative compounds from *Centaurea jacea* L. Fitoterapia. 2012;**83**:921-925

[41] Caesar LK, Cech NB. Synergy and antagonism in natural product extracts: When 1 + 1 does not equal 2. Nature Product Reports. 2019;**36**:869-888

[42] Salla M, Fakhoury I, Saliba N, Darwiche N, Gali Muhtasib H. Synergistic anticancer activi-ties of the plant-derived sesquiterpene lactokines salograviolide A andiso-seco-tanapartholide. Journal of Natural Medicine. 2013;**67**:468-479

[43] He M, Xia L, Li J. Potential mechanisms of plant-derived natural products in the treatment of cervical cancer. Biomolecules. 2021;**11**:1539

[44] Tanaka F, Kawano Y, Li M, et al. Prognostic significance of apoptotic index in completely resected non-small-cell lung cancer. Journal of Clinical Oncology. 1999;**17**:2728-2736

[45] Su M, Mei Y, Sinha S. Role of the crosstalk between autophagy and apoptosis in cancer. Journal of Oncology. 2013; Article ID 102735

[46] Mariño G, Niso-Santano M, Baehrecke EH, et al. Self-consumption: The interplay of autophagy and apoptosis. Molecular Cell Biology. 2014;**15**:81-94

[47] Prasad R, Rana NK, Koch B. Dendrobium chrysanthum ethanolic extract induces apoptosis via p 53 up-regulation in HeLa cells and inhibits tumor progression in mice. Journal of Complementary and Integrative Medicine. 2017;**14**:14

[48] Kim MS, Bak Y, Park YS, Lee DH, Kim JH, Kang JW, et al. Wogonin induces apoptosis by suppressing E6 and E7 expressions and activating intrinsic signaling pathways in HPV-16 cervical cancer cells. Cell Biological Toxicology. 2013;**29**:259-272

Chapter 10

Non-Encapsulated Trichinella Species: *T. pseudo spiralis*, *T. papuae* and *T. zimbawensis*

Devyani Sharma, Upninder Kaur and Rakesh Sehgal

Abstract

Trichinellosis is a meat-borne zoonotic disease caused by nematode worms of the genus *Trichinella* in humans. Sylvatic animals are the main reservoir hosts of this helminth but domesticated animals, mainly swine, can also acquire the infection when they are fed with scraps of game meat. The genus used to have only one species; however, it has subsequently evolved into a multispecies genus. Due to its broad host range, it has been able to establish itself in both domestic and sylvatic cycles, allowing it to maintain a vast host reservoir. Infection has been documented in a variety of experimental species, showing that it could potentially happen in natural settings as well. Due to the considerable genetic differences among the isolates, researchers predict that the number of species and genotypes discovered within *Trichinella* will increase. Outbreaks caused by various species in different parts of the world have also been reported therefore prevention and control are critical in order to limit the parasite's transmission to humans. Although molecular methods are used to identify the *Trichinella* species but these methods are not appropriate for the diagnosis of the infection in animals.

Keywords: *Trichinella*, non-encapsulated, *T. pseudo spiralis*, *T. papuae*, *T. zimbawensis*

1. Introduction

Trichinellosis, often referred to as Trichinosis, is a meat-borne zoonotic disease, which is spread by helminths belonging to the genus *Trichinella*. It is caused by the consumption of raw or undercooked meat of domestic or sylvatic animals infected with the larvae of *the* parasite [1]. The effective establishment of this parasite in nature was due to the availability of a diverse variety of animal reservoirs. With the exception of Antarctica, they are found all over the world [2, 3]. The discovery of *T. spiralis* was by serendipity when James Paget in 1835 discovered the cysts in the muscles of a patient who had succumbed to tuberculosis. They were then further described by Richard Owen in the same year [4]. The establishment of a multispecies genus concept within the *Trichinella* genus can be attributed to the scientific findings on the biological variety of *T. spiralis* isolates gathered from various geographical regions and wildlife [5]. Till date ten species and three genotypes have been identified.

These are further classified according to whether a collagen capsule surrounds the larvae in the host muscle, later forming a nurse cell complex [6]. The encapsulated clade consists of six *Trichinella* species and three genotypes, which include *T. spiralis, T. nativa, T. britovi, T. murrelli, T. nelsoni, T. patagoniensis, T. chanchalensis*, T6, T8 and T9, respectively. Another peculiar feature is that only mammals have been infected by them [7]. Till now only three species have been defined in the non-encapsulated clade infecting mammals along with birds or reptiles [8]. The study results of SSCP demonstrate that nonencapsulated species form a complex group that is distinguishable from encapsulated species, and support the current hypothesis that the encapsulated *Trichinella* group is present external to non-encapsulated forms, based on the independent biological and biochemical data sets [9]. The three non-encapsulated species revealed significant variation in four gene loci (cytochrome oxidase, P450, cyanate lyase and SB147D), indicating that they are distinct species [1]. A lot of experiments are still being carried out to determine the possible hosts of non-encapsulated species as well as their infectivity to humans, which will further widen our knowledge about them.

2. Classification and general life cycle of *Trichinella* species

Class – Enoplea
Subclass – Dorylaimia
Order – Trichinellida
Family – Trichinellidae [9]

2.1 Life cycle

The life cycle of *Trichinella* plays an important role in its establishment in domestic as well as wildlife settings. **Figure 1** represents the diagrammatic life cycle of the parasite. In many aspects, the life cycle is peculiar. To begin with, only one host serves as both the definitive and intermediate host. There are two types of cycles: domestic and sylvatic. The infection is transferred to humans in the domestic cycle by eating undercooked meat infected with the encysted larvae. The only ones who develop clinical symptoms are humans, also forming the dead end in the parasite's life cycle. The larva is released in the intestine after the digestion of the cyst, where they disrupt the columnar epithelium and moult four times before growing into male and female worms within 30 hours. After fertilising the female, the male dies, but the viviparous female continues to produce thousands of larvae. The larvae will only encyst in striated muscles, where they will modify muscle cells to ensure their own survival, establishing a nurse cell complex. A collagen capsule is produced around the nurse cell as a result of the host immune cell reaction, with a capillary network around it for nutrition. The larva experiences a developmental halt at this point, before this it becomes infective within five weeks and calcifying after months or years according to the *Trichinella* species and host species. The parasite is passed on to domestic pigs, rats and boars through cannibalism and the ingestion of meat scraps. Predation and scavenging habits among wildlife species, such as those of carnivores and omnivores, are pronounced in the sylvatic cycle. Humans become infected after consuming raw undercooked meat (**Table 1**) [1, 9, 10].

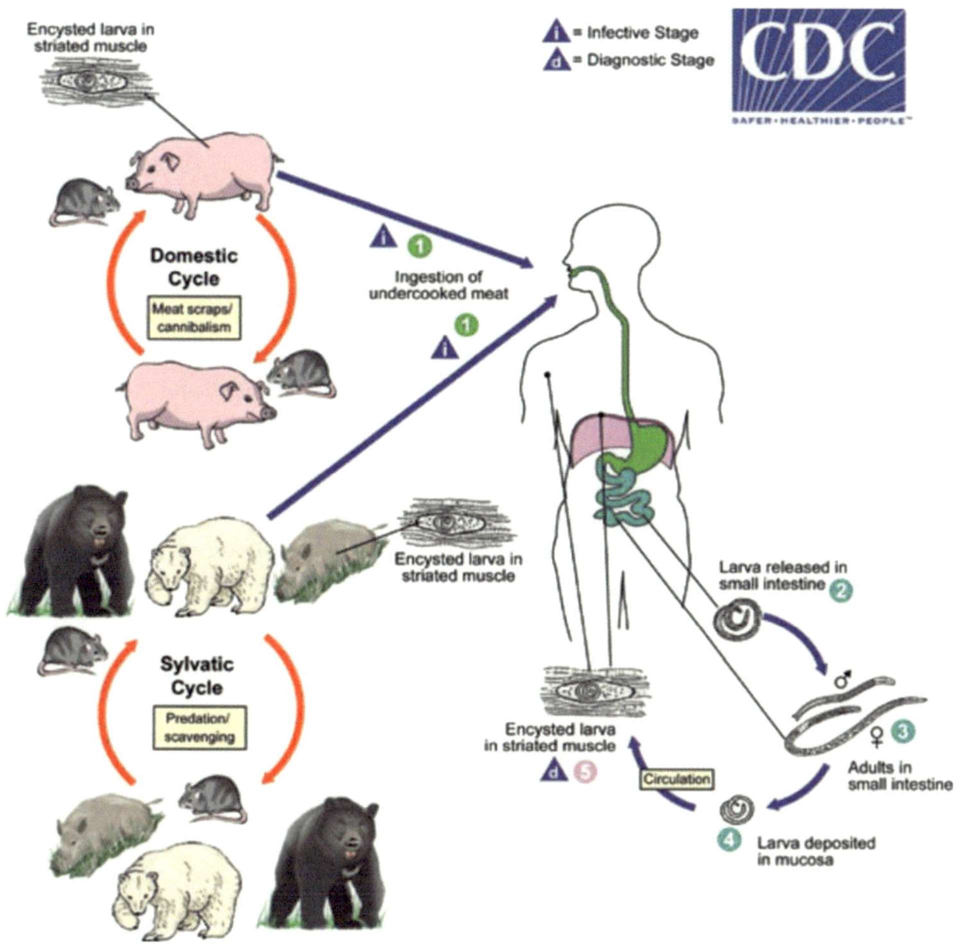

Figure 1.
Life cycle of Trichinella *species [10–13]. Image from the Centers for Disease Control and Prevention Image Library.*

Species	Geographical distribution	Cycle	Resistance to freezing	Pathogenicity to humans	Major hosts	Experimental transmission	Reference
T. pseudospiralis	Cosmopolitan	Sylvatic	No	High	Mammals and birds	Yes (Ferrets, guinea pigs and mice)	[14]
T. papuae	Papua New Guinea, Thailand, Cambodia	Domestic, Sylvatic	No	Moderate	Mammals and reptiles	NA	[12]
T. zimbawensis	Zimbabwe, Mozambique Ethiopia, South Africa	Sylvatic	No	Unknown	Mammals and reptiles	Rodents (*Rattus norvegicus*)	[11, 15–17]

Table 1.
Geographical distribution and general characteristics of the non-encapsulated species.

3. Non-encapsulated *Trichinella* species

3.1 *Trichinella pseudospiralis*

The first species to be discovered in the non-encapsulated clade was *T. pseudo spiralis* andsubsequently the fourth one in the genus *Trichinella*. **Figure 2** represents the timeline of discovery of the three non-encapsulated species. *T. pseudospiralis* was isolated from *Procyon lotor*, a raccoon caught in Krasnodar region. Significant differences were observed from the other species, which has been recorded earlier. Firstly, it can infect both mammals as well as birds, secondly, it was devoid of any collagen capsule and lastly, the adult worms and larvae had a smaller size from other species [8, 13]. It exhibits a cosmopolitan distribution in America, Asia, Australia and 20 different European countries, highlighting the importance of birds in carrying and spreading the parasite to new areas [14, 18]. The larvae of *T. pseudospiralis* is shown in **Figure 3.** In this case, the striated muscle fibres that surround the nurse cells are dense and branching longitudinally, but there is no netting pattern as seen in *T. spiralis* [20].

Since its discovery in 1972, this species has been detected in 249 animals, 237 of which were single infections and 11 of which were mixed infections with other *Trichinella* species. It has been discovered in 18 mammalian and eight avian species [14]. The parasite

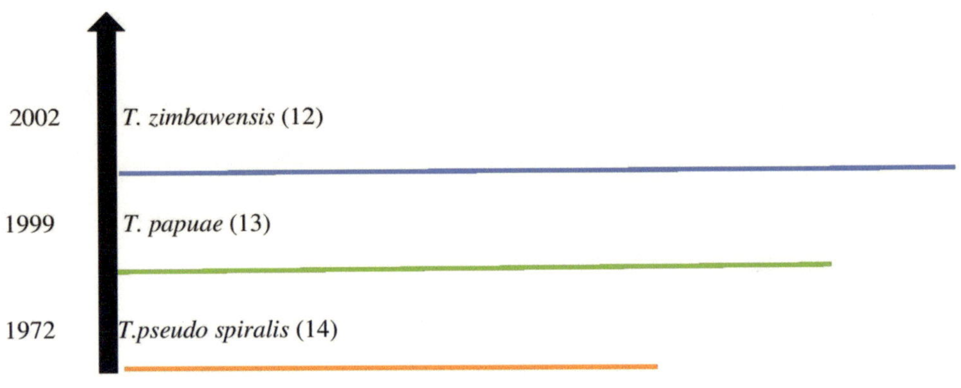

Figure 2.
Timeline of discovery of Trichinella *species.*

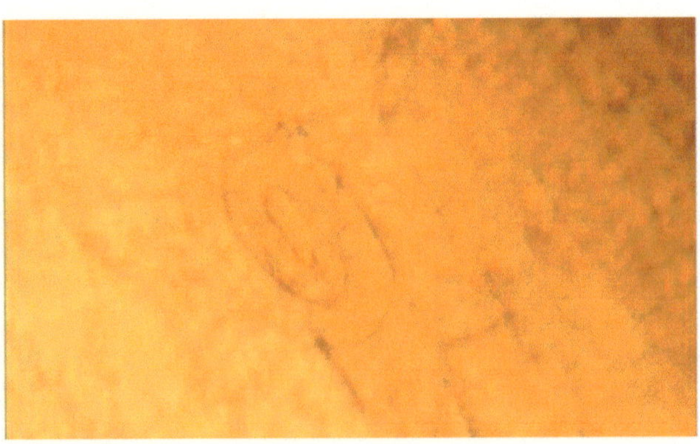

Figure 3.
T. pseudospiralis *larvae in the muscle of domestic pig [19].*

has also been detected in raccoon dogs in Germany [21–23], American mink in Poland [24], cougars from Colorado, United States [25], red foxes in Poland [26], raccoon dogs in Central Europe [27], wolverines from the Canadian north [28], Eurasian blackbird from Armenia [29], wild boars in Estonia [30], red kite from Italy [31], wolf from Central Italy [32], bobcats from Oklahoma [33]. *T. pseudospiralis* infection in red-eared sliders was found to be influenced by environmental temperature, as the infection was successful in turtles maintained at 38°C compared to those reared at 32°C and 28°C [32]. In order to identify the larvae, artificial digestion tests are preferred over trichinoscopy [1, 18].

Various studies have been carried out to elucidate the immune response of *T. pseudospiralis*. Matrix mettaloproteinases 9 and 2 have been identified as the markers for inflammatory response of both *T. spiralis* and *T. pseudospiralis* infection [33]. Infection with *T. pseudospiralis* also resulted in a reduction of follicular T helper cell differentiation [34]. A serpin gene, on the other hand, was discovered to play a key role in infection by activating the M2-polarised signaling pathway [35]. The parasite's excretory-secretory proteins can be used for early detection and the development of a vaccine candidate [36].

3.2 *Trichinella papuae*

An examination of domestic and wild swine from Papuae New Guinea, along with eighty-three wild animals, was conducted after the detection of non-encapsulated *Trichinella* larvae in five domestic female swine in the settlement of Balamuk in 1988. Six wild pigs were then tested positive in the Bula Plain in the years 1988 to 1998. The larvae detected in the diaphragm muscles of pigs can be seen in **Figure 4**. However, none of the 83 wild animals tested, including domestic pigs, had any larvae. The larvae from one of the wild pigs were then characterised and classified by Edoardo Pozio [12]. In Western Province, near Indonesia, 8.8% of the wild pig population was shown to be infected. Intake of infected wild pig meat was the source of infection [37].

It was also found in PNG's saltwater crocodiles and the source of infection was improper feeding of wild pigs to them [38, 39]. Varans, caimans, pythons and turtles have also been infected with *T. papuae* and *T. zimbawensis* in an experimental setting where varans were found to have the highest reproductive capacity rating of all the species. Despite receiving a high infection dose, just a small number of larvae were found in pythons and turtles. Furthermore, no clinical indications of the infection have been reported indicating that they do not play a substantial role in epidemiology. Only these two *Trichinella* species can complete their life cycle in both cold- and warm-blooded animals. As a result, they could trigger distinct physiological processes depending on the host they are infecting [40]. Further infection was investigated in the equatorial freshwater fishes *Serrasalmus nattereri* and *Serrasalmus rhombeus*, but no larvae or adult worms were found in any organ, implying that, despite being a food source for reptiles like crocodiles, they have no role to play in the epidemiology due to the entozoic habitat of these fishes, which is not suitable for these two *Trichinella* species [41]. **Table 2** lists the natural and experimental hosts of *T. papuae*.

Despite the absence of a collagen capsule, the larvae can thrive in a tropical climate, making them more likely to be e transmitted to a new host [45]. Infection with *T. papuae* was reported to reduce the severity of dextran sulphate sodium-induced colitis in mice. The absence of 57% of *T. papuae* lipids in humans indicates variations in lipid metabolism, which could aid in the development of innovative treatments [46].

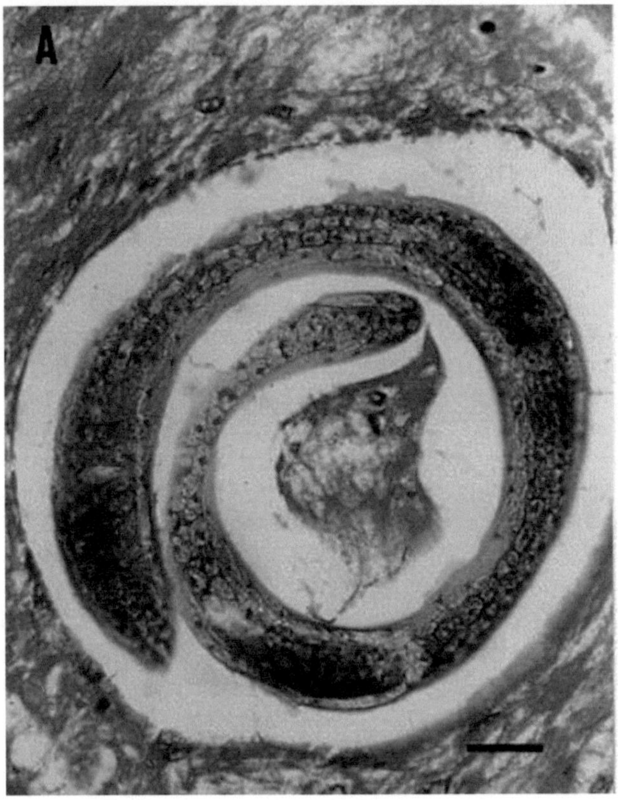

Figure 4.
In the village of Balamuk, larvae of Trichinella papuae were discovered in the diaphragm of an infected female pig (PNG), 1988 [12].

Natural host	*Crocodylus porosus* Papua New Guinea Wild pig Australian island in the Torres Strait region	[12, 38, 42]
Experimental host	*Ceratina sclerops*; Varanus *exanthematicus* *Python molurus* and *P. subrufa* Mice, rats, hamsters and gerbils Red foxes	[40, 43, 44]

Table 2.
List of natural and experimental hosts of T. papuae.

3.3 *Trichinella zimbawensis*

Trichinella larvae were identified in crocodile muscles in Zimbabwe in 1995. This was the first time *Trichinella* was found to naturally infect a reptile [47]. In an epidemiological survey, a farm near Victoria Falls was revealed to be the source of infection. The larvae isolated from crocodiles were able to infect domestic pigs and laboratory rats [48]. In the year 2002, Edoardo Pozio was the first to characterise and describe the larvae. This species has been found to infect both mammals and reptiles [11, 40]. Morphology of adults and larvae was determined to be comparable to that of *T. papuae. T. Zimbabwensis* males and females can procreate in both ways with *T. papuae* adults. As a result, the F1 offspring produces less viable F2 larvae [11]. **Figure 5** shows the *T.zimbawensis* larvae in the muscles of mice after four months of infection.

The parasite was then discovered in monitor lizards and Nile crocodiles in Zimbabwe and Mozambique, marking the first time to be found in wild reptiles [49]. Later, 38.5% prevalence rate was also been detected in wild Nile crocodiles in South Africa [15]. Natural infection has also been seen in mammals. In another experimental set-up baboons and vervet monkeys were also infected with the larvae. The most prevalent symptoms were fever, diarrhoea and muscular soreness. The infection was treated with ivermectin, but two baboons and two monkeys died as a result of the trial [50]. **Table 3** lists the experimental and natural hosts of *T. zimbawensis.*

Host age was found to have no effect on the distribution of parasites in various segments of intestine in case of golden hamsters and Balb C mice [51]. Also, increased progesterone levels in pregnant mice had a parasiticidal effect on the newly born larvae [53]. Co-infection of *Plasmodium berghei* with *T. zimbawensis* resulted in increased parasitemia in mice, which could further lead to severe malaria infection [54]. Various experiments have been carried out to study the immune response of *T. zimbawensis* infection. Non-encapsulated species have shown reduced inflammation and nitrosylation levels [55]. In an ELISA devised to detect the humoral response, *T. zimbawensis* was shown not to elicit a substantial immunological response in Nile crocodiles in terms of antibody titres and antibody persistence [56]. Another study revealed the same results, with the infection intensity not correlating with the amplitude of the humoral immune response [57]. The Th1, Th2 and T regulatory responses that are induced

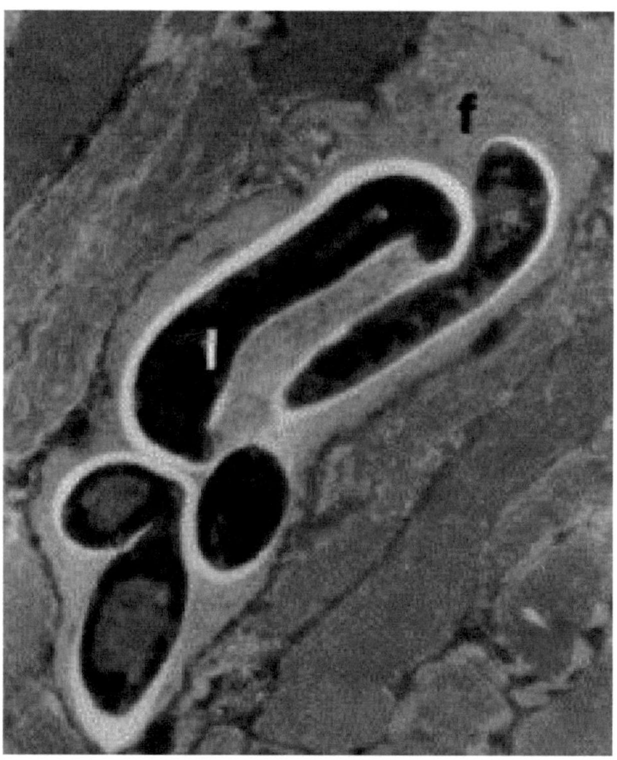

Figure 5.
T. Zimbawensis larvae in muscles of mice after four months of infection [11].

Hosts	Species	Region	References
Natural hosts	**Reptiles** *Nile crocodiles; Monitor lizards*	KNP of South Africa (Zimbabwe, Mozambique and Ethiopia) and Limpopo and Mpumalanga provinces of South Africa	[11, 15, 17, 47, 49]
	Mammals *Lion* (*Panthera leo*); Leopard *(Panthera pardus);* Spotted hyena (*Crocuta crocuta)* and Small spotted Genet *(Genetta genetta)*	KNP, South Africa	[16]
Experimental hosts	**Reptiles** Caimans; Varans; Pythons; Turtles **Mammals** (*Papio sp;* Vervet monkeys (*Cercopithecus aethiops)* Golden hamsters; Balb C mice		[40] [50] [51] [52]

Table 3.
List of natural and experimental hosts of T. zimbawensis.

during the different stages of infection were also shown to have significant variations [58]. This species has also been observed to affect metabolic parameters by inducing compensatory feeding in the host. During chronic infection, it was found to influence the host's Th1/Th17 immunological response [59]. The larvae were observed to invade the predilection muscles nearest to their release point in the small intestine first. The parasite load was found to be the highest in the fore and hind limb muscles. The use of biopsy samples from the dorso-lateral areas of the tail has also been recommended for surveillance purposes [60].

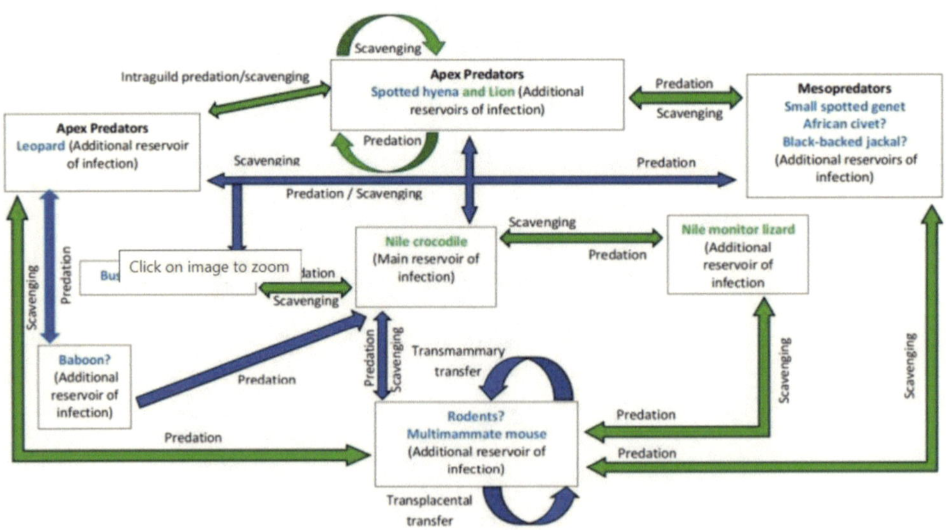

Figure 6.
Hypothetical transmission cycle of T. zimbawensis in *Kruger National Park [61].*

DOI: http://dx.doi.org/10.5772/intechopen.105680

A hypothetical transmission cycle for *T. zimbawensis* has been presented by Louis J. La Grange and Samson Mukaratirwa as shown in **Figure 6**. Recently, leopard and hyaena have been added as the apex predators along with few mesopredators [61].

The green arrows indicate the original hypothesised mode of transmission, while the blue arrows represent the modified way of transmission [61].

4. Human outbreaks due to non-encapsulated *Trichinella* species

Only *T. pseudospiralis* and *T. papuae* have been related to human outbreaks to date, while *T. zimbawensis* has not been associated to any such outbreaks. **Figure 7** represents the different outbreaks. Before the first outbreak in Kamchatka, only one case of *T. pseudospiralis* had been discovered in New Zealand [46, 63]. According to Edoardo Pozio, there were no species-specific primers available at the time of the outbreaks, therefore the species could have been *T. papuae* in the case of the Thai outbreak [47] but it had not been discovered yet [11]. The larvae were first isolated in a laboratory and then identified using cross-breeding procedures in the context of the Kamchatka epidemic [63]. Species-specific primers were utilised in the case of the French outbreak [11, 48]. The majority of outbreaks caused by *Trichinella Papuae* have occurred in Thailand, usually caused by the consumption of wild raw pig flesh [49, 50, 52]. In Taiwan, one outbreak was caused by the ingestion of soft-shelled turtles [51]. The most common clinical signs in these epidemics were myalgia, facial oedema and fever. The levels of creatine phosphokinase and aspartate aminotransferase were likewise higher. A few muscle biopsy specimens also included larvae [53]. These outbreaks suggest that there is a strong link between parasites and human behaviour, particularly the eating habits and certain rituals, such as the 'mumu' cooking method in Morehead District, PNG, which could have been a source of *T. papuae* infection [46, 64]. Albendazole, glucocorticosteroids and various supportive medications such as painkillers are generally used in the treatment [65]. A study found that the maslinic acid's efficiency in rats

Figure 7.
Map representing the outbreaks caused by non-encapsulated Trichinella *species [62].*

Species	Region	No of cases	Meat source	Reference
T. pseudo spiralis	New Zealand (1995)	1	Raw pork	[66]
	Kamchatka (1997)	28	Raw pork	[67]
	Thailand (1998)	59 (1 died)	Raw pork	[68]
	France (2000)	4	Wild boar	[69]
	Italy (2015)	36	Beef tartare mixed with wild boar meat	[70]
T. papuae	Thailand (2006)	28	Raw wild boar meat	[71]
	Thailand (2007)	34	Wild pig	[72]
	Taiwan (2008)	28	Soft-shelled turtles	[73]
	Thailand (2011)	1(imported case)	Raw wild pig meat	[74]
	Central Kampong Thom Province, Cambodia(2017)	3 persons were infected and 8 died	Raw mild pig meat	[65]
T. zimbawensis	No Human infection reported to date	NA	NA	[75]

Table 4.
Outbreaks caused due to different non-encapsulated Trichinella *species.*

was comparable to that of fenbendazole, with no side effects, indicating that it could be a promising anthelminthic drug against *Trichinella* larvae (**Table 4**) [76].

4.1 Species identification methods

A polymerase chain reaction-based on the mitochondrial large subunit ribosomal RNA gene was paired with a pyrosequencing technique to distinguish the four *Trichinella* species, this was successfully found to be sensitive enough to identify the individual larvae [77]. PCR based on the ITS1, ITS2 and ESV regions has also been utilised for the molecular identification of the species among wildlife in South Africa [78]. FRET-PCR and a melting curve analysis have also been utilised for the differential detection of the species [79]. Using Western blot, *T. pseudospiralis* infection can be differentiated from *T. spiralis* or *T. britovi* infection. When the source is unknown, this technique may be beneficial in epidemiological investigations [80].

5. Conclusions

The genus *Trichinella* used to have only one species; however, it has subsequently evolved into a multispecies genus. Due to its broad host range, it has been able to establish itself in both domestic and sylvatic cycles, allowing it to maintain a vast host reservoir. The infection has been documented in a variety of experimental species, showing that it could potentially happen in natural settings as well. Due to the considerable genetic differences among the isolates, researchers predict that the number of species and genotypes discovered within *Trichinella* will increase. Although molecular methods are used to identify the *Trichinella* species but these methods are not appropriate for the diagnosis of the infection in animals. Outbreaks caused by various species in different parts of the world have also been reported, therefore prevention and control are critical in order to limit the parasite's transmission to humans.

Acknowledgements

Professor Samson Mukaratirwara, Professor Edoardo Pozio and Professor Violeta Santrac graciously allowed us to use their valuable images of *Trichinella* species larvae. Special thanks to Professor Edoardo Pozio for revising the manuscript and providing valuable insights. Financial support was provided by Department of Science and Technology, New Delhi, India.

Conflict of interest

None

Acronyms and abbreviations

KNP	Kruger National Park
PNG	Papuae New Guinea
SSCP	Single strand conformation polymorphism analysis
FRET	Florescence resonance energy transfer

Author details

Devyani Sharma, Upninder Kaur and Rakesh Sehgal*
Department of Medical Parasitology, Postgraduate Institute of Medical Research and Research, Chandigarh, India

*Address all correspondence to: sehgalpgi@gmail.com

References

[1] Gottstein B, Pozio E, Nöckler K. Epidemiology, diagnosis, treatment, and control of *Trichinellosis*. Clinical Microbiology Reviews. 2009;**22**(1):127-145

[2] Pozio E. The broad spectrum of Trichinella hosts: From cold- to warm-blooded animals. Veterinary Parasitology. 2005;**132**(1-2):3-11

[3] Pozio E. Trichinella and trichinellosis in Europe. Veterinary Glass. 2022;**73**(2):65-84

[4] Owen R. Description of a Microscophc entozoon infesting the muscles of the human body. The Transactions of the Zoological Society of London. 1835;**1**(4):315-324

[5] Pozio E. Scientific achievements of the last 60 years: From a single to a multispecies concept of the genus Trichinella. Veterinary Parasitology. 2021;**297**:109042

[6] Wu Z, Sofronic-Milosavljevic L, Nagano I, Takahashi Y. *Trichinella spiralis*: Nurse cell formation with emphasis on analogy to muscle cell repair. Parasites & Vectors. 2008;**1**(1):27

[7] Zarlenga D, Thompson P, Pozio E. Trichinella species and genotypes. Research in Veterinary Science. 2020;**133**(289):296

[8] Pozio E, Zarlenga DS. New pieces of the Trichinella puzzle. International Journal for Parasitology. 2013;**43**(12):983-997

[9] Takahashi Y. Chapter 4 - Biology of Trichinella. In: Bruschi F, editor. Trichinella and Trichinellosis. Academic Press; 22 July 2021. p. 554. eBook ISBN: 9780128212677. Paperback ISBN: 9780128212097

[10] CDC. Trichinellosis – Biology. 2019

[11] Pozio E, Foggin CM, Marucci G, La Rosa G, Sacchi L, Corona S, et al. Trichinella zimbabwensis n.sp. (Nematoda), a new non-encapsulated species from crocodiles (Crocodylus niloticus) in Zimbabwe also infecting mammals. International Journal for Parasitology. 2002;**32**(14):1787-1799

[12] Pozio E, Owen IL, La Rosa G, Sacchi L, Rossi P, Corona S. Trichinella papuae n.sp. (Nematoda), a new non-encapsulated species from domestic and sylvatic swine of Papua New Guinea. International Journal for Parasitology. 1999;**29**(14):1825-1839

[13] Garkavi BL. The species of Trichinella isolated from wild carnivores. Veterinariia. 1972;**10**:90-91

[14] Pozio E. Trichinella pseudospiralis an elusive nematode. Veterinary Parasitology. 2016;**231**:97-101

[15] La Grange LJ, Marucci G, Pozio E. Trichinella zimbabwensis in wild Nile crocodiles (Crocodylus niloticus) of South Africa. Veterinary Parasitology. 2009;**161**(1-2):88-91

[16] La Grange LJ, Marucci G, Pozio E. Trichinella zimbabwensis in a naturally infected mammal. Journal of Helminthology. 2010;**84**(1):35-38

[17] La Grange LJ, Govender D, Mukaratirwa S. The occurrence of Trichinella zimbabwensis in naturally infected wild crocodiles (Crocodylus niloticus) from the Kruger National Park, South Africa. Journal of Helminthology. 2013;**87**(1):91-96

[18] La Rosa G, Marucci G, Zarlenga DS, Pozio E. Trichinella pseudospiralis populations of the palearctic region and their relationship with populations of the Nearctic and Australian regions. International Journal for Parasitology [Internet]. 2001;**31**(3):297-305

[19] Santrac V, Nedic D, Maric J, Nikolic S, Stevanović O, Vasilev S, et al. The first report of *Trichinella pseudospiralis* presence in domestic swine and *T. britovi* in wild boar in Bosnia and Herzegovina. Acta Parasitologica. 2015;**60**:471-475

[20] Khositharattanakool P, Morakote N, Uparanukraw P. The vasculature of nurse cells infected with non-encapsulated Trichinella species. The Southeast Asian Journal of Tropical Medicine and Public Health. 2013;**44**(4):561-567

[21] Mayer-Scholl A, Reckinger S, Schulze C, Nöckler K. Study on the occurrence of Trichinella spp. in raccoon dogs in Brandenburg, Germany. Veterinary Parasitology. 2016;**231**:102-105

[22] Hurníková Z, Kołodziej-Sobocińska M, Dvorožňáková E, Niemczynowicz A, Zalewski A. An invasive species as an additional parasite reservoir: Trichinella in introduced American mink (Neovison vison). Veterinary Parasitology. 2016;**231**:106-109

[23] Reichard MV, Logan K, Criffield M, Thomas JE, Paritte JM, Messerly DM, et al. The occurrence of Trichinella species in the cougar Puma concolor couguar from the state of Colorado and other regions of North and South America. Journal of Helminthology. 2017;**91**(3):320-325

[24] Cybulska A, Kornacka A, Bień J, Goździk K, Kalisińska E, Łanocha-Arendarczyk N, et al. The occurrence of Trichinella spp. in Red foxes (Vulpes vulpes) in different regions of Poland: Current data. Vector Borne and Zoonotic Diseases. 2016;**16**(11):717-721

[25] Cybulska A, Skopek R, Kornacka A, Popiołek M, Piróg A, Laskowski Z, et al. First detection of Trichinella pseudospiralis infection in raccoon (Procyon lotor) in Central Europe. Veterinary Parasitology. 2018;**254**:114-119

[26] Sharma R, Thompson P, Elkin B, Mulders R, Branigan M, Pongracz J, et al. Trichinella pseudospiralis in a wolverine (Gulo gulo) from the Canadian North. International Journal of Parasitology and Parasites Wildlife. 2019;**9**:274-280

[27] Movsesyan SO, Vlasov EA, Nikoghosian MA, Petrosian RA, GhasabyanMG,KuznetsovDN.Nematodes of birds of Armenia. Annals of Parasitology. 2020;**66**(4):447-455

[28] Kärssin A, Häkkinen L, Vilem A, Jokelainen P, Lassen B. Trichinella spp. in Wild Boars (Sus scrofa), Brown Bears (Ursus arctos), Eurasian Lynxes (Lynx lynx) and Badgers (Meles meles). Animals. 2021;**11**(1):183

[29] Marucci G, Romano AC, Interisano M, Toce M, Pietragalla I, Collazzo GP, et al. Trichinella pseudospiralis in a red kite (Milvus milvus) from Italy. Parasitology Research. 2021;**120**(6):2287-2290

[30] Ricchiuti L, Petrini A, Interisano M, Ruberto A, Salucci S, Marino L, et al. First report of Trichinella pseudospiralis in a wolf (Canis lupus italicus). International Journal of Parasitology Parasites Wildlife. 2021, 2021;**15**:195-198

[31] Reichard MV, Sanders TL, Prentiss NL, Cotey SR, Koch RW,

Fairbanks WS, et al. Detection of Trichinella murrelli and Trichinella pseudospiralis in bobcats (Lynx rufus) from Oklahoma. Veterinary Parasitology. 2021;**25**:100609

[32] Taira K, Nishiyama K, Ooi HK. Trichinella pseudospiralis infection in red-eared slider, Trachemys scripta, influenced by environmental temperature. Parasitology International. 2021;**84**:102413

[33] Bruschi F, Bianchi C, Fornaro M, Naccarato G, Menicagli M, Gomez-Morales MA, et al. Matrix metalloproteinase (MMP)-2 and MMP-9 as inflammation markers of Trichinella spiralis and Trichinella pseudospiralis infections in mice. Parasite Immunology. 2014;**36**(10):540-549

[34] Asano K, Wu Z, Srinontong P, Ikeda T, Nagano I, Morita H, et al. Nonencapsulated Trichinella pseudospiralis infection impairs follicular helper T cell differentiation with subclass-selective decreases in antibody responses. Infection and Immunity. 2016;**84**(12):3550-3556

[35] Xu N, Liu X, Tang B, Wang L, Shi HN, Boireau P, et al. Recombinant Trichinella pseudospiralis serine protease inhibitors Alter macrophage polarization In vitro. Frontiers in Microbiology. 2017;**8**:1834

[36] Wang Y, Bai X, Zhu H, Wang X, Shi H, Tang B, et al. Immunoproteomic analysis of the excretory-secretory products of Trichinella pseudospiralis adult worms and newborn larvae. Parasites & Vectors. 2017;**10**(1):579

[37] Owen IL, Sims LD, Wigglesworth MC, Puana I. Trichinellosis in Papua New Guinea. Australian Veterinary Journal. 2000;**78**(10):698-701

[38] Pozio E, Owen IL, Marucci G, La Rosa G. Trichinella papuae in saltwater crocodiles (Crocodylus porosus) of Papua New Guinea. Emerging Infectious Diseases. 2004;**10**(8):1507-1509

[39] Pozio E, Owen IL, Marucci G, La Rosa G. Inappropriate feeding practice favors the transmission of Trichinella papuae from wild pigs to saltwater crocodiles in Papua New Guinea. Veterinary Parasitology. 2005;**127**(3-4):245-251

[40] Pozio E, Marucci G, Casulli A, Sacchi L, Mukaratirwa S, Foggin CM, et al. Trichinella papuae and Trichinella zimbabwensis induce infection in experimentally infected varans, caimans, pythons and turtles. Parasitology. 2004;**128**(Pt 3):333-342

[41] Pozio E, Rosa GL. Evaluation of the infectivity of Trichinella papuae and Trichinella zimbabwensis for equatorial freshwater fishes. Veterinary Parasitology. 2005;**132**(1-2):113-114

[42] Cuttell L, Cookson B, Jackson LA, Gray C, Traub RJ. First report of a Trichinella papuae infection in a wild pig (Sus scrofa) from an Australian island in the Torres Strait region. Veterinary Parasitology. 2012;**185**(2-4):343-345

[43] Sadaow L, Intapan PM, Boonmars T, Morakote N, Maleewong W. Susceptibility of laboratory rodents to Trichinella papuae. The Korean Journal of Parasitology. 2013;**51**(6):629-632

[44] Webster P, Malakauskas A, Kapel CMO. Infectivity of Trichinella papuae for experimentally infected red foxes (Vulpes vulpes). Veterinary Parasitology. 2002;**105**(3):215-218

[45] Owen IL, Reid SA. Survival of *Trichinella papuae* muscle larvae in a pig carcass maintained under simulated natural conditions in Papua New

Guinea. Journal of Helminthology. 2007;**81**(4):429-432

[46] Mangmee S, Adisakwattana P, Tipthara P, Simanon N, Sonthayanon P, Reamtong O. Lipid profile of Trichinella papuae muscle-stage larvae. Scientific Reports. 2020;**10**(1):10125

[47] Foggin CM, Vassilev GD, Widdowson MA. Infection with Trichinella in farmed crocodiles (Crocodylus niloticus) in Zimbabwe. In: Abstract book on the 16th International Conference of The World Association for the Advancement of Veterinary Parasitology. 1997

[48] Mukaratirwa S, Foggin CM. Infectivity of Trichinella sp. isolated from *Crocodylus niloticus* to the indigenous Zimbabwean pig (Mukota). International Journal for Parasitology. 1999;**29**(7):1129-1131

[49] Pozio E, Foggin CM, Gelanew T, Marucci G, Hailu A, Rossi P, et al. Trichinella zimbabwensis in wild reptiles of Zimbabwe and Mozambique and farmed reptiles of Ethiopia. Veterinary Parasitology. 2007;**143**(3-4):305-310

[50] Mukaratirwa S, Dzoma BM, Matenga E, Ruziwa SD, Sacchi L, Pozio E. Experimental infections of baboons (Papio spp.) and vervet monkeys (*Cercopithecus aethiops*) with Trichinella zimbabwensis and successful treatment with ivermectin. The Onderstepoort Journal of Veterinary Research. 2008;**75**(2):173-180

[51] Mukaratirwa S, Nkulungo E, Matenga E, Bhebhe E. Effect of host age in the distribution of adult Trichinella zimbabwensis in the small intestines of golden hamsters (*Mesocricetus auratus*) and Balb C mice. The Onderstepoort Journal of Veterinary Research. 2003;**70**(2):169-173

[52] La Grange LJ, Mukaratirwa S. Experimental infection of tigerfish (Hydrocynus vittatus) and African sharp tooth catfish (Clarias gariepinus) with Trichinella zimbabwensis. The Onderstepoort Journal of Veterinary Research. 2020;**87**(1):e1-e5

[53] Hlaka L, Chitanga S, Masola B, Mukaratirwa S. Host pregnancy influences the establishment of Trichinella zimbabwensis in Balb C mice. Journal of Parasitic Diseases. 2017;**41**(3):799-804

[54] Murambiwa P, Silas E, Mdleleni Y, Mukaratirwa S. Chemokine, cytokine and haematological profiles in Sprague-Dawley rats co-infected with Plasmodium berghei ANKA and Trichinella zimbabwensis-A laboratory animal model for malaria and tissue-dwelling nematodes co-infection. Heliyon. 2020;**6**(2):e03475

[55] Bruschi F, Marucci G, Pozio E, Masetti M. Evaluation of inflammatory responses against muscle larvae of different Trichinella species by an image analysis system. Veterinary Parasitology. 2009;**159**(3-4):258-262

[56] Ludovisi A, La Grange LJ, Gómez Morales MA, Pozio E. Development of an ELISA to detect the humoral immune response to Trichinella zimbabwensis in Nile crocodiles (*Crocodylus niloticus*). Veterinary Parasitology. 2013;**194**(2):189-192

[57] La Grange LJ, Mukaratirwa S. Assessment of selected biochemical parameters and humoral immune response of Nile crocodiles (*Crocodylus niloticus*) experimentally infected with Trichinella zimbabwensis. Journal of the South African Veterinary Association. 2014;**85**(1):10. Art. no 1085. DOI: 10.4102/jsava.v85i1.1085

[58] Onkoba WN, Chimbari MJ, Kamau JM, Mukaratirwa S. Differential immune responses in mice infected with the tissue-dwelling nematode *Trichinella zimbabwensis*. Journal of Helminthology. 2016;**90**(5):547-554

[59] Onkoba N, Chimbari MJ, Kamau JM, Mukaratirwa S. Metabolic and adaptive immune responses induced in mice infected with tissue-dwelling nematode Trichinella zimbabwensis. Open Veterinary Journal. 2016;**6**(3): 178-184

[60] La Grange LJ, Mukaratirwa S. Distribution patterns and predilection muscles of *Trichinella zimbabwensis* larvae in experimentally infected Nile crocodiles (*Crocodylus niloticus* Laurenti). The Onderstepoort Journal of Veterinary Research. 2014;**81**(1):e1-e7

[61] La Grange LJ, Mukaratirwa S. Epidemiology and hypothetical transmission cycles of Trichinella infections in the Greater Kruger National Park of South Africa: An example of host-parasite interactions in an environment with minimal human interactions. Parasite. 2022;**9**:13

[62] Free World Maps. Atlas of the World. 2022

[63] Adisakwattana P, Nuamtanong S, Kusolsuk T, Chairoj M, Yenchitsomanas PT, Chaisri U. Non-encapsulated Trichinella spp., T. papuae, diminishes severity of DSS-induced colitis in mice. Asian Pacific Journal of Allergy and Immunology. 2013;**31**(2):106-114

[64] Owen IL, Muke L, Davies HL. Trichinellosis: A possible link between human infection and the traditional earth-oven or "mumu" method of cooking in Morehead District, Western Province, Papua New Guinea. Anthropological Medicine. 2008;**15**(3):189-197

[65] Caron Y, Bory S, Pluot M, Nheb M, Chan S, Prum SH, et al. Human outbreak of Trichinellosis caused by Trichinella papuae Nematodes, Central Kampong Thom Province, Cambodia. Emergency Infectious Diseases. 2020;**26**(8):1759-1766

[66] Andrews: Identification of Trichinella pseudospiralis... - Google Scholar [Internet]. [cited 2022 May 9]. Available from: https://scholar.google.com/scholar_lookup?title=Identification%20of%20Trichinella%20pseudospiralis%20from%20a%20human%20case%20using%20random%20amplified%20polymorphic%20DNA&publication_year=1995&author=J.R.%20Andrews&author=C.%20Bandi&author=E.%20Pozio&author=M.A.%20Gomez%20Morales&author=R.%20Ainsworth&author=D.%20Abernethy

[67] Britov: Trichinellosis in Kamchatka - Google Scholar [Internet]. [cited 2022 May 8]. Available from: https://scholar.google.com/scholar_lookup?title=Trichinellosis%20in%20Kamchatka&author=V.A.%20Britov&publication_year=1997&pages=287-288

[68] Jongwutiwes S, Chantachum N, Kraivichian P, Siriyasatien P, Putaporntip C, Tamburrini A, et al. First outbreak of human trichinellosis caused by *Trichinella pseudospiralis*. Clinical Infectious Diseases. 1998;**26**(1):111-115

[69] Ranque S, Faugère B, Pozio E, La Rosa G, Tamburrini A, Pellissier JF, et al. Trichinella pseudospiralis outbreak in France. Emerging Infectious Diseases. 2000;**6**(5):543-547

[70] Gómez-Morales MA, Mazzarello G, Bondi E, Arenare L, Bisso MC, Ludovisi A, et al. Second outbreak of Trichinella pseudospiralis in Europe: Clinical patterns, epidemiological investigation and identification of the etiological agent based on the western blot patterns of the patients' serum. Zoonoses Public Health. Feb 2021;**68**(1):29-37

[71] Khumjui C, Choomkasien P, Dekumyoy P, Kusolsuk T, Kongkaew W, Chalamaat M, et al. Outbreak of trichinellosis caused by Trichinella papuae, Thailand, 2006. Emerging Infectious Diseases. 2008;**14**(12):1913-1915

[72] Kusolsuk T, Kamonrattanakun S, Wesanonthawech A, Dekumyoy P, Thaenkham U, Yoonuan T, et al. The second outbreak of trichinellosis caused by Trichinella papuae in Thailand. Transactions of the Royal Society of Tropical Medicine and Hygiene. 2010;**104**(6):433-437

[73] Lo YC, Hung CC, Lai CS, Wu Z, Nagano I, Maeda T, et al. Human trichinosis after consumption of soft-shelled turtles, Taiwan. Emergency Infectectious Diseases. 2009;**15**(12):2056-2058

[74] Intapan PM, Chotmongkol V, Tantrawatpan C, Sanpool O, Morakote N, Maleewong W. Molecular identification of Trichinella papuae from a Thai patient with imported trichinellosis. The American Journal of Tropical Medicine and Hygiene. 2011;**84**(6):994-997

[75] La Grange LJ, Mukaratirwa S. Epidemiology and hypothetical transmission cycles of Trichinella infections in the Greater Kruger National Park of South Africa: An example of host-parasite interactions in an environment with minimal human interactions. Parasite. 2020;**27**:13

[76] Mukaratirwa S, Gcanga L, Kamau J. Efficacy of maslinic acid and fenbendazole on muscle larvae of *Trichinella zimbabwensis* in laboratory rats. Journal of Helminthology. 2016;**90**(1):86-90

[77] Sadaow L, Tantrawatpan C, Intapan PM, Lulitanond V, Boonmars T, MorakoteN,etal.Moleculardifferentiation of *Trichinella spiralis*, *T. pseudospiralis*, *T. papuae* and *T. zimbabwensis* by pyrosequencing. Journal of Helminthology. 2015;**89**(1):118-123

[78] Mukaratirwa S, La Grange LJ, Malatji MP, Reininghaus B, Lamb J. Prevalence and molecular identification of Trichinella species isolated from wildlife originating from Limpopo and Mpumalanga provinces of South Africa. Journal of Helminthology. 2019;**93**(1):50-56

[79] Tantrawatpan C, Intapan PM, Thanchomnang T, Lulitanond V, Boonmars T, Wu Z, et al. Differential detection of *Trichinella papuae*, *T. spiralis* and *T. pseudospiralis* by real-time fluorescence resonance energy transfer PCR and melting curve analysis. Veterinary Parasitology. 2012;**185**(2-4):210-215

[80] Gómez-Morales MA, Ludovisi A, Amati M, Cherchi S, Tonanzi D, Pozio E. Differentiation of Trichinella species (*Trichinella spiralis*/*Trichinella britovi* versus *Trichinella pseudospiralis*) using western blot. Parasites & Vectors. 2018;**11**(1):631

Chapter 11

Anticancer Functions of Pyridine Heterocycles

Kereyagalahally H. Narasimhamurthy, Nichhapurada Kallesha, Chakrabhavi D. Mohan and Kanchugarakoppal S. Rangappa

Abstract

Pyridine is a heterocyclic molecule with a nitrogen atom that is often found in nature. As a prosthetic group taking part in redox processes in the biological system, it plays an important function in many enzymes of the living system. Pyridine is an important pharmacophore, a privileged scaffold, and a superior heterocyclic system in drug development, with various applications in anticancer research because of its ability to work on significant receptors. Typically, it is the core of several currently available medicines. In the fight against cancer, many pyridine derivatives have been shown to inhibit kinases, androgen receptors, tubulin polymerization, topoisomerase enzyme, human carbonic anhydrase, and several other targets. Researchers are now concentrating on developing pyridine novel entities with other moieties for cancer therapy. This section presents pyridine derivative synthesis and biological expansions, as well as their target receptor sites.

Keywords: synthesis, leukemia, tumor, anticancer, *N*-containing heterocycles, pyridine

1. Introduction

The name pyridine is derived from the Greek word and is a combination of two words: pyr means fire and idine is used for aromatic bases [1]. It is nitrogen-containing heterocycle [2], six-membered and aromatic, and it plays a vital role in the field of medicinal chemistry [3]. Cancer is a group of more than 100 different diseases. It can progress almost anywhere in the body. The causes of cancers are host variables such as genetics, epigenetics, microbiome, age, gender, metabolic state, inflammatory state, and immune function [4]. Environmental factors such as food contamination, viruses, UV radiation, carcinogens from the environment, and diet/lifestyle factors such as nutrients, energy consumption, phytochemicals, other food ingredients, alcohol, physical activity, and smoking [5]. Some of the derivatives of pyridine nucleus containing molecules that are potential drug candidates are streptonigrin, streptonigrone, and lavendamycin, which were reported in the literature [6]. Some of the reported pyridine molecules are selective toward topoisomerase inhibitors [7]. Some of the pyridine-conjugated derivatives were PIM-1 kinase inhibitors [8], human carbonic anhydrase inhibitors [9], proto-oncogene tyrosine-protein kinase (ROS) [10],

ALK/ROS1 dual inhibitors, receptor tyrosine kinase (RTK) c-Met, epidermal growth factor receptor [11], EGFR and HER-2 kinase inhibitors, cyclin-dependent kinase (CDK) inhibitors [12], VEGFR-2 inhibitors [12], topoisomerases, phosphoinositide 3-kinase, maternal embryonic leucine zipper kinase (MELK), NF-κB inhibitors [13], etc. Considering all this information, exploration of these heterocycles is very important for the development of potential anticancer drug candidates. Hence, we covered all the reports related to the pyridine moiety in this book chapter.

2. Anticancer efficacy of diverse pyridine derivatives

Shudo and group developed pyridine derivatives [14] and these were tested for their reverse drug resistance in a multidrug-resistant human carcinoma cell line, KB-C2. Among the synthesized derivatives, compound **1** was the most active in reversing multidrug resistance. Its activity is higher than that of verapamil, cepharanthine, nimodipine, and nicardipine. Few of the synthesized pyridine derivatives displayed lower calcium channel blocking activity and more potent resistance-reversing activity than other calcium channel blockers.

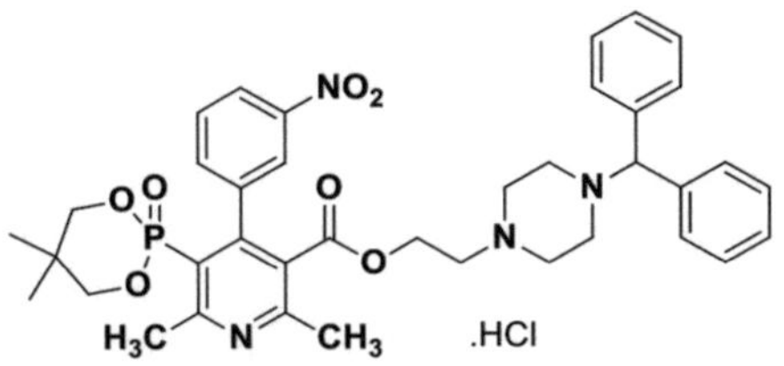

1

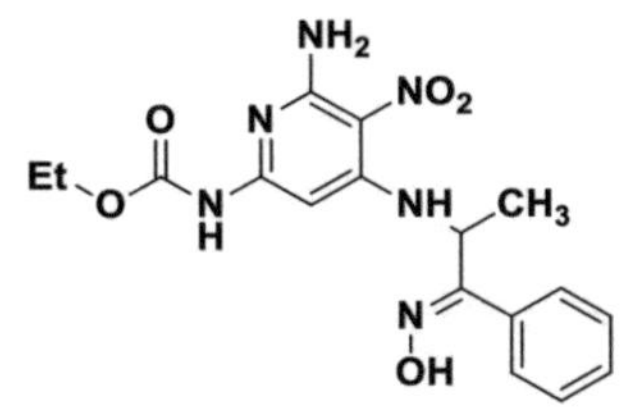

2

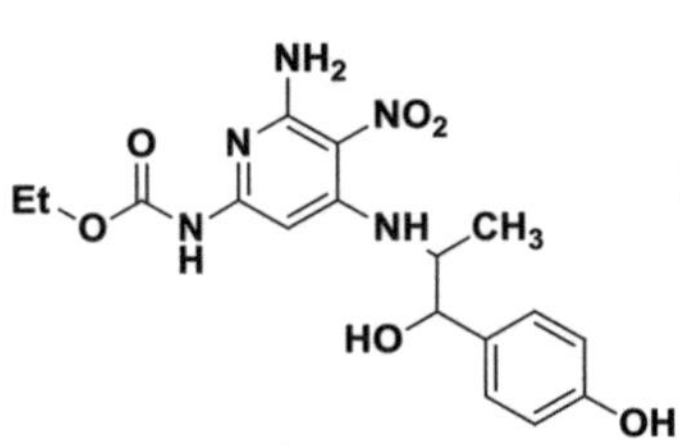

3

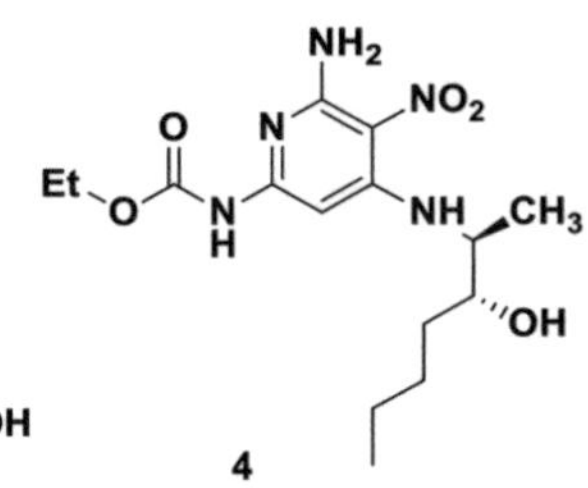

4

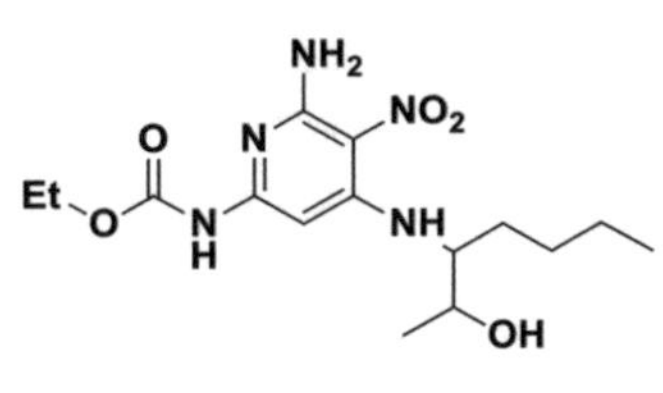

5

Temple Jr. and his group reported [15] the synthesis and structure-activity studies of some pyridine derivatives and cytotoxic activity against lymphoid leukemia L1210 cells. From the results, it was confirmed that compounds act by multiple modes of action. The primary mode of action of compound **3** might be through the inhibition of the incorporation of pyrimidine nucleosides into DNA and RNA. However, there were two other compounds, **4** and **5**, whose primary mode of action was tubulin polymerization inhibition.

Liu and coworkers [16] reported the synthesis of pyridine-2-carboxaldehyde thiosemicarbozone derivatives. The above molecules were evaluated for anti-neoplastic activity in mice containing L1210 leukemia. The 3-amino derivative compounds **6** and **7** were comparable in their antitumor efficacy against L1210 leukemia. The 5-amino derivatives **8**, **9**, and 5-hydroxy amino derivatives **10** were comparable to the 5-HP anti-neoplastic agents. The above lead molecules containing only amino groups were selected for further studies and screened against L1210 leukemia-bearing CD_2F_1 female mice. Compounds **6–7** and **8–9** were found to be the most active against L1210 leukemia.

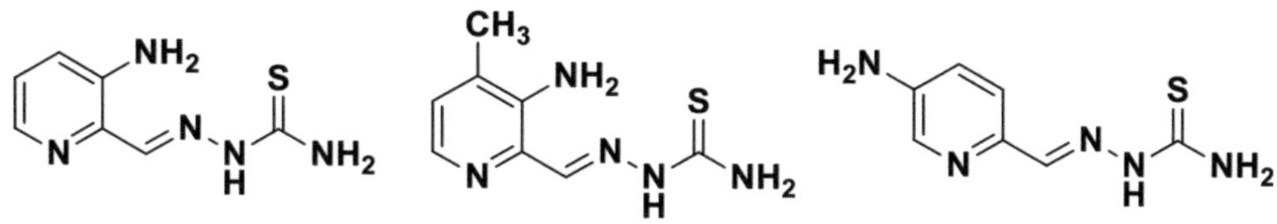

6 **7** **8**

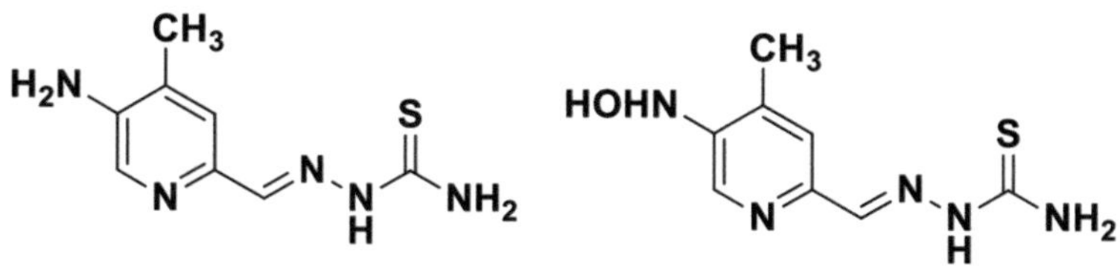

9 **10**

One more research group tried to improve the efficacy of pyridine-2-carboxaldehyde thiosemicarbozone derivatives. However, the incorporation of cytidine into DNA *via* ribonucleotide reductase was inhibited markedly. Thus, a pronounced decrease in the formation of [^{14}C] deoxy ribonucleotides from radioactive cytidine occurred in the acid-soluble fraction of compounds **6** and **7** treated L1210 cells. And it is consistent with DNA replication that at lower concentrations, cells generally accumulate in the S-phase of the cell cycle at higher concentrations of compounds **6** and **7**. Cells at the G0/G1 phase of the cell cycle are observed with a loss of S phase population. The above results provide more support for the development of HCTs. Specifically, compounds **6** and **7** are potential drug candidates for clinical use in the treatment of cancer [17].

Jew and coworkers synthesized [18] a 6-formyl-pyridine-2-carboxylate derivative, and these molecules were tested for telomerase inhibitory activity. Among the series, compound **11** was identified as the lead molecule, and most of the thioester derivatives showed higher activity than the reference compound. A wide variation in the activity was observed based on the position of the halide on the aromatic ring. From the results, it was evident that the *para* chloro derivative showed higher potency than the *meta*- and *ortho*-substituted derivatives. The number of chlorides on the ring is not directly affected by activity. Following the *in vivo* assay, the authors investigated the *in vitro* activity using cancer cell lines HT-29, Caki-2, A549, HEC-1-B, and HL-60, with camptothecin serving as a positive control. From the *in vitro* results, it was confirmed that lead molecules are not that effective in *in vitro* assays. Hence, the antitumor mechanism is different from cytotoxicity.

In the year 2006, Amr and group reported pyridine, [19] pyran, and pyrimidine derivatives. *In vitro* activity of the above-synthesized molecules was performed using 59 different cancer cell lines. Among the series, several active molecules showed higher activity, but our topic of interest is pyridine; hence, compound **12** is identified as a lead molecule and it is selective toward leukemia cell lines. Structure-activity relationship studies of the above series indicate that the presence of nitrile in the molecules enhances the activity. Onnis and group reported [20] the synthesis of trifluoromethyl pyridine derivatives and their *in vitro* cytotoxicity assays using diverse cancer cell lines. Among the series, compound **13** emerged as a potent molecule and gave activity at nanomolar concentration. And because it is free from animal toxicity, the given lead molecule was further evaluated for *in vivo* assay.

Amin and his group reported [21] a series of tetralin-6-ylpyridines. These molecules were evaluated for *in vitro* antiproliferative activity using two cell lines, HepG2 and MCF-7 cell lines. From the results, it was confirmed that compound **14** was selected for liver cancer and compound **15** was selective for breast cancer. Elgemeie and his group reported [22] the pyridine thioglycosides as anticancer agents. The *in vitro* antiproliferative activity was conducted using four different cancer cell lines such as HepG2, H460, MCF-7, U251, and the animal cell line EAC cells. These molecules displayed good cytotoxicity against four cell lines and the animal cell line EAC. Flow cytometric analysis of the aforementioned derivatives against U251 and HepG2 cell lines later revealed that cell cycle arrest occurred in the S phase. This mechanism is almost the same as the antimetabolite cell cycle arrest.

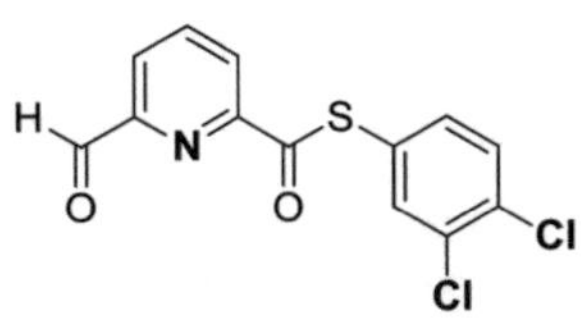

11

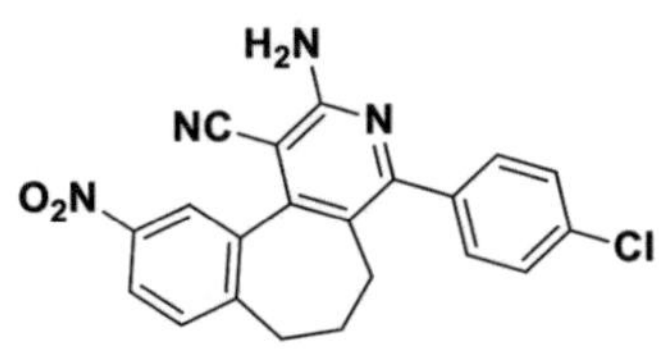

12

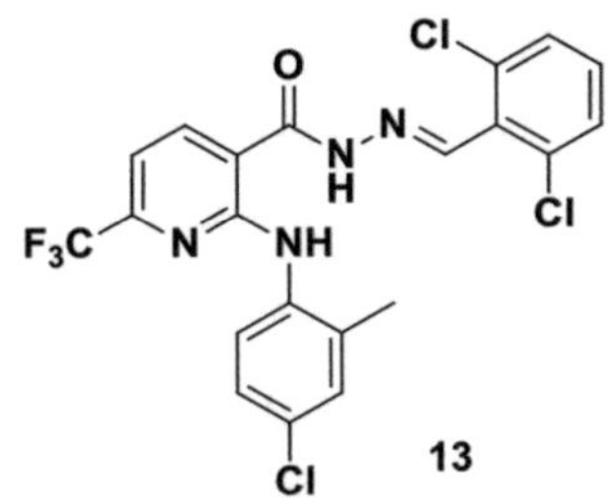

13

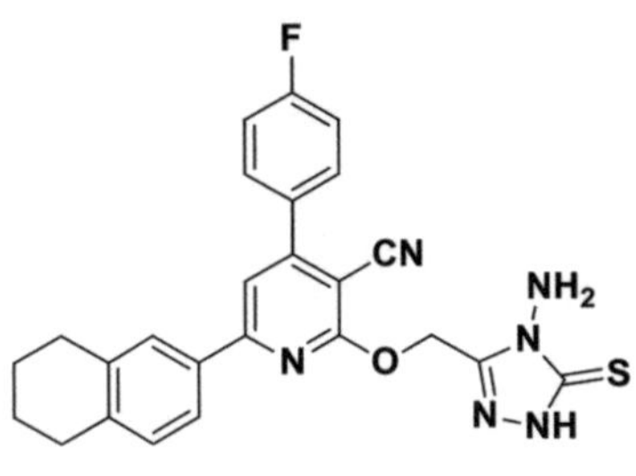

14

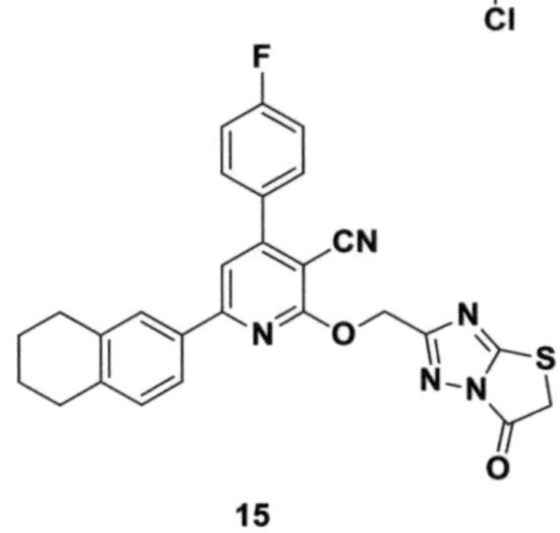

15

Elzahabi reported [23] the pyridine-conjugated benzimidazoles as anticancer agents. These were tested for *in vitro* antiproliferative activity using 41 different panel cancer cell lines. It was confirmed that compounds **16** and **17** are lead molecules in most of the cell lines they tested. The structure-activity relationship of the above synthesized derivatives gave some information. The *para*-substituted chloro group and methoxy group greatly enhanced the activity.

Liu and coworkers developed a series of benzo[5,6]cyclohepta[1,2-*b*]pyridine containing thiourea derivatives as anticancer agents. *In vitro* activity is performed using MCF-7, MDA-MB 231, and HT-29 cancer cell lines using 5-fluorouracil as a positive control. In an *in vitro* assay, the results showed that the activities of the molecules were comparable to those of 5-fluorouracil [24].

Bassyouni and coworkers [25] developed a series of pyridine conjugates, and after synthesis, the above derivatives were tested for anticancer activity. *In vitro* activity of the above compounds was performed using the liver cancer cell line HepG2 and

5-fluorouracil and doxorubicin as positive controls. Among the synthesized derivatives, compounds **18–23** displayed better activity than the positive control.

In the year 2014, one more group reported [26] the quinoline pyrazole pyridine hybrids as anticancer agents. All the synthesized derivatives were tested for EGFR kinase and antiproliferative activity against cell lines such as A549 and HepG2 cell lines using erlotinib as a positive control. From the results, it was confirmed that compounds **24** and **25** were identified as lead molecules against EGFR and other cell lines. Zheng and group reported [27] the synthesis of a series of pyridine bridged analogs of combretastatin-A4. The above-synthesized derivatives were tested for *in vitro* antiproliferative activity using three different cell lines: MDA-MB-231, A549, and HeLa. The three-atom linker containing nitrogen emerged as the more favorable structure. Among the synthesized molecules, compounds **26**, **27,** and **28** were identified as lead molecules. These molecules inhibit cell survival and growth and arrest the cell cycle. The competitive binding assay confirms the binding posture of CA-4, **26**, and **27**, which is very similar to CA-4.

Lu and coworkers developed [28] a series of sulfonyl groups containing pyridine derivatives as potential anticancer agents. The above-synthesized derivatives were tested for their *in vitro* activity using A2780, MCF-7, and HCT-116, and here, ON01910 is used as a positive control. After *in vitro* analysis, lead molecules were identified, that is, compounds **29–32**. Later, these molecules are tested on a panel of cancer cell lines. Of the four compounds, **30** and **31** gave better antitumor activity in an *in vivo* assay.

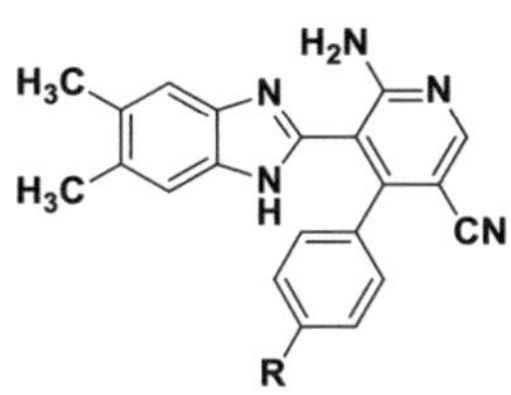

R = Cl, **16**
R = OCH_3, **17**

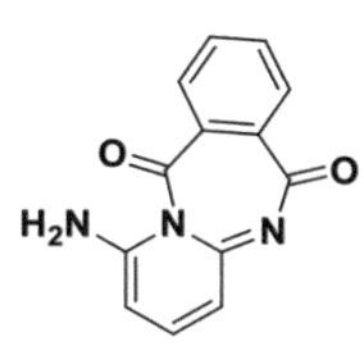

18

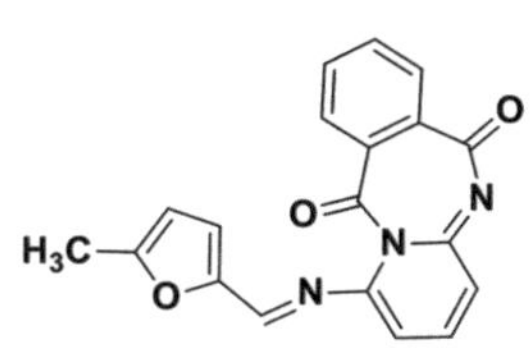

19

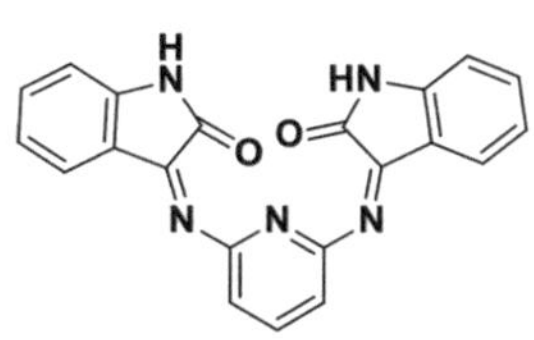

20

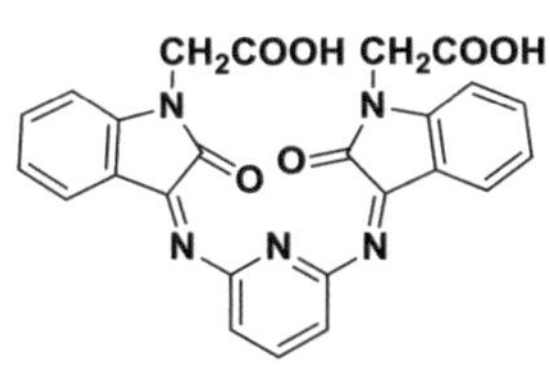

21

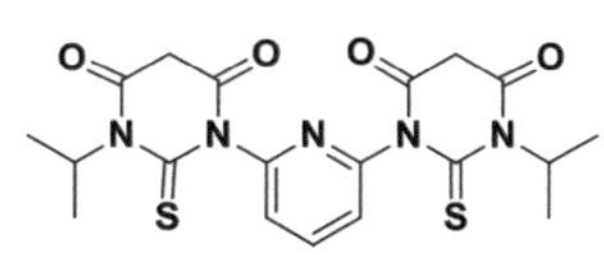

22

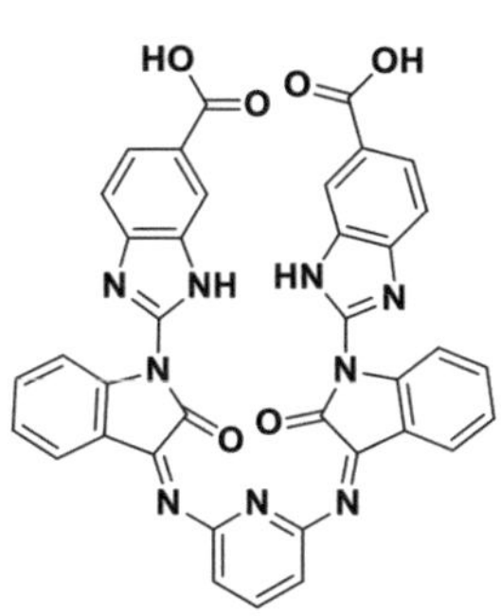

23

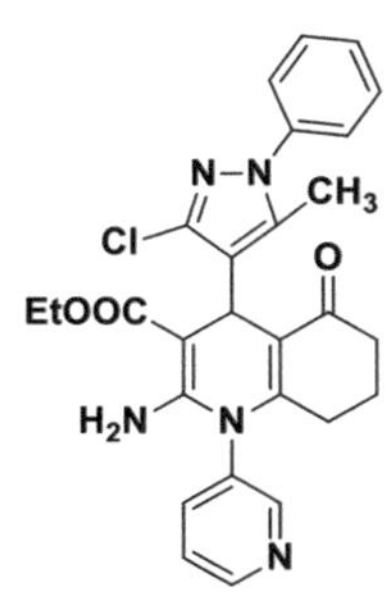

24

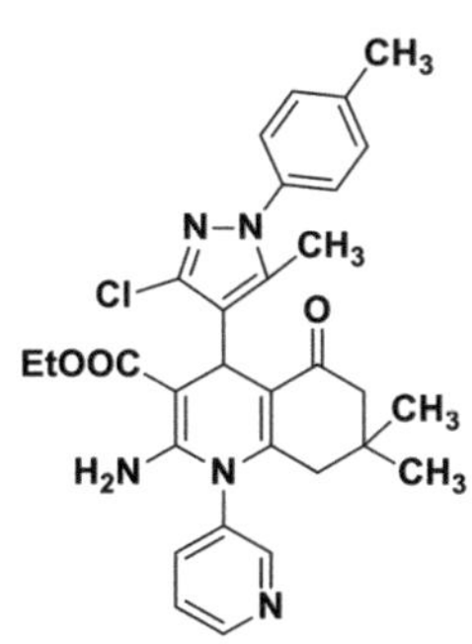

25

Eldehna and his group reported [29] the isatin-pyridine derivatives as antiproliferative agents, with *in vitro* activity being performed using HepG2, A549, and MCF-7 cancer cell lines. Among the isatin derivatives, compound **33** was found to be more active against HepG2 cancer cell lines than the reference compound doxorubicin, while compound **34** was found to be active against A549 and MCF-7 cell lines.

Previously reported tetraindole derivatives had some disadvantages; hence, to overcome that, Fu and coworkers [30] reported a series of tetraindole derivatives. The synthesized derivatives were evaluated against triple-negative breast cancer cell lines and adenocarcinoma cell lines. Among the synthesized derivatives, compound **35** displayed selective cytotoxicity against breast cancer cell lines over normal cell lines. In addition, its mode of action is shown to involve the G2/M phase of cell cycle arrest and also blocks cancer cell metastasis effectively.

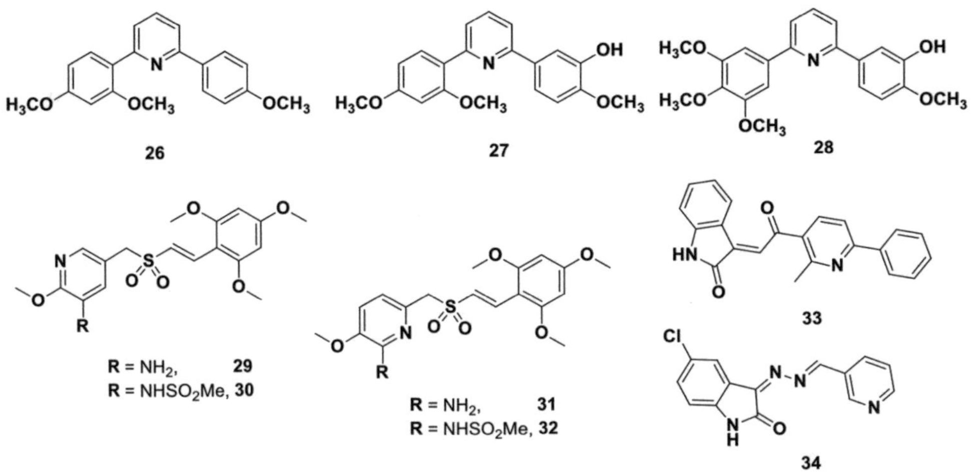

One more research group [31], in the same year, reported a 1,2,4-triazine group containing derivatives. Synthesized derivatives were tested for *in vitro* antiproliferative activity using different cancer cell lines. Among the series, compound **36** was identified as the lead molecule. This compound also showed prominent activity in *in vivo* activity. Abbas and his group reported [32] the synthesis of an imidazole group containing pyridine derivatives. The above-synthesized derivatives were evaluated for anticancer activity. *In vitro* antiproliferative studies were conducted using MCF-7 and HepG2 cell lines using doxorubicin as a positive control. Among the synthesized derivatives, compound **37** is identified as the lead molecule in both cell lines.

Abdelazem and coworkers reported [33] a series of diary amides with the pyrimidinyl pyridine group. These were evaluated for *in vitro* antiproliferative activity using 60 different cancer cell lines. Among all the prepared derivatives, compound **38** gave very good results. This compound showed activity in micromolar concentration in all nine cancer types, but it was the highest against melanoma cell lines.

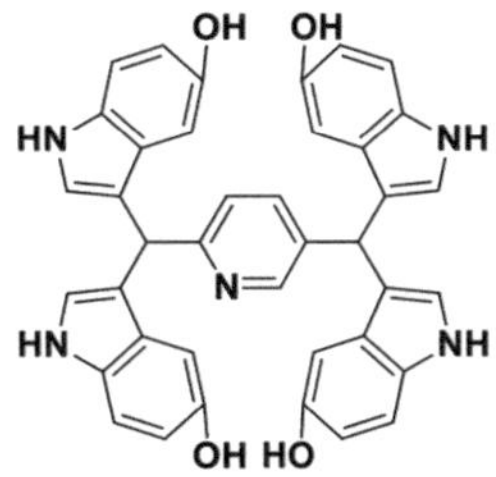

35

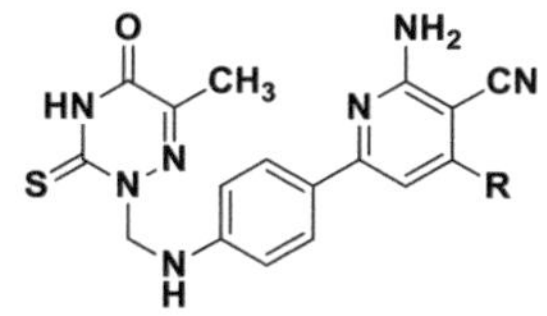

R = 2,4-dimethoxy-5H-furo[3,2g] chromen-5-one

36

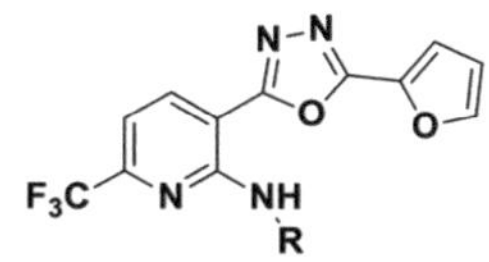

R = 2-methyl furyl **39**
R = C_3H_7 **40**

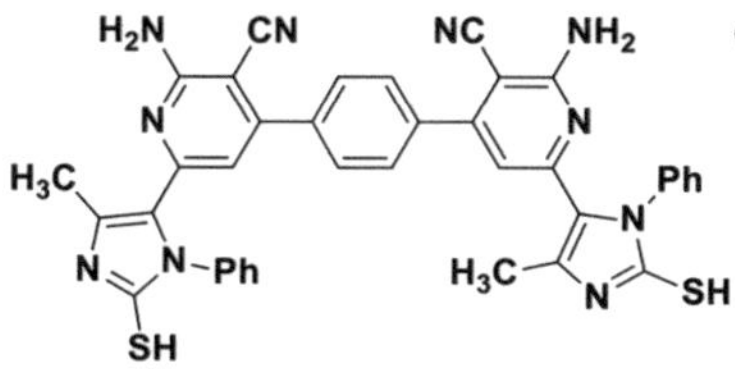

37

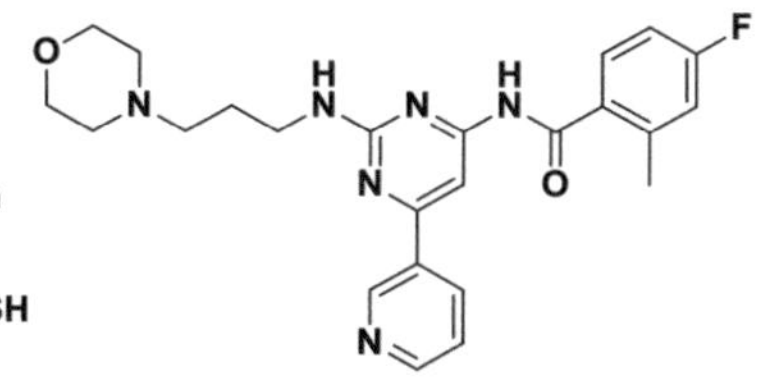

38

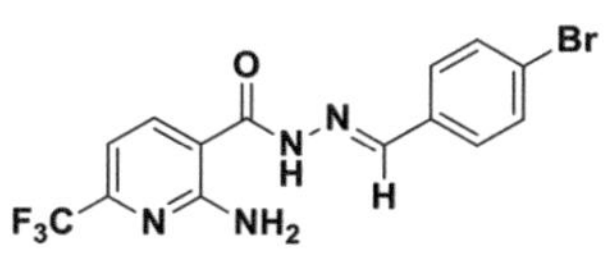

41

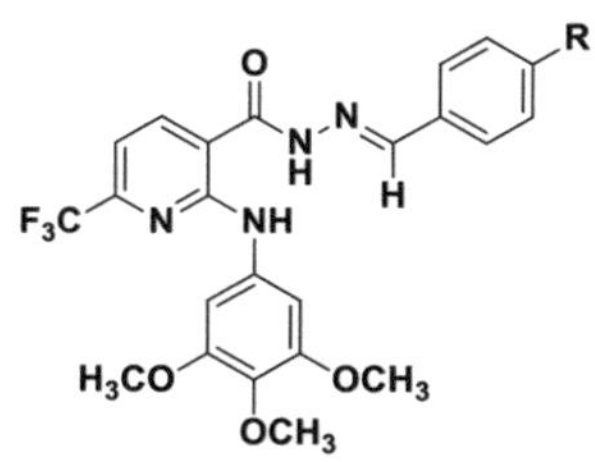

R = 4-OH **42**
R = H, **43**

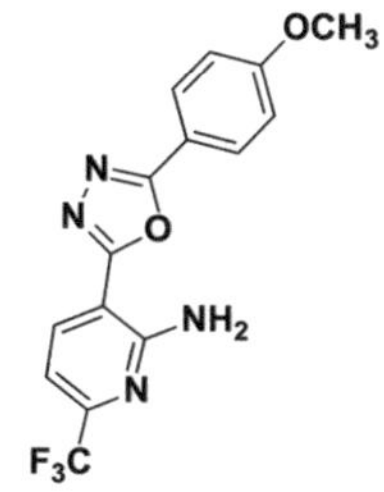

44

Naresh kumar and group designed [34] and synthesized oxadiazolo pyridine derivatives. *In vitro* antiproliferative assay was carried out for the cell lines such as HeLa, DU145, HepG2, and MBA-MB-231 cell lines; here, 5-fluoro uracil is used as a positive control. Among the synthesized derivatives, compounds **36–41** showed better activity against DU145 and HepG2 cancer cell lines only.

Wu and coworkers developed [35] benzimidazole propyl ketone derivatives; after synthesis, these molecules are evaluate for *in vitro* cytotoxicity assay, using cell lines such as HCT-116, MCF-7, and HepG2 cell lines. Here, 5-fluoro uracil and paclitaxel are used as positive control. From the *in vitro* studies, it was evident that some of the molecules exhibited better activity than others, but the topic of interest was pyridine. Hence, compound **42** is the only molecule containing a pyridine ring that displayed good activity. *In vivo* studies of these compounds showed promising activity. Compound **42** displayed better activity; hence, this heterocyclic core is considered a promising drug candidate.

Zhou and group designed [36] and prepared a series of pyridine analogs of curcumin as human prostate cancer inhibitors. Effects of curcumin analogs are on the human prostate cancer cell line CWR-22Rvl. Among the synthesized derivatives, compounds **45–48** were identified as lead molecules. The inhibitory effects of these compounds were tested by using an androgen receptor-linked luciferase assay. Results suggest that compounds **46–48** had the strongest inhibitory effect.

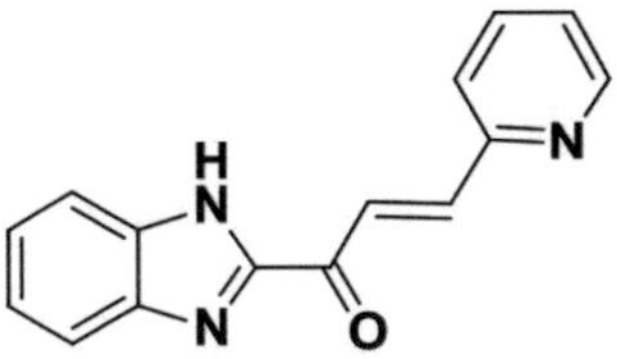

45

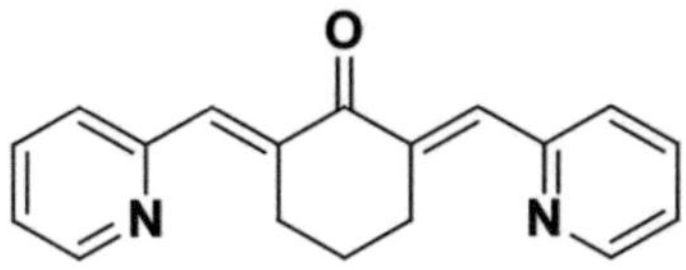

46

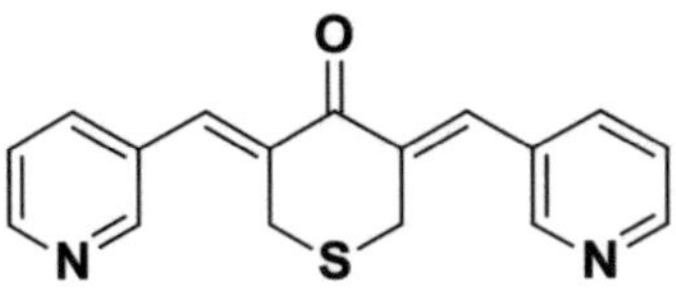

47

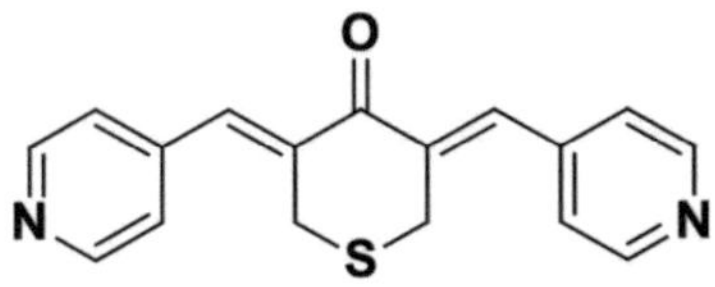

48

Gu and coworkers reported [37] the synthesis of fluoro phenoxy pyridine derivatives. The above-synthesized derivatives were checked for dual c-Met/VEGFR-2 targets. Initially, the above-synthesized derivatives were tested in *in vitro* assays on both c-Met and VEGFR-2. From the results, it was evident that compounds **49–51** showed very high inhibitory potency. Furthermore, an *in vitro* enzyme assay confirmed that compound **51** is the lead molecule. Molecular docking studies also confirmed that compound **51** was a potential compound for cancer treatment.

Abdelaziz and his group designed [38] and synthesized a series of pyridine analogs, and these synthesized derivatives were evaluated for anticancer PIM-I kinase activity. All the synthesized derivatives were evaluated by using 60 different cancer cell lines. From the results, it was confirmed that compounds **52–55** were identified as lead molecules. The active molecules were selected for PIM-1 kinase inhibitory activity. Those molecules that were active in *in vitro* activity displayed very good activity in PIM-1 kinase inhibitory activity.

Ansari and coworkers reported [9] the pyridine thiazolidinones as anticancer agents. Specifically, the above-synthesized derivatives were evaluated for the human carbonic anhydrase IX target. Among the synthesized derivatives, **56** and **57** showed very good enzyme inhibitory activity. Docking studies also supported their findings that the above identified active molecules showed good interaction and hydrogen bonding in the active pocket site. After that, the authors tested these molecules for *in vitro* activity against three cancer cell lines: HEK-293, MCF-7, and HepG2. Compounds **56** and **57** outperformed the reference doxorubicin *in vitro* activity with the cell lines MCF-7 and HepG2.

Durgapal and his group designed [39] and synthesized the 3-amino methyl pyridine derivatives. The above-synthesized derivatives were tested for their *in vitro* antiproliferative studies and DNA binding activity. *In vitro* activity was conducted using two cancer cell lines, that is, A549 and MCF-7, where 5-fluorouracil was used as a positive control. Among the series, compound **58** is identified as a lead molecule, and it is more active than 5-fluorouracil. Then, the further evaluation of this compound toward a DNA binding assay showed that compound **58** is twofold more active than compound **59**. Further evaluation of compound **59** by different tests proved to be efficient.

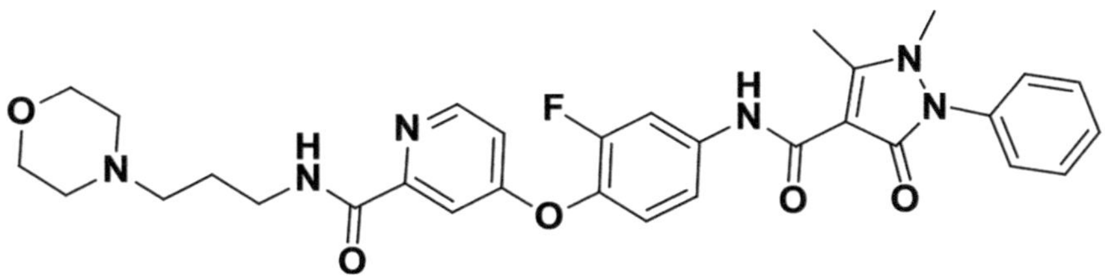

49

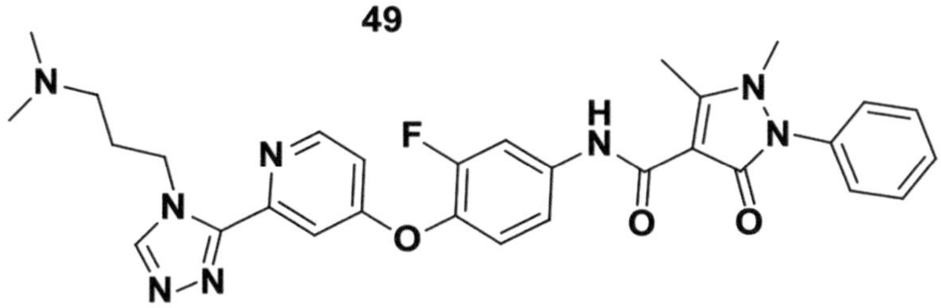

50

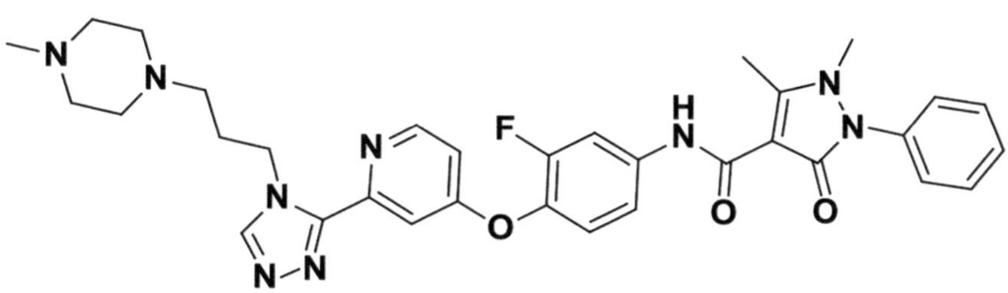

51

Gomha and his group developed [40] a series of thiadiazolo pyridine derivatives. The above-synthesized derivatives were tested for their anticancer activity using two cancer cell lines, that is, A549 and HepG2 cell lines, using cisplatin as a reference. Among the synthesized molecules, compound **60** emerged as the lead molecule in the HepG2 cell line and the most active molecule in the A549 cell line.

Another group reported some pyridine analogues for anticancer activity targeting G-Quadruplex [41]. Through the FLET melting assay, it was confirmed that compounds were selective for G-4 over duplexes. Most active G-4 ligands were tested for antiproliferative activity by using HL60 and K562 cell lines. Compound **61** is identified as the lead molecule from this assay. In the year 2018, another research group developed pyridine urea derivatives as anticancer agents [42]. Initially, the authors investigated *in vitro* activity against only MCF-7 cancer cell lines. Later, selected molecules were tested for *in vitro* activity against several panels of cell lines. According to the results of the studies, compounds **62** and **63** are potent molecules. Later, active molecules were tested against VEGFR-2. Both compounds exhibited good activity at micromolar concentrations.

Androutsopoulos and his group reported [43] the synthesis and biological evaluation of pyridine molecules. Initially, the authors tested these molecules in *in vitro* assays using HepG2 and MCF-7 cell lines. From the studies, it was confirmed that compound **65** is more active than compound **64** and in these compounds, HepG2 cells showed more sensitivity than other cell lines. From the cell cycle analysis, induction of G2/M phase arrest was observed. Down-regulation of the cell cycle associated protein cyclin D1 was also induced, as was up-regulation of the cell cycle inhibitors p53 and p21. These results indicate that these molecules are promising drug candidates for cancer.

Fayed and group reported [44] that the coumarin group contains pyridine analogs. The above-synthesized derivatives were tested in an *in vitro* antiproliferative assay using four different cancer cell lines, that is, HCT-116, MCF-7, HepG2, and A549 cell lines. From the results, it was evident that compounds **66–68** were identified as lead molecules. Further study of these molecules toward flow cytometric analysis revealed that cell cycle arrest in the G2/M phase is followed by apoptosis. In addition, the caspase-3 activity of lead molecules was confirmed; these compounds increased the caspase-3 activity more than the control group.

Eldehna and coworkers reported [45] a series of pyridine phenyl urea derivatives. The above-prepared derivatives were evaluated for *in vivo* activity. Cancer cell lines such as A549 and HCT-116 are used, and doxorubicin is used as a positive control. Among the phenyl area derivatives, compound **69** is identified as the lead molecule in both cell lines. Later, the activity of this lead molecule was tested against subpanels. The above-mentioned lead molecule causes apoptosis in HCT-116 cells, as evidenced by decreased expression of the anti-apoptotic Bcl-2 protein and increased levels of pro-apoptotic proteins. In addition, active molecules interrupted the cell cycle by arresting the G2/M phase. Later, an annexin V-FITC/propedium iodide assay was performed by treating the lead molecule with HCT-116 cells. An increase in positive annexin V-FITC apoptotic cells was observed, a nearly eightfold increase in comparison with control.

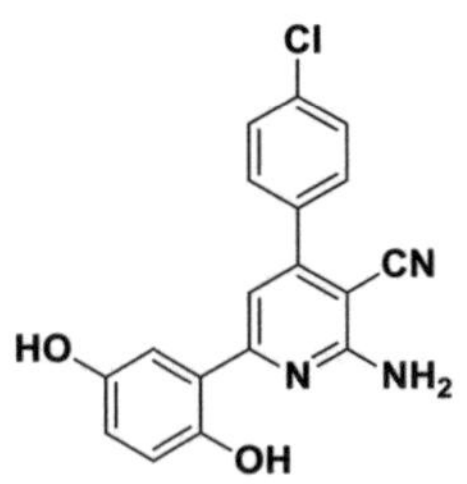

52

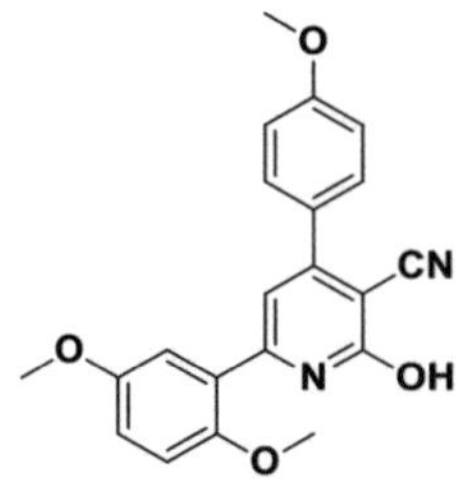

53

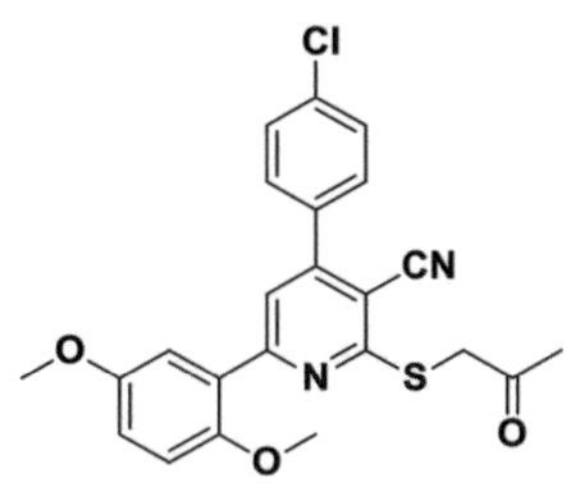

54

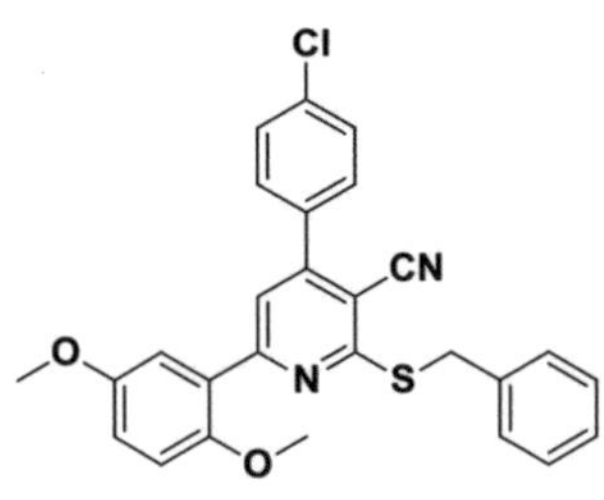

55

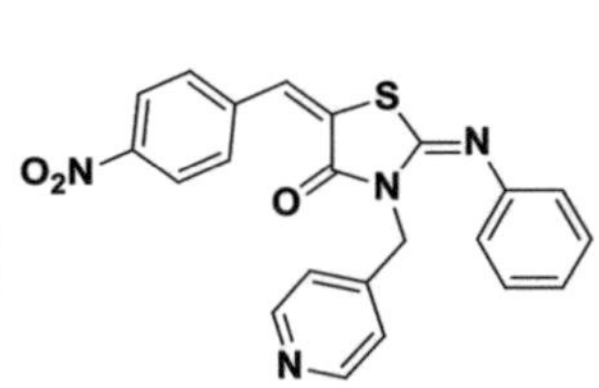

56

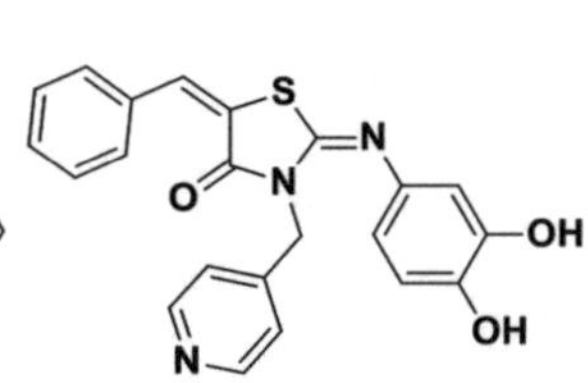

57

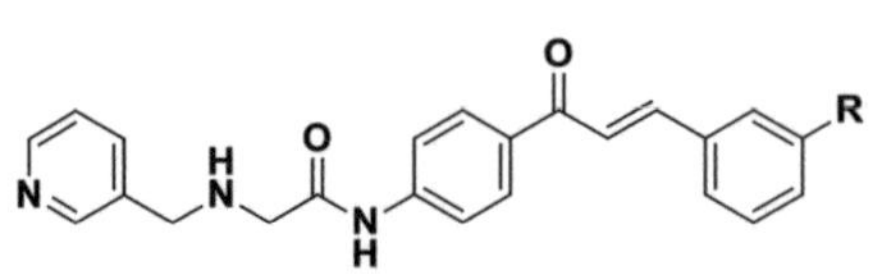

R = OMe **58**
R = Cl **59**

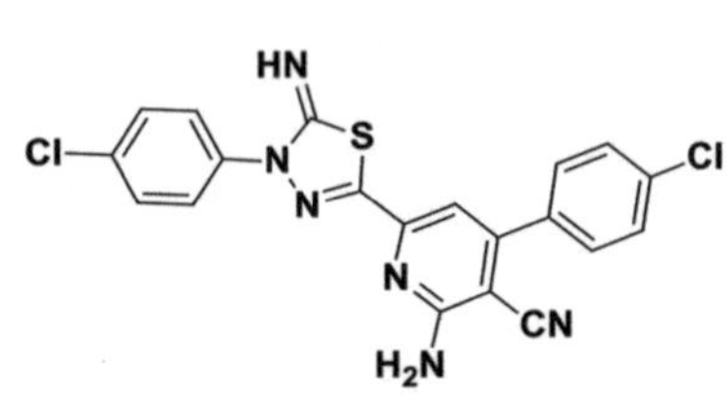

60

Muruguvel and team reported [46] the biological evaluation of the thiophene containing triazole and pyridine structures. Initially, *in vitro* activity of the above moiety was done by using human cancer cell lines such as A549, PC-3, and MDAMB-231 using doxorubicin as a positive control. The above pyridine derivative **70** showed better activity against breast cancer (MDAMB-231) cell lines than others. Xu and his group developed [47] chalcone pyridine analogues as anti-tubulin agents. Compound **71** displayed the most potent activity, and it effectively inhibited tubulin polymerization reactions by binding at colchine site of tubulin. In addition, cellular mechanism studies revealed that cell cycle arrest occurs at the G2/M phase. Notably, the *in vivo* efficacy of compound **71** was more potent than that of CA-4.

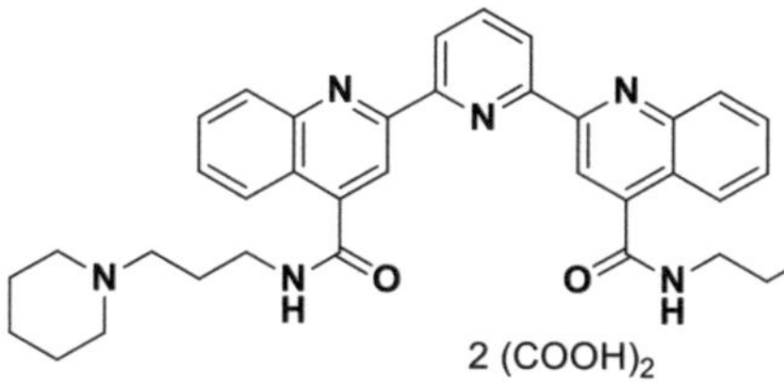

61

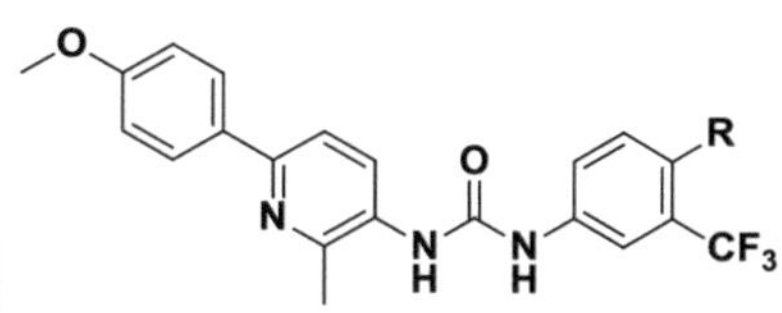

R = H, **62**
R = Cl, **63**

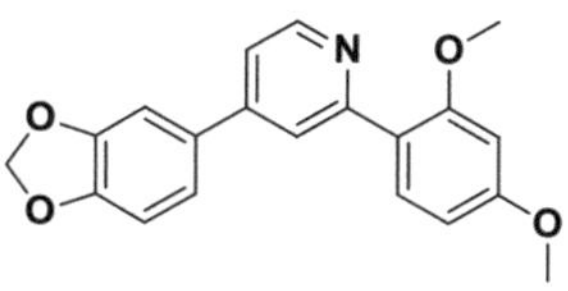

64

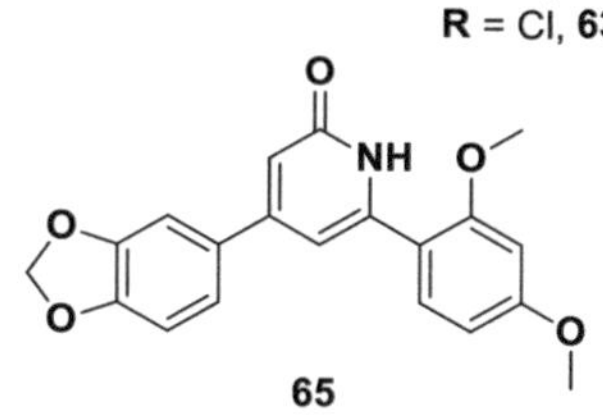

65

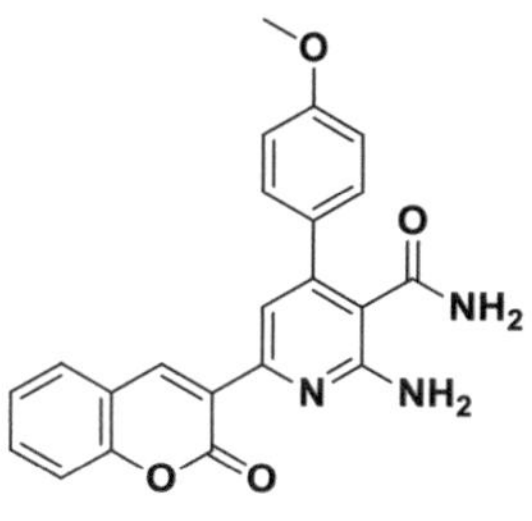

66

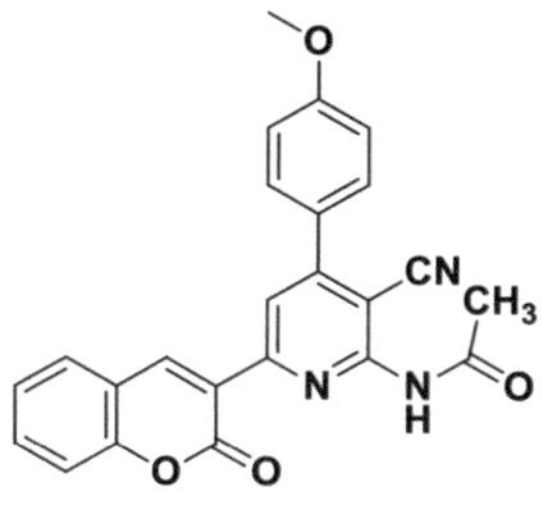

67

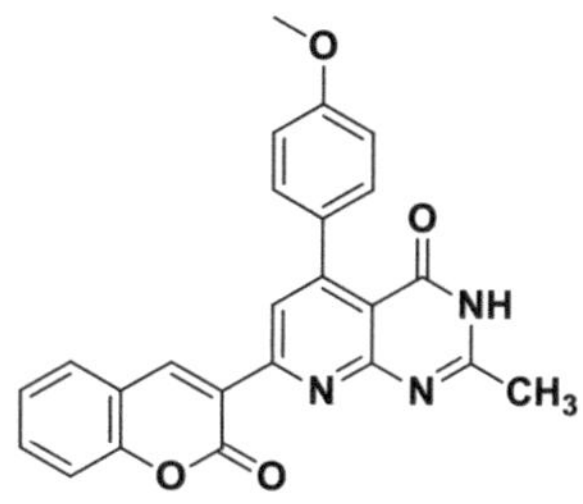

68

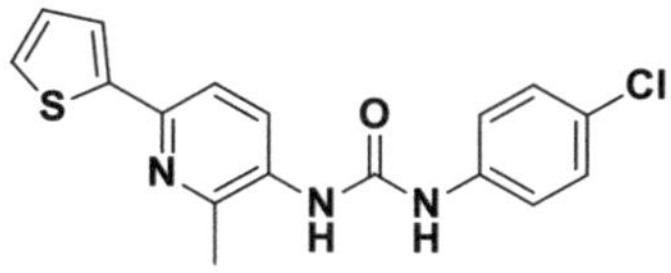

69

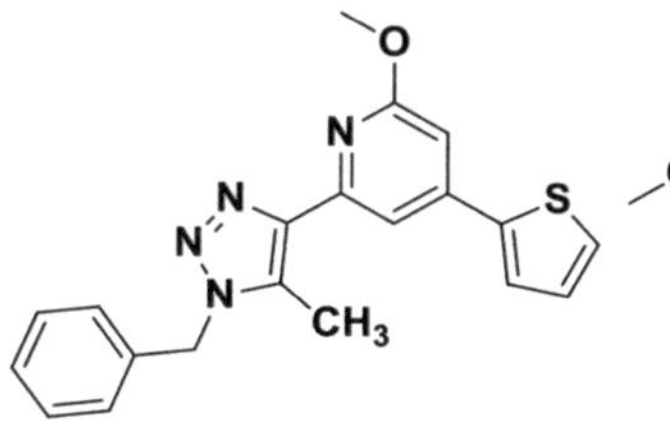

70

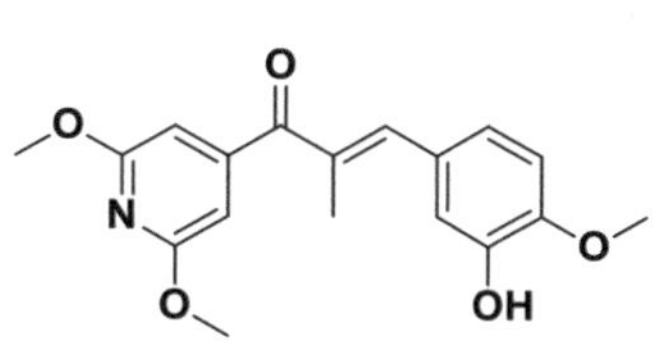

71

3. Conclusion

The results of several investigations into pyridine with anticancer qualities are listed in this review, as well as the prospective function of the pyridine nucleus in the creation of anticancer drugs. As can be seen from the biological actions of pyridine derivatives, the pyridine nucleus is a very versatile nucleus in the pharmacological sector. Anticancer drugs are frequently utilized with their derivatives. As a result, the pyridine nucleus could be thought of as a cure-all for a variety of ailments.

Acknowledgements

K.S.R. thanks the Council of Scientific & Industrial Research, New Delhi, and the Indian Science Congress Association, Kolkata, for providing the Emeritus Scientist and Asutosh Mookerjee Fellowship, respectively. K.S.R. and C.D.M. thank DST-Promotion of University Research and Scientific (PURSE), Institution of Excellence, and the University of Mysore for providing laboratory facilities.

Conflict of interest

The authors declare no conflict of interest.

Author details

Kereyagalahally H. Narasimhamurthy[1], Nichhapurada Kallesha[2], Chakrabhavi D. Mohan[3*] and Kanchugarakoppal S. Rangappa[4]

1 Department of Studies in Organic Chemistry, University of Mysore, Mysuru, India

2 The Himalaya Drug Company, Makali, Bengaluru, India

3 Department of Studies in Molecular Biology, University of Mysore, Mysuru, India

4 Vignana Bhavan, Institute of Excellence (IOE), University of Mysore, Mysuru, India

*Address all correspondence to: cd.mohan@yahoo.com

References

[1] Allais C, Grassot J-M, Rodriguez J, Constantieux T. Metal-free multicomponent syntheses of pyridines. Chemical Reviews. 2014;**114**(21):10829-10868. DOI: 10.1021/cr500099b

[2] Narasimhamurthy KH, Sajith AM, Joy MN, Rangappa KS. An overview of recent developments in the synthesis of substituted thiazoles. Chemistry Select. 2020;**5**(19):5629-5656. DOI: 10.1002/slct.202001133

[3] Oehninger L, Rubbiani R, Ott I. N-Heterocyclic carbene metal complexes in medicinal chemistry. Dalton Transactions. 2013;**42**(10):3269-3284

[4] Lynn DJ, Benson SC, Lynn MA, Pulendran B. Modulation of immune responses to vaccination by the microbiota: Implications and potential mechanisms. Nature Reviews Immunology. 2022;**22**(1):33-46. DOI: 10.1038/s41577-021-00554-7

[5] Hoffman R, Gerber M. The Mediterranean diet: Health and science. Wiley-Blackwell. 2013. https://www.wiley.com/en-gb/9781118713389

[6] Rashdan HR, Shehadi IA, Abdelmonsef AH. Synthesis, anticancer evaluation, computer-aided docking studies, and ADMET prediction of 1, 2, 3-triazolyl-pyridine hybrids as human aurora B kinase inhibitors. ACS Omega. 2021;**6**(2):1445-1455

[7] Liang X, Wu Q, Luan S, Yin Z, He C, Yin L, et al. A comprehensive review of topoisomerase inhibitors as anticancer agents in the past decade. European Journal of Medicinal Chemistry. 2019;**171**:129-168

[8] Naguib BH, El-Nassan HB. Synthesis of new thieno [2, 3-b] pyridine derivatives as pim-1 inhibitors. Journal of Enzyme Inhibition and Medicinal Chemistry. 2016;**31**(6):1718-1725

[9] Ansari MF, Idrees D, Hassan MI, Ahmad K, Avecilla F, Azam A. Design, synthesis and biological evaluation of novel pyridine-thiazolidinone derivatives as anticancer agents: Targeting human carbonic anhydrase IX. European Journal of Medicinal Chemistry. 2018;**144**:544-556. DOI: 10.1016/j.ejmech.2017.12.049

[10] Lawson M, Rodrigo J, Baratte B, Robert T, Delehouze C, Lozach O, et al. Synthesis, biological evaluation and molecular modeling studies of imidazo [1, 2-a] pyridines derivatives as protein kinase inhibitors. European Journal of Medicinal Chemistry. 2016;**123**:105-114

[11] Lin R, Johnson SG, Connolly PJ, Wetter SK, Binnun E, Hughes TV, et al. Synthesis and evaluation of 2, 7-diamino-thiazolo [4, 5-d] pyrimidine analogues as anti-tumor epidermal growth factor receptor (EGFR) tyrosine kinase inhibitors. Bioorganic & Medicinal Chemistry Letters. 2009;**19**(8):2333-2337

[12] Misra RN, Rawlins DB, Xiao H-Y, Shan W, Bursuker I, Kellar KA, et al. 1H-Pyrazolo [3, 4-b] pyridine inhibitors of cyclin-dependent kinases. Bioorganic & Medicinal Chemistry Letters. 2003;**13**(6):1133-1136

[13] Han JG, Gupta SC, Prasad S, Aggarwal BB. Piperlongumine chemosensitizes tumor cells through interaction with cysteine 179 of IκBα kinase, leading to suppression of NF-κB–regulated gene products. Molecular Cancer Therapeutics. 2014;**13**(10):2422-2435

[14] Shudo N, Mizoguchi T, Kiyosue T, Arita M, Yoshimura A, Seto K, et al. Two pyridine analogues with more effective ability to reverse multidrug resistance and with lower calcium channel blocking activity than their dihydropyridine counterparts. Cancer Research. 1990;**50**(10):3055-3061

[15] Temple C, Rener GA, Waud WR, Noker PE. Antimitotic agents: Structure-activity studies with some pyridine derivatives. Journal of Medicinal Chemistry. 1992;**35**(20):3686-3690. DOI: 10.1021/jm00098a014

[16] Liu MC, Lin TC, Sartorelli AC. Synthesis and antitumor activity of amino derivatives of pyridine-2-carboxaldehyde thiose micarbazone. Journal of Medicinal Chemistry. 1992;**35**(20):3672-3677. DOI: 10.1021/jm00098a012

[17] Cory JG, Cory AH, Rappa G, Lorico A, Mao-Chin L, Tai-Shun L, Sartorelli AC. Inhibitors of ribonucleotide reductase: Comparative effects of amino- and hydroxy-substituted pyridine-2-carboxaldehydethiosemicarbazones. Biochemical Pharmacology. 1994;**48**(2): 335-344. DOI: 10.1016/0006-2952(94)90105-8

[18] Jew S-s, Park B-s, Lim D-y, Kim MG, Chung IK, Kim JH, et al. Synthesis of 6-formyl-pyridine-2-carboxylate derivatives and their telomerase inhibitory activities. Bioorganic & Medicinal Chemistry Letters. 2003;**13**(4):609-612. DOI: 10.1016/S0960-894X(02)01041-7

[19] Amr A-GE, Mohamed AM, Mohamed SF, Abdel-Hafez NA, Hammam AE-FG. Anticancer activities of some newly synthesized pyridine, pyrane, and pyrimidine derivatives. Bioorganic & Medicinal Chemistry. 2006;**14**(16):5481-5488. DOI: 10.1016/j.bmc.2006.04.045

[20] Onnis V, Cocco MT, Fadda R, Congiu C. Synthesis and evaluation of anticancer activity of 2-arylamino-6-trifluoromethyl-3-(hydrazonocarbonyl) pyridines. Bioorganic & Medicinal Chemistry. 2009;**17**(17):6158-6165. DOI: 10.1016/j.bmc.2009.07.066

[21] Amin KM, El-Zahar MI, Anwar MM, Kamel MM, Mohamed MH. Synthesis and anticancer activity of novel tetralin-6-ylpyridine and tetralin-6-ylpyridine derivatives. Acta Poloniae Pharmaceutica-Drug Research. 2009;**66**(3):279-291

[22] Elgemeie GH, Mahdy EM, Elgawish MA, Ahmed MM, Shousha WG, Eldin ME. A new class of antimetabolites: Pyridine thioglycosides as potential anticancer agents. Zeitschrift für Naturforschung C. 2010;**65**(9-10):577-587. DOI: 10.1515/znc-2010-9-1008

[23] Elzahabi HSA. Synthesis, characterization of some benzazoles bearing pyridine moiety: Search for novel anticancer agents. European Journal of Medicinal Chemistry. 2011;**46**(9):4025-4034. DOI: 10.1016/j.ejmech.2011.05.075

[24] Liu W, Zhou J, Zhang T, Zhu H, Qian H, Zhang H, et al. Design and synthesis of thiourea derivatives containing a benzo[5,6]cyclohepta[1,2-b] pyridine moiety as potential antitumor and anti-inflammatory agents. Bioorganic & Medicinal Chemistry Letters 2012;**22**(8):2701-2704. DOI: 10.1016/j.bmcl.2012.03.002

[25] Bassyouni FA, Tawfik HA, Soliman AM, Rehim MA. Synthesis and anticancer activity of some new pyridine derivatives. Research on Chemical Intermediates. 2012;**38**(7):1291-1310. DOI: 10.1007/s11164-011-0413-9

[26] Sangani CB, Makawana JA, Zhang X, Teraiya SB, Lin L, Zhu H-L. Design, synthesis and molecular modeling of pyrazole–quinoline–pyridine hybrids as a new class of antimicrobial and anticancer agents. European Journal of Medicinal Chemistry. 2014;**76**:549-557. DOI: 10.1016/j.ejmech.2014.01.018

[27] Zheng S, Zhong Q, Mottamal M, Zhang Q, Zhang C, LeMelle E, et al. Design, synthesis, and biological evaluation of novel pyridine-bridged analogues of combretastatin-A4 as anticancer agents. Journal of Medicinal Chemistry. 2014;**57**(8):3369-3381. DOI: 10.1021/jm500002k

[28] Lu T, Goh AW, Yu M, Adams J, Lam F, Teo T, et al. Discovery of (E)-3-((Styrylsulfonyl)methyl)pyridine and (E)-2-((Styrylsulfonyl)methyl)pyridine derivatives as anticancer agents: Synthesis, structure–Activity relationships, and biological activities. Journal of Medicinal Chemistry. 2014;**57**(6):2275-2291. DOI: 10.1021/jm4019614

[29] Eldehna WM, Altoukhy A, Mahrous H, Abdel-Aziz HA. Design, synthesis and QSAR study of certain isatin-pyridine hybrids as potential anti-proliferative agents. European Journal of Medicinal Chemistry. 2015;**90**:684-694. DOI: 10.1016/j.ejmech.2014.12.010

[30] Fu C-W, Hsieh Y-J, Chang TT, Chen C-L, Yang C-Y, Liao A, et al. Anticancer efficacy of unique pyridine-based tetraindoles. European Journal of Medicinal Chemistry. 2015;**104**:165-176. DOI: 10.1016/j.ejmech.2015.09.032

[31] Abd El-All AS, Osman SA, Roaiah HMF, Abdalla MM, Abd El Aty AA, AbdEl-Hady WH. Potent anticancer and antimicrobial activities of pyrazole, oxazole and pyridine derivatives containing 1,2,4-triazine moiety. Medicinal Chemistry Research. 2015;**24**(12):4093-4104. DOI: 10.1007/s00044-015-1460-3

[32] Abbas I, Gomha S, Elaasser M, Bauomi M. Synthesis and biological evaluation of new pyridines containing imidazole moiety as antimicrobial and anticancer agents. Turkish Journal of Chemistry. 2015;**39**:334-346

[33] Abdelazem AZ, Al-Sanea MM, Park H-M, Lee SH. Synthesis of new diarylamides with pyrimidinyl pyridine scaffold and evaluation of their anti-proliferative effect on cancer cell lines. Bioorganic & Medicinal Chemistry Letters. 2016;**26**(4):1301-1304. DOI: 10.1016/j.bmcl.2016.01.014

[34] Naresh Kumar R, Poornachandra Y, Nagender P, Santhosh Kumar G, Krishna Swaroop D, Ganesh Kumar C, et al. Synthesis of novel nicotinohydrazide and (1,3,4-oxadiazol-2-yl)-6-(trifluoromethyl)pyridine derivatives as potential anticancer agents. Bioorganic & Medicinal Chemistry Letters. 2016;**26**(19):4829-4831. DOI: 10.1016/j.bmcl.2016.08.020

[35] Wu L-t, Jiang Z, Shen J-j, Yi H, Zhan Y-c, Sha M-q, et al. Design, synthesis and biological evaluation of novel benzimidazole-2-substituted phenyl or pyridine propyl ketene derivatives as antitumour agents. European Journal of Medicinal Chemistry. 2016;**114**:328-336. DOI: 10.1016/j.ejmech.2016.03.029

[36] Zhou DY, Zhao SQ, Du ZY, Zheng X, Zhang K. Pyridine analogues of curcumin exhibit high activity for inhibiting CWR-22Rv1 human prostate cancer cell growth and androgen receptor activation. Oncology Letters. 2016;**11**(6):4160-4166. DOI: 10.3892/ol.2016.4536

[37] Gu W, Dai Y, Qiang H, Shi W, Liao C, Zhao F, et al. Discovery of novel 2-substituted-4-(2-fluorophenoxy)

pyridine derivatives possessing pyrazolone and triazole moieties as dual c-Met/VEGFR-2 receptor tyrosine kinase inhibitors. Bioorganic Chemistry. 2017;**72**:116-122. DOI: 10.1016/j.bioorg.2017.04.001

[38] Abdelaziz ME, El-Miligy MMM, Fahmy SM, Mahran MA, Hazzaa AA. Design, synthesis and docking study of pyridine and thieno[2,3-b] pyridine derivatives as anticancer PIM-1 kinase inhibitors. Bioorganic Chemistry. 2018;**80**:674-692. DOI: 10.1016/j.bioorg.2018.07.024

[39] Durgapal SD, Soni R, Umar S, Suresh B, Soman SS. 3-Aminomethyl pyridine chalcone derivatives: Design, synthesis, DNA binding and cytotoxic studies. Chemical Biology & Drug Design. 2018;**92**(1):1279-1287. DOI: 10.1111/cbdd.13189

[40] Gomha SM, Muhammad ZA, Abdel-aziz MR, Abdel-aziz HM, Gaber HM, Elaasser MM. One-pot synthesis of new thiadiazolyl-pyridines as anticancer and antioxidant agents. Journal of Heterocyclic Chemistry. 2018;**55**(2):530-536. DOI: 10.1002/jhet.3088

[41] Das RN, Chevret E, Desplat V, Rubio S, Mergny J-L, Guillon J. Design, synthesis and biological evaluation of new substituted diquinolinyl-pyridine ligands as anticancer agents by targeting G-quadruplex. Molecules. 2018;**23**(1):81

[42] El-Naggar M, Almahli H, Ibrahim HS, Eldehna WM, Abdel-Aziz HA. Pyridine-ureas as potential anticancer agents: Synthesis and in vitro biological evaluation. Molecules. 2018;**23**(6):1459

[43] Androutsopoulos VP, Spandidos DA. Anticancer pyridines induce G2/M arrest and apoptosis via p53 and JNK upregulation in liver and breast cancer cells. Oncology Reports. 2018;**39**(2):519-524. DOI: 10.3892/or.2017.6116

[44] Fayed EA, Sabour R, Harras MF, Mehany ABM. Design, synthesis, biological evaluation and molecular modeling of new coumarin derivatives as potent anticancer agents. Medicinal Chemistry Research. 2019;**28**(8):1284-1297. DOI: 10.1007/s00044-019-02373-x

[45] Eldehna WM, Hassan GS, Al-Rashood ST, Al-Warhi T, Altyar AE, Alkahtani HM, et al. Synthesis and in vitro anticancer activity of certain novel 1-(2-methyl-6-arylpyridin-3-yl)-3-phenylureas as apoptosis-inducing agents. Journal of Enzyme Inhibition and Medicinal Chemistry. 2019;**34**(1):322-332. DOI: 10.1080/14756366.2018.1547286

[46] Murugavel S, Ravikumar C, Jaabil G, Alagusundaram P. Synthesis, computational quantum chemical study, in silico ADMET and molecular docking analysis, in vitro biological evaluation of a novel sulfur heterocyclic thiophene derivative containing 1,2,3-triazole and pyridine moieties as a potential human topoisomerase IIα inhibiting anticancer agent. Computational Biology and Chemistry. 2019;**79**:73-82. DOI: 10.1016/j.compbiolchem.2019.01.013

[47] Xu F, Li W, Shuai W, Yang L, Bi Y, Ma C, et al. Design, synthesis and biological evaluation of pyridine-chalcone derivatives as novel microtubule-destabilizing agents. European Journal of Medicinal Chemistry. 2019;**173**:1-14. DOI: 10.1016/j.ejmech.2019.04.008

Chapter 12

Comparison of the Pharmacokinetics of Eflornithine after Application of Eflornithine Cream and "Eflornithine: Armenicum" Composition in Rates

Hovhannes Ghazaryan and Areg Hovhannisyan

Abstract

This chapter reports the study results to determine percutaneous absorption and pharmacokinetics of eflornithine following topical treatment with eflornithine hydrochloride 13.9% cream and "eflornithine–armenicum" composition in rats. The model of aerobic wounds was developed. Eflornithine hydrochloride cream (dose of 460 mg/kg) was applied in group I, and "eflornithine–armenicum" composition was applied in group II at a same dose of Eflornithine. The plasma concentration-time profile of racemic eflornithine following frequent sampling was determined by the HPLC method (LLOQ, 1.5 ng/ml). Eflornithine concentrations were measurable at 24 h, with peak concentrations in plasma 5.3 ng/ml after cream and 3.8 ng/ml after composition application ($p < 0.001$) and the average time to reach the maximum concentration of eflornithine increases from 2 h to 3.3 h. The area under the pharmacokinetic curve was decreased after composition application by 25%. Eflornithine was eliminated from plasma with a mean terminal half-life of 11.6 hours. It can be assumed that the use of "eflornithine–armenicum" composition allows for maintaining the optimal concentration of two anti-inflammatory compounds at the site of application for a long time, which can improve their pharmacological effect compared to separate use of eflornithine cream.

Keywords: Eflornithine, "Eflornithine–Armenicum," composition, pharmacokinetics, percutaneous absorption

1. Introduction

One of the actual problems of modern medicine is the views and approaches to wound healing, depending on the cause of wounds.

They are numerous, especially depending on the effect of bacteria, such as aerobic and anaerobic, viruses, fungi, injuries to the skin due to burns, and various chemical and physical factors, such as low temperature [1].

In recent years, difficulties in wound healing have been observed mainly due to the ineffectiveness of various widely used antimicrobials, especially antibiotics. It is more severe if it develops against some existing somatic diseases, such as diabetes and non-diabetes mellitus, some acquired connective tissue diseases, congenital and acquired immune deficiencies, etc. The search and development for new effective means of symptomatic-pathogenic treatment of wounds are one of the priorities of modern medicine.

In connection with this, it becomes expedient to carry out pharmacokinetic research to find out the ability of the components of the given composition to remain in the wound.

Studies show that the pharmacokinetics of Eflornithine rats are characterized by a slow systemic blood flow after co-administration of Eflornithine paste to rats, providing a pronounced restorative effect on the wound surface than with Eflornithine alone.

Wound healing has always been fraught with difficulties, as the use of a wide range of antibacterial drugs, including antibiotics, has often been ineffective due to the polymorphic nature of the bacteria found in the wound.

The search for new effective antimicrobials for wound healing, including the development of new formulations the ingredients of which may have a multifaceted inhibitory effect on the growth of bacteria in the wound, is still a very promising direction in modern medicine.

In this regard, Armenicum (ointment and paste) exhibits significant cytotoxic effects on a number of resident conditionally pathogenic bacteria [1, 2].

The authors attribute the effects of "Armenicum" quite well to iodine, which has a direct and/or mediated bactericidal effect by stimulating the activity of radicals in the wounds of diseased tissues [1, 3–6].

The new composition "Eflornithine–Armenicum" has been developed by taking into account the new scientific interpretations that have appeared in the inflammatory processes of wounds during the last 10 years, which are given below:

- "Armenicum" shows the spectrum of antibacterial and anti-inflammatory effects in wound processes caused by aerobic bacteria.

- Eflornithine is endowed with a spectrum of antimicrobial activity against pathogenic and conditionally pathogenic resident bacteria containing polyamines.

Using the results of modern scientific research for the treatment of wounds, we found it expedient to develop the drug composition "Eflornithine–Armenicum", which will be characterized by access restriction of polyamines to microorganisms.

The inclusion of Eflornithine in the composition of the Armenicum will inhibit diabetes and non-diabetes mellitus bacterial persistence in the earliest stages of the local inflammatory process, which will probably lead to a faster and earlier wound-healing process.

It should be noted that pharmacological examinations should be mandatory for the approbation of the therapeutic efficacy of the "Eflornithine–Armenicum" composition on the wound model.

Eflornithine human racemic pharmacokinetics are characterized by an oral bioavailability of approximately 50%, mainly renal elimination (>80%) of low extraction, no reported metabolites, negligible plasma protein binding, and a multiphasic plasma concentration-time profile, probably contributing to highly varying estimates of half-life ranging from 2 to 30 h and percutaneous absorption of topical applications of Eflornithine hydrochloride (HCl) cream is very low (<1%). Absorbed Eflornithine is rapidly eliminated from plasma and predominantly excreted in urine without being metabolized. While the percutaneous absorption of Eflornithine increased after twice-daily topical application, relative to the first application, absorption reached a steady state within four days of twice-daily application and remained low. Trough plasma concentrations also remained constant after four days of twice-daily application. Systemic exposure was low, with steady-state peak and trough plasma concentrations during twice-daily treatment of 10 and 5 ng/ml, respectively, and the terminal half-life averaged 8 to 11 hours. The low degree of percutaneous absorption and low systemic exposure to Eflornithine offer a favorable clinical safety profile of Eflornithine HCl 13.9% cream [7–9].

The results of the pharmacokinetics study of Armenicum paste following its application on wound surface in two doses showed that the pharmacokinetics of the major active ingredient of Armenicum paste was characterized by slow absorption into the blood and remained on the wound surface for a long time providing more pronounced pharmacological effect [6].

It should be noted that pharmacological studies should serve as an obligatory stage in testing the therapeutic efficacy of the "Eflornithine–Armenicum" composition on a wound model. For this purpose, we carried out pharmacokinetic studies of the medicinal composition created by us on the basis of the pharmacological company "Arpimed" (Armenia) in order to determine the duration of its stay on the wound surface and the rate of absorption from the wounds into the bloodstream.

2. Materials and methods

2.1 Drugs and reagents

Vaniqa 11.5% cream (11.5%, Vaniqa, Batch: 838225): Each gram of cream containing 115 mg of Eflornithine hydrochloride monohydrate as the active ingredient, 47.2 mg of cetostearyl alcohol, 14.2 mg of stearyl alcohol, 0.8 mg of methyl parahydroxybenzoate, and 0.32 mg of propyl parahydroxybenzoate (Almirall Limited, Harman House, UB8 1QQ, UK).

Armenicum paste (Arpimed CJS, Batch: 0019): Each gram of paste containing iodide anion (J2) – 5.93%, calcium iodide (KJ) – 8.89%, lithium chloride (LiCl) – 0.15%, dextrin ($C_{18}H_{32}O_{16}$) – 74.13%, polyvinyl alcohol (C_2H_4O) – 2.23%, sodium chloride (NaCl) – 6.67%, and water – 2%.

To get the 100 g "Eflornithine–Armenicum" composition, add 85 g of Armenicum paste and 15 g of Eflornithine hydrochloride monohydrate (Eflornithine hydrochloride is 13.9 g) to the composition of the container. Stir the mixture at room temperature for 10–15 minutes before using.

The standard used for the Eflornithine hydrochloride monohydrate was the RS USP standard (series R06840). For analytic method validation, the L-ornithine hydrochloride was obtained from Sigma (St. Louis, MO, USA; product A-4571, Batch: 19/L), methanol, high-performance liquid chromatography (HPLC) gradient was obtained from grade Carl Toth GmbH + Co (Art. 7342.1), acetic acid HPLC class was obtained from Carl Roth GmbH (Art. 3738.3), high purity (+99%) acetone was obtained from Carl Roth GmbH (Art. No. 7328.2), and high purity (+99%) dansyl chloride was obtained from Sigma-Aldrich (product –d2625).

1-Heptansulfonic acid sodium salt monohydrate was obtained from Carlo Erba (France Batch No: V8A553138I), distilled water for HPLC (Part No. 5062–8578), Hewlett-Packard, and 0.45-μm nylon filter was obtained from Carl Roth Germany (Carl Roth Germany, Art.7330.1). All chemicals were of analytical grade, and all solvents were of HPLC grade.

2.2 Study animals

A total of 72 non-breed white male rats, bred by the Institute of Fine Organic Chemistry of the National Academy of Science, Yerevan, Armenia, were used in the study. All animals were clinically examined upon arrival and those that showed signs of abnormality or disease were excluded. Animals were kept in the animal house for 10–15 days prior to the commencement of the study under a 12 h/12 h light/dark cycle at 25 to 27°C and 60 to 65% humidity and were offered humidity and were offered standard rat chow ad libitum. Unfit animals were replaced prior to the start of the study, and no animals were replaced after the study began. The initial weights of the study animals were in the range of 160–200 g.

Care of rats and all interventions were performed according to the PHS Guide for the laboratory animals [10] in accordance with requirements of the YSMU Ethics Committee. All experiments were performed during the light phase of the cycle.

2.3 Experimental design

A 4 cm^2 wound was incised on the inner surface of the hind leg of animals by the model of aerobic wounds created according to Oganesyan S.S. methodology [11]. After this trauma, the musculoskeletal tissues were compressed twice by the Kocher clamp. The preparations were investigated 5–7 minutes after the onset of the injury: Eflornithine paste or "Eflornithine–Armenicum" composition was applied to the entire wound surface of the animals, which covered all the upper wounds.

Development of aerobic wounds in animals and further operations were performed under ether anesthesia. The animals were kept in isolation without bandages after the wound was healed.

The animals were divided into 2 groups given below:

- Group I: 36 rats (mean weight 162 ± 39 g) were treated using a single dose of Eflornithine paste at a dose of 460 mg/kg.
- Group II: 36 rats (mean weight 167 ± 37 g) were treated using a single dose of "Eflornithine–Armenicum" composition (Eflornithine: 460 mg/kg, Armenicum: 5.1 mg/kg).

At 0, 2.0, 4.0, 8.0, 12.0, and 24.0 h after the application of investigated preparations, six animals from the both group were sacrificed and blood samples taken.

Following decapitation with a laboratory guillotine, blood was collected from each animal in separate heparinized centrifuge tubes that were centrifuged at 600 g for 15 min in order to obtain blood plasma. An aliquot of plasma (1.0 ml) was taken for Eflornithine assay and the remainder of the sample was conserved by closing the centrifuge tube with a cap and storing it in a freezer for 1–3 days prior to further assay. On the day of the assay, plasma was removed from the refrigerator and stored at room temperature for 1 hour.

2.4 Racemic quantification of Eflornithine

Racemic Eflornithine was quantified using precolumn derivatization, followed by high-performance liquid chromatography (HPLC) and ultraviolet (UV) detection according to published methods [8, 9], modified as described below. The HPLC system consisted Shimadzu LC/UV/MS instrument (LC-20 AD/T, Shimadzu Corporation, Kyoto, Japan) consisting of an autosampler (LC-20 AD/SIL-20A), a UV-diode array detector (SPD-M20A IVDD), and data acquisition and analysis software (Lab Solutions, Version 3.40.299. Shimadzu) set at 330 nm.

Plasma samples (0.5 ml) were precipitated with ice-cold methanol (3 ml) containing an internal standard (DL-4-amino-3-hydroxybutyric acid) at a concentration of 20 ng/ml. The samples were placed on a vortex mixer for approximately 10 s, centrifuged for 10 min at 12,000 g, and thereafter kept at 37°C for 10 minutes. The supernatants were transferred to new tubes and evaporated to dryness at 65°C under a gentle stream of air. The dried samples were dissolved in 100 μl phosphate buffer (0.1 M; pH 7.5). The derivatization mixture was prepared daily by mixing o-phthalaldehyde (20 mg), ethanol (1 ml), nitriloacetic acid (4 mg), mercaptoethanol (100 μl), and 10 ml phosphate buffer (0.1 M; pH 7.5).

Prior to injection, the samples were mixed by adding two 50-μl volumes of derivatization mixture (i.e., o-phthalaldehyde) to the 48-well-plate autoinjector.

The temperature for the autoinjector was kept constant at 20°C. The plasma samples containing the derivatization mixture were programmed to stand in the autoinjector for 2.00 min prior to injection. Eflornithine was separated on a Chromolith Performance RP-18e 100-mm by 4.6-mm-ID column (VWR International, Darmstadt, Germany) protected by a ChromSep Guard SS 10-mm by 2-mm-ID column (Varian, Palo Alto, CA), using a gradient program. Mobile phase A consisted of 92% phosphate buffer (0.1 M; pH 7.5), 5% methanol, and 3% acetonitrile. Mobile phase B consisted of 80% methanol, 10% acetonitrile, and 10% water. The flow rate was set to 2 ml/min, using the following gradient program: t = 0 to 6.75 min, linear decrease of A from 80–40%; t = 6.75 to 8.0 min, 40% A; and t = 8.0 to 10.0 min, linear increase of A from 40–80%. The typical retention times for the internal standard and Eflornithine were 5.5 ± 0.3 min and 7.3 ± 0.2 min, respectively.

Calibration curves, constructed using Eflornithine standards, were linear in the range 1.00–10.0 ng/ml with correlation coefficients (r) of 0.9991, for the target signals at 330 nm. The limit of detection (at a signal/noise ratio of 3) was 0.80 ng/ml and the lower limit of quantification (LLOQ) was 1.5 ng/ml. The LLOQ for racemic Eflornithine was set to 1.5 ng/ml at which level precision and accuracy were < 12%. The experimental plasma samples for method validation were prepared at three concentrations (1.5, 3, and 6 ng/ml) and were analyzed in duplicate at each level during the analytical runs to ensure that experimental samples were accurately and precisely determined. Validation experiments demonstrated that the accuracy was 96.9 ± 3.5%, the precision (coefficient of variation) 3.6%, and the recovery 93.94 ± 2.9%.

2.5 Pharmacokinetic analysis

The pharmacokinetic parameters were calculated using Kinetica 4.4.1 software (Thermo Electron Corporation, 2004, USA).

The measured parameters were as follows:
maximum plasma concentration (C_{max}, ng/ml);
elimination rate constant (k_{el}, h^{-1});
elimination half-life ($t_{1/2}$, h);
area under concentration versus time curve extrapolated to infinity ($AUC_{0\text{-}\infty}$, hng/ml); absorption rate constant (k_a, h^{-1});
the time to C_{max} (t_{max}, h).

2.6 Statistical analysis

Statistical analyses were performed using GraphPad PRISM software (version 2.0, 1996; GraphPad Software, San Diego, USA). The statistical significance of differences between the pharmacokinetics profiles of Eflornithine released from Eflornithine cream and "Eflornithine–Armenicum" composition over time were assessed using two-way between/within ANOVA wherein an interaction effect indicates a different response over time between the two dosage forms.

3. Results

The pharmacokinetic profiles of Eflornithine determined in control rats (Group I) and in those (Group II) that had been treated with "Eflornithine -Armenicum" composition are shown in **Figure 1**.

Maximum plasma concentrations of Eflornithine (C_{max}) averaged 5.193 ± 0.74 ng/ml after Eflornithine cream application in Group I and decreased following "Eflornithine–Armenicum" composition 3.716 ± 1.198 ng/ml.

A comparison of the two pharmacokinetic profiles of Eflornithine indicates that the concentration of Eflornithine in the blood plasma of animals treated with Eflornithine cream was slightly higher than in that of the "Eflornithine–Armenicum" composition group during the 2–24 h following application and attained its maximum value (C_{max}) at 2.00 h.

For Group II treated with "Eflornithine–Armenicum" composition, the observed average time to reach maximum concentration in blood plasma (t_{max}) was slightly

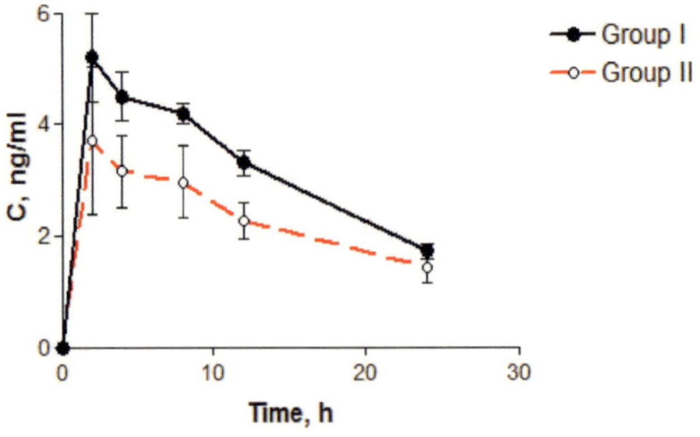

Figure 1.
Mean plasma profiles of Eflornithine following topical application of Eflornithine cream (group I) formulation of HCl and "Eflornithine–Armenicum" composition (group II) (n = 6, mean ± SD).

lower (3.33 ± 0.74 h), but the difference between the mean values was not statistically significant (**Table 1**).

After reaching the maximum, the pharmacokinetic profile of the Eflornithine concentration in plasma is almost the same in both groups (**Table 2** and **Figure 1**).

The values for the average elimination half-life ($t_{1/2}$) and of Eflornithine were very similar for both groups: (11.600 ± 0.961 h and 11.761 ± 0.956 h for groups I and group II, respectively), and the difference between the mean values was not statistically significant (**Table 1**).

Notable differences were found; however, in the absorption rate constant (k_a) of Eflornithine, which was higher in the Eflornithine cream-treated group and averaged 4.595 ± 0,53 h^{-1} following Eflornithine cream application against 3.244 ± 0,54 h^{-1} for "Eflornithine–Armenicum" composition treated group and the difference between the mean values was statistically significant (**Table 1**).

The areas under the concentration-time curve $AUC_{0-\infty}$ were also different between the control group and the group treated with "Eflornithine–Armenicum" composition. The difference between the mean values was statistically significant (**Table 1**).

The results showed that when Eflornithine was used concomitantly with Armenicum and the maximum plasma concentrations of Eflornithine were approximately 30% lower than those of Eflornithine cream. The total amount of Eflornithine in blood plasma ($AUC_{0-\infty}$) also decreased by 23%. This indicates that Eflornithine enters the bloodstream from the application site significantly more slowly and to a lesser extent when used in combination with Armenicum.

Pharmacokinetics Parameters	Eflornithine HCI cream	"Eflornithine–Armenicum" composition	p
C_{max}(ng/ml)	5.362 ± 0.53	3.822 ± 0.42	0.005***
t_{max}, (h)	2.00 ± 0.00	3.333 ± 0.74	0.363^{ns}
$K_a(h^{-1})$	4.595 ± 0.53	3.244 ± 0.54	0.0011**
$AUC_{0-\infty}$ (hng/ml)	76.872 ± 4.72	57.913 ± 3.57	0.0001***
$t_{1/2}$(h)	11.588 ± 0.96	11.761 ± 0.95	0.8131^{ns}

Table 1.
Pharmacokinetic parameter estimates for eflornithine in the rat after application of eflornithine HCI cream and "Eflornithine–Armenicum" composition.

Time, h	Eflornithine cream, (Eflornithine HCl 460 mg/kg)	"Eflornithine–Armenicum" composition, (Eflornithine HCl 460 mg/kg)
0	0	0
2	5.193 ± 0.74	3.716 ± 1.198
4	4.620 ± 0.40	3.160 ± 0.60
8	4.341 ± 0.21	2.973 ± 0.597
12	3.307 ± 0.18	2.27 ± 0.299
24	1.625 ± 0.30	1.146 ± 0.37

Table 2.
Mean plasma profiles of Eflornithine following topical application of Eflornithine cream and formulation of Eflornithine HCl and "Eflornithine–Armenicum" composition (n = 6, mean ± SD).

Thus, it is possible that Eflornithine can remain on the surface of the site of application for a longer time, which probably contributes to the longer manifestation of its local anti-inflammatory effect.

If the theoretical data obtained are correct, then in practice, we can calculate the relative bioavailability of Eflornithine using the method of Ritschel and Kearns (1998), after its co-administration with Armenicum [12, 13].

According to generally accepted pharmacokinetic approaches, if two dosage forms contain the same substance in the same dosage and are used in combination with another substance, it is possible to assess its relative bioavailability by setting the bioavailability of the first dosage form at 100%, if this ingredient has been used alone, and has the same dosage in the investigated pharmaceutical form [12, 13].

The percent of relative bioavailability of Eflornithine was calculated as a ratio of Eflornithine's $AUC_{0-\infty}$ after applying "Eflornithine–Armenicum" and $AUC_{0-\infty}$ of Eflornithine after applying Eflornithine cream.

The results of the calculation showed that the relative bioavailability of Eflornithine after application of "Eflornithine–Armenicum" composition is $77.4 \pm 6.04\%$ of the bioavailability obtained following the application of Eflornithine cream alone.

Given the fact that the elimination rate of the drug after reaching the maximum concentration in blood is practically the same, it can be assumed that decreased relative bioavailability of Eflornithine is probably associated primarily with its lower and slower absorption of Eflornithine to the systemic circulation from the application site, following the application of "Eflornithine–Armenicum" composition [14].

As known from the literature data, in particular, the percutaneous absorption of Eflornithine remains low (less than 1% of the dose, based on excretion of radioactivity in urine and feces) after single and multiple doses applied under conditions of clinical use. The obtained data are largely similar to those obtained by Malhotra B. and coauthors [7], who confirmed that the average transdermal absorption of Eflornithine to blood plasma after Eflornithine cream application is approximately 1–2% of the radioactive dose. The authors concluded that the percutaneous absorption may correspond to a zero-order kinetic absorption.

The latter presupposition allows concluding that Eflornithine is probably absorbed into the bloodstream after percutaneous application by nonlinear absorption kinetics, which is in line with the previous conclusion by several authors [15–18].

The data do not exclude the possibility that the transfer of Eflornithine from the application surface to the general bloodstream occurs through active carriers. The kinetics of this process was very different from ordinary passive transport.

A system that performs such special transport of compounds through biomembranes (involving carriers) usually has a limited capacity [14, 19–21].

It can be assumed that when the "Eflornithine–Armenicum" composition was used, two biologically active substances of Armenicum, iodide-dextrin and eflornithine, can compete with each other for space on the carrier since it has previously been shown that the iodide anion can also be transported through biomembranes using carriers [1, 4, 5].

Consequently, as the density of the Eflornithine–iodine–dextrin complex increases at the surface of application, their penetration rate into the bloodstream reaches a constant value at some point from the site of application, leading to a slow absorption process. Thus, it can be assumed that the combined use of Eflornithine and Armenicum in the form of a composition allows for maintaining the optimal concentration of two anti-inflammatory compounds at the site of application for a long time, which can improve their pharmacological effect compared to the separate use of Eflornithine cream.

4. Conclusion

Eflornithine concentrations were measurable at 24 h, with peak concentrations in plasma of 5.3 ng/ml after cream and 3.8 ng/ml after composition application ($p < 0.001$). After combined use of Eflornithine and Armenicum, the average time to reach the maximum concentration of Eflornithine is observed, which increased from 2 h to 3.3 h. The area under the Eflornithine plasma concentration versus time curve was decreased after composition application by 25%. Eflornithine was eliminated from plasma with a mean terminal half-life of 11.6 hours without statistically significant difference after cream and composition application.

Thus, it can be assumed that the combined use of Eflornithine and Armenicum in the form of a composition allows for maintaining the optimal concentration of two anti-inflammatory compounds at the site of application for a long time, which can improve their pharmacological effect compared to the separate use of Eflornithine cream.

Conflict of interest

The authors declare no conflict of interest.

Author details

Hovhannes Ghazaryan[1] and Areg Hovhannisyan[2*]

1 Arpimed Pharmaceutical Company, SCJ, Abovyan, Armenia

2 Institute of Fine Organic Chemistry of the National Academy of Science of the Republic of Armenia, Yerevan, Armenia

*Address all correspondence to: dopingareg@gmail.com

References

[1] Zilfyan A, Avagyan S, Ghazaryan A. The Effect of Armenicum Paste on the Course of Wound Process. Germany: Monograf LAB LAMBERT Academic Published; 2016

[2] Aleksanyan YT, Kamalyan LA, Melik-Andresyan GG. Experimental Study of the Antiviral Effect of the Preparation "Armenicum". Yerevan: "Armenicum" - Experimental research, Gitutyun; 2000. pp. 49-69

[3] Abrahamyan H, Muradyan R, Ghazaryan A, Hovhannisyan A. The pharmacokinetics of Armenicum paste on the rats. Vesnik, S-P. 2009;**14**:75-78

[4] Ghazaryan A, Topchyan H, Hovhannisyan A, Abrahamyan H, Muradyan R. The pharmacokinetics of Armenicum paste tested on the rats. The new Armenian Medical Journal. 2010;**4**:48-49

[5] Abrahamyan H, Hovhannisyan A. Comparative pharmacokinetics of iodide anion after administration of iodine preparations. The new Armenian Medical Journal. 2011;**5**:26-32

[6] Ghazaryan A, Avagyan S, Topchyan H, Zilfyan A. Shifts in the content of samatostatin in blood serum of intact rats under experimentally induced wound process with the application of armenicum drug and paste. The New Armenian Medical Journal. 2016;**10**:53-56

[7] Malhotra B, Noveck R, Behr D, Palmisano M. Percutaneous absorption and pharmacokinetics of Efornithine HCl 13.9% cream in women with unwanted facial hair. Journal of Clinical Pharmacology. 2001;**41**:972-978

[8] Jansson R, Malm M, Roth C, Ashton M. Enantioselective and nonlinear intestinal absorption of Eflornithine in the rat. Antimicrobial Agents and Chemotherapy. 2008;**52**:2842-2848. DOI: 10.1128/AAC.00050-08

[9] Johansson CC, Gennemark P, Artursson P, Äbelö A, Ashton M, Jansson-Löfmark R. Population pharmacokinetic modeling and deconvolution of enantioselective absorption of eflornithine in the rat. Journal of Pharmacokinetics and Pharmacodynamics. 2013;**40**:117-128. DOI: 10.1007/s10928.9293-x.012

[10] Guide for the Care and Use of Laboratory Animals. Garber JC, editor. Eighth ed. Washington: The National Academic Press; 2011. p. 246

[11] Oganesyan SS, Tarverdyan NA, Zilfyan AV. Method of Simulating Abscess, Inventor's certificate number SU 1347089 A1, international number G09B 23/28, 1987 Date of registration: 28.02.1986; Date of publication: 23.10.1987 1987

[12] Ritschel WA. Absorption of drugs. Pharmaceutical International. 1974;**1**:343-346

[13] Ritschel WA, Kearns GL. Handbook of Basic Pharmacokinetics Including Clinical Application. 5th ed. Washington, D.C.: American Pharmaceutical Association; 1998. p. 563

[14] Gibaldi M. Limitation of classical theories of drug absorption. In: Nimmo LE, editor. Prescott, W.S. Book Drug Absorption. Lancaster: MTP Press Limited Falcon House; 1981. p. 458

[15] Griffin CA, Slavik M, Chien J, Hermann G, Thompson O, Blanc GD, et al. Phase I trial and pharmacokinetic study of intravenous and oral alpha-difluoromethylornithine. Investigational New Drugs. 1987;**5**:177-186

[16] Rillema JA, Hill MA. Prolactin regulation of the pendrin-iodide transporter in the mammary gland. American Journal of Physiology. Endocrinology and Metabolism. 2003;**284**:E25-E28

[17] Burri C, Brun R. Eflornithine for the treatment of human African trypanosomiasis. Parasitology Research. 2003;**90**:S49-S52

[18] Na-Bangchang K, Doua F, Konsil J, Hanpitakpong W, Kamanikom B, Kuzoe F. The pharmacokinetics of eflornithine (alphadifluoromethylornithine) in patients with late-stage T.b. gambiense sleeping sickness. European Journal of Clinical Pharmacology. 2004;**60**:269-278. DOI: 10.1007/s00228-004-0759-7

[19] Hayton WL. Rate-limiting barriers to intestinal drug absorption: a review. Journal of Pharmacokinetics and Biopharmaceutics. 1980;**8**:321

[20] Lacroix L, Pourcher T, Magnon C, Bellon N, Talbot M, Intaraphairot T, et al. Expression of the apical iodide transporter in human thyroid tissues: A comparison study with other iodide transporters. The Journal of Clinical Endocrinology and Metabolism. 2004;**89**:1423-1428

[21] Volkenstein MV. Biophysics: Textbook. Allowance. 3rd ed. St. Petersburg; Moscow-Krasnodar: Nauka; 2008. p. 594

Chapter 13

Trending Advancements in Technologies Pertinent to Therapeutical Pharmacodynamics

Saravanan Priyadharshini, Muthusamy Velusamy Nivedha, Molabanti Vasupriya, Santhanakrishnan Jayasree, Subramaniam Mahalakshmi, Gunasekaran Sinduja and Usha Singaravelu

Abstract

It is omniscient that pharmacodynamics of a drug is understood as its effect on body by interacting with the structure of targets to either activate or inhibit their function/action. Based on the activity of the drug to the target binding site they have been classified into different types. Physiological, cellular, molecular, biochemical, and toxicological effects individually have a significant role in drug's effect and response. The mechanism of action, dosage and its response, therapeutic index are some of the noteworthy parameters to be considered in drug pharmacodynamics. This chapter comprehends the above-mentioned concepts, their importance in pharmacodynamics of drug, and the impact of recently developed methods like genome-wide or transcriptome-wide sequencing, chronopharmacodynamics, systems biology, pharmacometabolomics, etc., in different stages of drug discovery process, and how the digitization of therapeutics and healthcare direct the path to personalized medicine. The integration of bioinformatics, systems biology, and big data related approaches like ML & AI with the pharmacological (PK/PD) study highly benefits the patients' therapeutics.

Keywords: drug response, pharmacodynamic parameters, integrated approaches, pharmacology, digital therapeutics

1. Introduction

Pharmacodynamics (PD) elucidates the relationship of drug's concentration with the body (target), the way it affects and the impact it creates. It can be an ultimate characteristic feature to determine the effects exhibited by the concentration of the drug so as to assure its efficacy and safety. This feature in turn helps to avoid the adverse and toxic reactions caused by the drug that is administered to the body. PD forms a part of pharmacology that deals with the concepts of drug's interaction with any one of the four primary groups of target proteins, namely, ion channels, enzymes, membrane carriers, and receptors, and concentration of drug with respect to time, and the effect caused by the dosage of the drug. Since, the concentration of drug is significantly related to the drug's effect, there are some parameters to confine its efficiency and some advanced computer-based techniques to highlight the concerns in this regard. These kinds of techniques enable PD to unlock the problems occurring while designing a novel drug which could be rectified by the development of precision making medicines through digital therapeutics/treatment. This in-turn paves way for PD to place its remarkable contribution in the phase of personalized medicine design based on the individual's requirement.

This chapter mainly focuses on the advanced technologies in today's trend in order to exemplify the crucial part of drug's PD characteristics in drug discovery process in consideration to human health and also, some of the basic concepts need to be taken into mind, like its importance in drug discovery process and the technologies which escalate the identification of PD of a drug are discussed in brief.

2. Parameters of pharamacodynamics

Parameters of PD illustrate the mode of action/interaction of the drug in the active site of the target. It is important that the drug should meet the PD parameters of concentration to inhibit the function of the target [1]. The drug responses can be manifested by the PD parameters.

2.1 Mechanism of action

Mechanism of action (MoA) is simply a biochemical interaction resulted in physiological response. This mechanism is enacted by the process called signal transduction which provides the description of the drug's (chemical) action (i.e. transmission of signal (chemical/drug) from the surface of the cell to the cytoplasm and/or nucleus of the cell) in the body/cell/receptor. At the beginning of this process, the cell recognizes the signal that bound to the receptor which is embedded either in the plasma membrane or present in the intracellular region. Transduction of signal takes place in the mid-way of the signal transduction process. Here the multi-step pathway utilized to carry out the rapid transmission of signal to the series of molecules. Protein kinases and protein phosphatases are the enzymes, trigger the phosphorylation cascade, and dephosphorylation processes in respect to activate and deactivate the protein. As a result of this, the secondary messengers transmit the signals brought from the cell surface receptor to the intracellular receptor to exhibit the cellular response by making changes in gene regulation processes.

Traditional methods (Direct approach) like affinity chromatography, and modern "omics" related high throughput methods (Indirect approach) are aid to identify the drug's mode of action by comparing the diseased cell (i. e., cell treated with the drug) with the normal cell [2]. The determination of MoA of a drug advantages in the categorization of the patients who are all exposing the similar kind of response to the treatment and estimation of accurate dosage of the drug can be done by monitoring the patients' target pathway [3].

2.1.1 Grouping of drugs based on mode of action

The signal transduction process allows us to group the drugs based on their binding mode and effects (either positive or negative). In general, drugs (ligands) can be grouped into two (i) Agonist, and (ii) Antagonist. *Agonist* is a kind of drug that imitates the ligand embedded within the receptor structure that exhibits positive effect and produces biological response by activating the receptor in the signal transduction process. In simple words, it tends to replicate the expected (positive) reaction. More or less agonists possess the ability to meet up the maximum response by partially engrossing the receptors.

There are four different kinds of agonists lie under a single umbrella. The first type is the *full agonist* that means, it causes maximum amount of biological effect at maximum concentration. The second type is the *partial agonist,* that causes less maximum effect and therefore it receives partial response from the activation of receptor. The third type is an *inverse agonist* that binds with the receptor which is already in active state to minimize its activity. The fourth type of agonist is the *biased agonist* which makes the receptor capable of giving signals with various efficacies in their multiple downstream pathways.

Antagonist is another kind of drug that has no intrinsic activity in the receptor to produce any response. It prevents the agonist to bind at the receptor binding site by its increased concentration and reduces the fractional occupancy of the drug [4]. It forms a number of irreversible covalent bonds that will elongate the pharmacological activity [5].

There are two types of pharmacologic antagonist viz. *Competitive antagonists*, as its name says battle with the agonist to bind in the same active site of the receptor. *Non-competitive antagonists* lower the capacity of an agonist in its response towards the irreversible binding of the antagonist to the receptor. It produces a large antagonistic effect as an outcome [6].

2.2 Dosage and its response

PD encourages the utilization of several mathematical models namely, fixed effect model, logarithmic model, E_{max} model, and sigmoid E_{max} model [6] to explain the PD parameters such as efficacy (E_{max}), potency (EC_{50}), equilibrium rate constant (K), and sigmoidicity constant (n). These PD parameters can feature the drug-target interaction and their effect in the magnitude and duration of drug response. Rashed [7] estimated the magnitude and duration of response by comparing the maximum observed effect (MOE) to a sequence of doses and he also estimated the time taken for the half-life effect at various concentrations.

Efficacy (E_{max}) is the maximum response caused by the drug that evaluates the relationship between concentration and effect. *Potency* (EC_{50}) is the concentration of the drug at a stationary state which shows the half of the maximum effect. When the potency increases, the response from the respective concentration will decrease. This depends on some properties such as the receptor density, efficiency of stimulus–response mechanisms which is in use, drug affinity, and drug efficacy denoted in terms of $EC_{50}/ED_{50}/Kd$ [6]. The hill coefficient (g) is an empiric parameter derived from the sigmoidicity of the effect-concentration correlation which defines the relationship between drug concentration and its effect. The hill coefficient value of >2 indicates a steep relationship, >3 indicates that it can be either all or none effect. The sigmoidicity constant (n) is a linear relationship of concentration-effect profile. The relationship between n and effect is inversely proportional (i.e.,) the increase in will decrease the concentration.

Affinity of a drug is also one of a parameter to be concern, defined by the time course and strength of binding of the drug with its target at certain degree of concentration and it falls under the factors which are essential to determine the potency of a drug [5].

Some other parameters to be considered in PD are drug dissociation (Kd), receptor occupancy, up and down regulation of receptor. The *drug dissociation* (Kd) value shows how strongly the drug binds to its receptor for example, if the Kd value obtain from analysis is low, it is the indication of strong binding and higher affinity of a drug. The fundamental of *receptor occupancy* lies in the law of mass action, which says that an excellent PD response can only be attained if the number of receptors covered by the drug is more. The *up and down regulations of receptor* have been happened due to the long-time subjection of a receptor to an antagonist and/or an agonist [8].

2.3 Therapeutic index

The therapeutic index (TI) is one of the parameters help to determine and optimize safety and efficacy of the drug. It can be referred as the ratio of the drug's dosage at which the drug attains the no toxicity state [9]. It differentiates the drug effect caused at a particular concentration of blood from the amount of the effect that causes toxicity [10]. TI is the composition of drug effect (ED_{50}), toxic effect (TD_{50}), and lethal effect (LD_{50}). ED_{50} can be define as a scale of the concentration at which the half percent (50%) of patients exhibiting a particular pharmacologic effect. TD_{50} the amount of dosage needed to exhibit a particular toxic effect. LD_{50} is the prediction of the amount of dosage used to kill 50% of patients.

3. Role of pharmacodynamics IN DRUG'S effect and response

In basic, PD can be denoted as the determination of the effect of pharmacology and toxicology of the drug that appears as a result of the interaction occurred between the drug and receptor [11]. An individual drug's response to the target could be knocked off by the activities such as activation, inhibition of the targets' disease mechanisms at the active site described quantitatively with the support of PD. Depending on the activity of target and its biochemical pathway, the PD effects can be classified as direct, indirect, immediate, and delayed effects. The direct effect takes place in biochemical interaction pathway and the indirect effect occurs far from the end of the biochemical pathway.

With the knowledge of the drug concentration at the active site of the receptor, we can quantify the intense of the drug's effect and the magnitude of the response that is differing for the drug with the same concentration due to the disease and/or pre-administered drug. The response could be interrupted by many criteria such as the density of receptor on cell surface, signal transmission, and some regulatory factors.

The apparent effects such as physiological, cellular, molecular, biochemical and toxicological effects are the products of the drug-receptor interaction that is significant to make decision about dosage/concentration of the drug. Respective to the response of the drug action, the distribution and dosage of the drug increases simultaneously [8].

3.1 Physiological effects

Physiological factors like body type, quantitative amount of drug action, time course of drug action to the target, produces the effect that could interrupt the actual behavior of the drug. Age is one of the significant physiological factors affecting the drug's action. When fixing the dosage of the drug, the age of the patients is necessary to be taken into account because, the absorption and distribution of the drug dosage requires more attention than the binding site of target. This is the reason why clinical trials mostly involve the middle and old age patients. Next to age, specificity becomes an important physiological factor affecting the drug's effect and response. The drug and the target should be very specific to provide successful results, otherwise it may lead to fatal results and hence, accuracy of monitoring is very much important. Another physiological factor to be considered is the time duration which can be defined as either at which time the drug enters the body or the time the drug reaches the target and at which time it's being excreted [12].

3.2 Cellular effects

Cell has been serves as a medium in which the interaction of drug and target takes place. There is a possibility for the occurrence of two different instances, such as, the bonding of drug with the target makes it function properly as a metabolite and the other one is, it forms a group of toxic compounds which exhibit toxicological effects as an end product. It is through the process of endocytosis, the drug enters into the cell surface area. The drug needs to be a lipophilic component, since the cellular membrane is made up of lipid bilayer, which permits only the lipophilic components and cause the cytochrome P450 to produce the inhibition action/to simulate the production of the factors which cause disease, thereby describing the effect and functioning of a drug.

3.3 Molecular effects

Drug by interacting with the macromolecules of the cellular organelles like, proteins, lipids, etc. causes some changes in the rate or magnitude of an intrinsic cellular response without producing any new responses. Many drugs that have been interacting with receptors (e.g., serum albumin) within the body do not cause any biochemical or physiological response directly even though some positive and negative effects like agonism and antagonism are stimulated by the action of the drug to the receptor. The understanding of drug interactions at the molecular level guides us to know about drug's effect in detail, and hence the dosage and distribution of the drug.

3.4 Biochemical effects

The biochemical effects, also an outcome of the drug-receptor interaction that is referred as therapeutic processing of drugs that aids the modification of the drug's conformation according to the environment of the receptor. Since the function of the molecule is associated with the structure of molecule, if the structure of the molecule is modified, then mechanism of receptor might be modified or inhibited by the instance of biochemical interaction which as a result exposes biochemical effects. The administration of oral drug creates a long lasting intense effect whereas the injections possesses and intense effects [13].

3.5 Toxicological effects

Toxicity referred as the production of toxic effect that is due to the entry of the xenobiotic components which are external chemical substances for humans/animals that alter the function of the administered drug. Instead, there is a theory in the name of 'Darwinian toxicology' that shows the contrast view on toxicological effects by stating that these effects are the body's protective response against the drug has been administered to the body [14].

The detection of toxicological effects remains as a complex issue. It is even yet difficult to solve the toxic effects of drugs which are being subjected to the clinical trials. Perhaps, the drug passes all the pre and post drug designing/discovery stages by exhibiting positive results till the end of the clinical trials, it may become a huge problematic situation if it shows toxic effects after marketing the drug. It is inevitable to repeat the process again from the start point. This shows the importance of understanding the behavior of the drug molecule at the molecular level to avoid such toxicological effects. One of the "omics" concepts, biomarker identification helps to screen drugs with its associated toxic effect.

There are five different toxicity conditions, the first one being "On-target" that is the mechanism-based toxicity caused due to the interaction of drug with the same target to produce required pharmacological response. The second one is "Hypersensitivity" and "Immune response". The third one is "Off-target" toxicity, in which the drug interacts with alternate target and causes toxicity. The fourth one is "Bioactivation", which converts multiple drugs as product of metabolites (reactive). These elements manipulate and being a starting point of toxicity of the proteins that has to be avoided. The fifth condition is an unfamiliar condition named as "Idiosyncratic reactions" known as troublesome responses [15].

4. The impact of recently developed methods over pharmacodynmics approaches

4.1 Genome-wide association studies (GWAS)

Genome-wide association studies (GWAS) act as a tool to identify the position of the gene that affects the drug susceptibility and response to the body. The genomic variants cause the PD-PGx effects with respective to the variations in patient response in the pathways of drug-target interaction. Whole genome sequencing helps to determine the PGx markers exhibiting distinct features by employing conventional genetic screening [16].

The genetic architecture of drug response can be described in deep by incorporating the PD process with GWAS. This permits to find out the genetic influences regarding the drug response and adverse drug reactions through computational tools using the mathematical equation and statistical method [17].

In this regard, candidate-gene studies offer data on the cases that may vary from adverse drug reactions to the single gene's alleles. Some victorious examples speak about and suggest this approach for the drug toxicity investigation, interpretation of drug-response. The response for a drug might be differing from patient to patient because they may not respond properly for a particular drug or it may need an error-free dosage to acquire good results. Phase III of clinical trials adopts GWAS to determine the best drug with a precise dose [18].

4.2 Chronopharmacodynamics

Chronopharmacology is a field which depends on the biological time and endogenous periodicities to discern the drugs' biochemical and physiological effects on body, drug action mechanisms, drug-concentration relationship, and impact of circadian clock to the effect of the drug. Circadian clock has established its role in regulatory metabolism, detoxification, and few other physiological processes.

Chronopharmacodynamics validates the changes of circadian in the drugs' mode of action to provide optimized pharmacodynamic response. Chronotherapeutics deals with the scientific approach of synchronizing drug delivery with the body's circadian rhythm, to maximize the therapeutic index and enhance effectiveness. Studies have revealed that a drug may have different effects depending on the time of dosage [19].

Circadian pharmacodynamics/Chronesthesy could be a variation associated with time effects that allow some modifications in treatment to improve the chances of efficacy and safety and to lower the side effects of drug. However, side effects of drugs are influenced by physio-chemical properties and PK/PD of the drug. Since, circadian variation affect the PD, it becomes essential to consider the circadian rhythm prior to drug administration so as to prevent the timely variations that occur in the drug's MoA [20].

4.3 Systems biology

Systems biology sets a trend of using data mining and statistical tools which aid to analyze the networks to locate the topology of biological systems and to build dynamic ordinary differential equation (ODE) models to represent the mechanisms of biochemical reaction. Systems pharmacology has evolved as a discipline that possesses the features of both systems biology and pharmacology that could be employed in all the phases of drug research and development. The enhanced PD (ePD) models highlight the importance of testing the drug effects in a multilevel network and the treatment of an individual patient to promote accuracy and benefits of clinical decision-making [21]. It integratively employs various domains such as systems engineering, systems biology, and PK/PD which enhances the understanding of the complex biological systems by iterating the computation and mathematical model construction, experiments and quantitatively analyzing different interactions that occur between drug and biological systems [22].

The Network-based Systems Pharmacology is an impactful way to understand the adverse effects of drug. It benefits by increasing the drug efficacy, regulating signaling pathway with multiple channels, and provide higher success rate of clinical trials, and also lowering the costs of drug discovery and development [6].

4.4 Pharmacometabolomics

Pharmacometabolomics (PMx) related studies have turned their concentration towards the areas such as biomarker identification and metabolic patterns. PMx also known as Pharmacometabonomics, is a recently developed discipline that puts in together different aspects of an individual's metabolite profile via metabolic approaches and predicts the individual's variation with respect to drug response phenotypes. It paves a way to identify endogenic metabolites and their associated pathways, individual drug PK/PD characteristic predictions and biomarkers for observing disease progression.

Metabotype (an individual metabolic profile) is one of a new strategy that consists of baseline metabotype and treatment metabotype. *Baseline metabotype* can be obtained from pre-dose sample which deals with the heterogenic data and subtypes of disease. *Treatment metabotype* are obtained from the samples which are collected while dosing in order to determine the effect/side effects of drug and changes in response exhibited in molecular pathways. These two metabotypes help to find the variation occurred due to the effect of drug and that is exposed by inter and intra-patients. The investigation of therapeutic responses depends on these two metabotypes and the metabolic signatures that help to build models of patient's response to drug.

Another strategy is the PMx based biomarker identification which involves the steps such as data collection, process, statistical analysis, and biochemical interpretation. The biomarkers identified as an end-product of the PMx procedure may act effectively for responders and safely for non-responders in addition to some other unfavorable action and they are used to fix the range of drug dosage based on the metabolites discovered from treatment samples. PMx also identifies biomarkers involved in the downstream effects of pathophysiology and PD/PK events [11].

4.5 Artificial intelligence (AI)

AI speeds up the validation and optimization of the target and the drug structure respectively. AI approaches have been employed to check the safety and efficacy of the drug molecules by developing and analyzing big data.

Recently, QSAR approaches have also been transformed as AI-dependent QSAR approaches, namely, linear discriminant analysis (LDA), support vector machines (SVM), random forest (RF), decision trees (DT) and so on. The nearest-neighbor classifier, RF, extreme learning machines, SVMS, and deep neural networks (DNNs) are able to predict the *in vivo* activity and toxicity, physicochemical properties and bioactivity of the developed drugs. With the usage of training data set (n number of compounds), predictive models have been developed and tested to predict the properties of the data (n number of compounds lesser than training data set) based on some parameters such as molecular surface area, molecular mass, total hydrogen count, refractivity, volume, logP, total polar surface area, sum of E-states indices, solubility index (log S), and rotatable bonds. By choosing anyone of the drug feature and its respective target, AI can predict the binding affinity of the drug to its target. Tools designed based on ML and DL approaches such as KronRLS, SimBoost, DeepDTA, and PADME are used to determine drug-target binding affinity. DL-based methods like DeepDTA, PADME, WideDTA and Deep Affinity produce better results when compared to the ML-based methods via the application of network-based methods [23].

Deep learning (DL) models surpassing the PK/PD methods are being followed in the clinical trials to predict the time course of the response for the respective dose of the administered therapeutic. This requires quite a lot man power to model the dynamic systems for the bulk amount of data to produce the best product (drug).

The integration of major concepts with deep learning workflow may provide an effective outcome. The models which are built with the combination of machine and human support gain a center role in the diagnostic procedures to monitor the real time treatment data. The deep learning and PK/PD approach together help to meet the necessities of automatic modeling to define the dosage by reducing the amount of time and human energy has been spent.

AI and ML system is used to detect and predict the data of individuals' symptom by means of feedback loop using digital biomarkers thereby leading to precision medicine. Certain companies involving the AI robots combined with psychological models in therapeutic areas. Some other companies design pills with an ingestible mini sized sensor coated with copper and magnesium on alternate side. Once the pill is administered to a person (patient), it will pair the two sides and start to generate signal for the wearable sensor patch. The person who worn the sensor patch may get a digital record via a mobile app with the permission of his/her doctor (healthcare provider). The sensor patch will monitor the persons' activity, heart rate, sleep quality, temperature and even monitors disease conditions like diabetes, hypertension, etc.

Although AI is a boon for the pharmacological area, it has some notable limitations in handling data, such as its range, growth, variability, and unpredictability. The future of AI in clinical pharmacology has been expected to predict the concentration and effect of the drug with the help of DL models which could be implemented in digital devices that aid to monitor the treatment data automatically [24].

5. Digitized therapeutics and healthcare IN personalized medicine

Personalized medication become robust by predicting the accurate treatment results for the drug which will administer in the live environment and the high-throughput measurement of bio-molecules (genes, proteins, and metabolites) present in a cell. To fulfill this criterion, currently emerging technologies of digital health provides the capability to evaluate the patient's response to the drug in the real world and time by converting therapeutics as a Digital Personalized Medicine (DPM).

The Internet of Things (IoT), a network-based technology provides support in this regard. The Internet of Pharmaceutical Things (IoPT) digitally quantifies the effects of drug in real-time PD and enables remote monitoring [25]. Sensors and wireless connections are used to capture drug interaction and phenotype data. For example, Digital twin concept is used to improve diagnostics and treatments. Digital twins are an assembly of unlimited copies of models of all phenotypic, molecular, and environmental factors related to disease mechanisms. Each twin is treated computationally with potential drugs and a drug with high effectiveness is selected for treatment [26].

Personalized drug delivery system (PPDS) is a constant measurement of dosage that incorporates the Active Pharmaceutical Ingredients (APIs) with patient-adopted precise dose and possesses different aspects that could be modified to enhance the drug identification, absorbability, excretion and monitoring the treatment. The incorporation of PDDS and Digital health is the current upcoming area that may aid to upgrade the healthcare systems, cost-efficient treatments and boost the result of overall healthcare [27].

Digital health is a merge of diverse platforms and systems which offer solutions based on technologies to improve healthcare. This comprises almost all the technologies that are responsible for patients' health and the concepts such as mobile health (mHealth), telehealth (telemedicine), smart devices, sensors and wearables, health information technology and personalized medicine.

Digital therapeutics (DTx) is one such category of digital health solutions that provides evidence-based, software-driven therapeutic interventions for the prevention and management of medical disorders or diseases that aims to reduce the consumption level of pharmaceuticals or to be drug-free under certain conditions. It serves as an indicator to analyze drug effectiveness and is also used to compute the response of a person to a drug which is under investigation in clinical trials.

Products (digital therapeutics) developed by the implementation of advanced technology have to be approved by regulators and they should follow certain principles which are categorized based on the medical purpose provided by Digital Therapeutic Alliance (DTA) to serve individuals with safe and effective treatment. Criteria like, growth of digital analytical tools, up-gradation of data visualization and machine learning encourage researchers to convert data as knowledge that will assist clinical decision-making [28].

Nowadays the term "E-patient", pioneered by Dr. Tom Ferguson in 1999, has been gaining attention, representing the patients who are all able to decide their own health and healthcare and capable of making health-related decision utilizing the technology. They are developed to treat/manage/prevent certain disease conditions including chronic diseases. With the help of DTx, physicians can treat their patients residing anywhere and anytime beyond the border of a clinic or a hospital. It helps them to make accurate diagnostic decisions at the right time with the concern about patient care. This ideology proceeds towards the personalized treatment of the patient through monitoring and tracking the response for the prescribed therapies [29].

6. Conclusion

Development of drug is a multistep process that deals with multiple concepts and domains. It requires a holistic knowledge of all related areas to identify a potential drug to provide an expected result. The drug begins its action from administration to excretion from the body. The prediction of drug's action in human body/cell is essential to develop a potent drug with no toxic effects. This can be accomplished by the already developed/developing computational approaches with the insight of the PD associated concepts like parameters to consider, and the relative effect and response of the drug.

This opens up the door of personalized medicines in the form of digital therapeutics, enable the physician to prescribe medication on the basis of patients' requirement. But it is still important to test and mining evidences in order to evaluate its efficacy and safety.

This chapter helped us to realize that the computational approaches discussed in this chapter are providing such good outcomes when combining them with one another. Even though, PD has attained a remarkable place in the drug discovery and marketing, an effective product (drug) can be obtain by unifying the PK/PD properties of a drug. In other words, as a whole pharmacology of a drug should be taken into concern in order to design a patient-friendly drug/medicine.

Author details

Saravanan Priyadharshini, Muthusamy Velusamy Nivedha, Molabanti Vasupriya, Santhanakrishnan Jayasree, Subramaniam Mahalakshmi, Gunasekaran Sinduja and Usha Singaravelu*
Integrated Bio-computing Laboratory, Department of Bioinformatics, Bharathiar University, Coimbatore, India

*Address all correspondence to: usha@buc.edu.in

References

[1] Gallagher JJ, Williams-Bouyer N, Villarreal C, Heggers JP, Herndon DN. Chapter 12 – Treatment of infection in burns. In: David NH, editor. Total Burn Care. Third ed. Saunders, W. B., Elsevier Inc; 2007. pp. 136-176

[2] Tulloch LB, Menzies SK, Coron RP, Roberts MD, Florence GJ, Smith TK. Direct and indirect approaches to identify drug modes of action. IUBMB Life. 2017;**70**(1):9-22

[3] Puneeth GR, Mahesh SR, Adithya H, Bhargava SJ, Chaya Kumari HA, Gururaj HL, et al. Analysis of drug classification using mechanism of action. Journal of Physics: Conference Series. 2021;**1914**:012034

[4] Reeves PT, Roesch C, Raghnaill MN. Drug action and pharmacodynamics. In: MSD Veterinary Manual. Rahway, NJ, USA: Merck & Co., Inc.; 2016

[5] Lees P, Cunningham FM, Elliott J. Principles of pharmacodynamics and their applications in veterinary pharmacology. Journal of Veterinary Pharmacology Therapeutics. 2004;**27**:397-414

[6] Salahudeen MS, Nishtala PS. An overview of pharmacodynamic modeling, ligand-binding approach and its application in clinical practice. Saudi Pharmaceutical Journal. 2016;**25**:165-175

[7] Rashed SF. Pharmacodynamic Parameters: Influence on Dose-Response Relationship and Estimation. Rhode Island: University of Rhode Island Digital Commons; 1999

[8] Marino M, Jamal Z, Zito PM. Pharmacodynamics. In: StatPearls [Internet]. Treasure Island (FL): StatPearls Publishing; 2022

[9] Muller PY, Milton MN. The determination and interpretation of the therapeutic index in drug development. Nature Reviews: Drug Discovery. 2012;**11**(10):751-761

[10] Tamargo J, Le Heuzey JY, Mabo P. Narrow therapeutic index drugs: A clinical pharmacological consideration to flecainide. European Journal of Clinical Pharmacology. 2015;**71**(5):549-567

[11] Beger RD, Schmidt MA, Kaddurah-Daouk R. Current concepts in Pharmacometabolomics, biomarker discovery, and precision medicine. Metabolite. 2020;**10**(4):129

[12] Martens CR, Denman BA, Mazzo MR, Armstrong ML, Reisdorph N, McQueen MB, et al. Chronic nicotinamide riboside supplementation is well-tolerated and elevates NAD+ in healthy middle-aged and older adults. Nature Communications. 2018;**9**(1):1286

[13] Burns JJ, Shore PA. Biochemical effects of drugs. Annual Review of Pharmacology. 1961;**1**:79-104

[14] Horii I. Toxic effect onset and evaluations of medicinal drugs horizon for Darwinian toxicological thought. The Journal of Toxicological Sciences. 2010;**35**(4):425-435

[15] Guengerich FP. Mechanisms of drug toxicity and relevance to pharmaceutical development. Drug Metabolism and Pharmacokinetics. 2011;**26**(1):3-14

[16] Katsila T, Patrinos GP. Whole genome sequencing in pharmacogenomics. Frontiers in Pharmacology. 2015;**6**:61

[17] Wu R, Tong C, Wang Z, Mauger D, Tantisira K, Szefler SJ, et al. A conceptual framework for pharmacodynamic genome-wide association studies in pharmacogenomics. Drug Discovery Today. 2011;**16**(19-20):884-890

[18] Daly A. Genome-wide association studies in pharmacogenomics. Nature Reviews Genetics. 2010;**11**:241-246

[19] Ozturk N, Ozturk D, Kavakli IH, Okyar A. Molecular aspects of circadian pharmacology and relevance for cancer chronotherapy. International Journal of Molecular Sciences. 2017;**18**(10):2168

[20] Erkekoglu P, Baydar T. Chronopharmacodynamics of drugs in toxicological aspects: A short review for clinical pharmacists and pharmacy practitioners. Journal of Research in Pharmacy Practice. 2012;**1**(2):41-47

[21] Iyengar R, Zhao S, Chung SW, Mager DE, Gallo JM. Merging systems biology with pharmacodynamics. Science Translational Medicine. 2012;**4**(126):126ps7

[22] Van der Graaf PH, Benson N. Systems pharmacology: Bridging systems biology and pharmacokinetics-pharmacodynamics (PKPD) in drug discovery and development. Pharmaceutical Research. 2011;**28**:1460-1464

[23] Paul D, Sanap G, Shenoy S, Kalyane D, Kalia K, Tekade RK. Artificial intelligence in drug discovery and development. Drug Discovery Today. 2021;**26**(1):80-93

[24] Lu J, Bender B, Jin YJ, Guan Y. Deep learning prediction of patient response time course from early data via neural-pharmacokinetic modeling. Nature Machine Intelligence. 2021;**3**(8):1-9

[25] Ozdemir V. The changing face of personalized medicine and clinical pharmacology with digital drugs and real-time pharmacodynamics. OMICS: A journal of. Integrative Biology. 2020;**24**(4):173-174

[26] Bjornsson B, Borrebaeck C, Elander N, Gasslander T, Gawel DR, Gustafsson M, et al. Digital twins to personalize medicine. Genome Medicine. 2020;**12**:4

[27] Raijada D, Wac K, Greisen E, Rantanen J, Genin N. Integration of personalized drug delivery systems into digital health. Advanced Drug Delivery Reviews. 2021;**176**:113857

[28] Sverdlov O, VanDam J, Hannesdottir K, Thornton WT. Digital therapeutics: An integral component of digital innovation in drug development. Clinical Pharmacology & Therapeutics. 2018;**104**(1):72-80

[29] Dang A, Arora D, Rane P. Role of digital therapeutics and the changing future of healthcare. Journal of Family Medicine and Primary Care. 2020;**9**(5):2207-2213